AGROSPHERE
Nutrient Dynamics, Ecology and Productivity

AGROSPHERE
Nutrient Dynamics, Ecology and Productivity

K.R. Krishna

Science Publishers, Inc.
Enfield (NH), USA   Plymouth, UK

SCIENCE PUBLISHERS, Inc.
Post Office Box 699
Enfield, New Hampshire 03748
United States of America

Internet site: *http://www.scipub.net*

sales@scipub.net (marketing department)
editor@scipub.net (editorial department)
info@scipub.net (for all other enquiries)

Library of Congress Cataloging-in-Publication Data
Krishna, K. R. (Kowligi R.)
   Agrosphere : nutrient dynamics, ecology, and productivity / K.R.
     Krishna. p. cm.
   Includes bibliographical references (p.).
   ISBN 1-57808-275-7
     1.Crops and soils. 2. Soil Fertility. 3. Agricultural ecology.
   4. Agriculture--Environmental aspects. I. Title.

S596.7.K75 2003
631.4`22--dc21

2003041639

ISBN 1-57808-275-7

© 2003, Copyright Reserved

All rights reserved. No part of this publication may be reproduced, stored in a retrieval system, or transmitted in any form or by any means, electronic, mechanical, photocopying or otherwise, without the prior permission from the publisher. The request to produce certain material should include a statement of the purpose and extent of the reproduction.

Published by Science Publishers Inc., NH, USA
Printed in India.

"Let me begin this book by saluting and expressing gratitude to the first, unknown yet most important farmer(s)/agriculturist(s) who invented seeding, devised farming procedures and cultivated food crops in the 'Fertile Crescent' and 'Indus Valley'. Then, to modern agriculturists world-wide, who developed present day, predominantly prairie based 'Agrosphere' that produces for us food grains, vegetables, fruits and other useful commodities"

—Kowligi R. Krishna

Preface

Nature has bestowed several bounties on this earth such as soils interlaced with nutrients, moisture and air; a wide range of crop species with adaptability to different environments, diverse genotypic, phenotypic and yield potential; ambient environment including atmosphere, radiation, precipitation and biotic component. All have been crucial in the agricultural evolution on this globe. The most remarkable is the fact that such an agricultural phenomenon was initiated, guided and endeavored through *human ingenuity*.

The agrosphere as defined in this book encompasses several agro-ecosystems and useful vegetation. Today, it extends into a third of the global land surface. It is interesting to note that *human ingenuity* and his *agricultural acumen* has preferred cereals, legumes, and oil-seed based *prairie* vegetation for generating food. Such prairies constitute 80% of the agrosphere, and extend into a wide spectrum of environments. In tune with this agricultural reality, I have provided detailed explanations on 'Agrosphere' in the first chapter. Then, discussed several field crop-based agroecosytems. A chapter on Citrus in Florida, and two on forestry represent plantation ecosystems. Indeed, there are several other cropping ecosystems that would have been appropriate for inclusion into this volume. For example, the Corn belt of United States of America, Coffee land in South America, Grape zones in Southern Europe, Rubber plantations in Malaysia and the cropping zones in Yangtze Kiang and Hwong Ho river basins in China and more. Constraints of space, concern for brevity of this volume, lack of access to relevant scientific information, and my own inability to rapidly accumulate knowledge have been the reasons for not discussing other cropping ecosystems. However, interesting discussions on some of these, and other agroecosystems are possible in the next edition of this volume.

Initial few paragraphs within each chapter explain the geographical setting, the agroenvironment, and cropping pattern. They should be useful in gaining better insight. Nutrients in soil, their physico-chemical transformations and dynamics in relation to productivity form the core of this book. Literature quoted emanate from various international agricultural centres, reputed universities world-wide, and other specialized organizations. However, discussions are confined to influence of nutrients and their dynamics on productivity of various cropping ecosystems. The ecological aspects addressed are confined to soil microbial transformations, nitrogen fixers, mycorrhiza and net gains achieved through such symbioses.

At this point I wish to inform that the 'Agrosphere' includes a range of other equally interesting aspects in addition to nutrients, their dynamics and productivity. For example, volumes titled 'Agrosphere: dynamics of pestilence' or Agro-entomosphere; 'Agrosphere: economics, productivity and human welfare' or 'Agro-econosphere', that deal with relevant aspects are a clear possibility. However, they are out of the purview of this book, and are not my expertise.

Overall I believe that this book is one of the best scholarly editions in the recent times depicting global agriculture and its uniqueness. There are innumerable advantages in studying 'Agrosphere' independently with all its specificities, peculiarities and splendors. I have used the word "Agrosphere" and request my colleagues in Science, Agriculture and other faculties to judge the usefulness of this concept and spread the idea.

I would like to thank my wife (Dr. Uma Krishna) and son (Sharath Kowligi) who kept up excellent progress in their professions without distracting me, and encouraging where feasible.

November, 2002   Dr. K.R. Krishna PhD

Acknowledgements

Quite a large number of specialists from different organizations have spared their recent publications, institutional reports and related information that were requested. I express my sincere appreciation to them all and to their institutions. Following is the list of researchers and their institutions:

Alison Magill, Complex Research Systems, Durham, University of New Hampshire, New Hampshire, USA; *Ardell Halvorson*, USDA-ARS, Northern Plains area, Fort Collins, Colorado, USA; *Ashok Alva*, USDA, Washington State University, Pullman, Washington State, USA; *Bill Raun*, Department of Crop Science, Oklahoma State University, USA; *Brain Weinhold*, USDA-ARS, Soil and Water Science, University of Nebraska, Lincoln, Nebraska, USA; *Catherine Watson*, Agricultural and Environmental Science Division, Department of Agriculture, The Queens University of Belfast, New Forgelane, Belfast, United Kingdom; *Clinton Shock*, Oregon State University Agricultural Experimental Station, Oregon, USA; *David Foster*, Harvard Forests, Harvard University, Massachusetts, USA; *David Powlson*, Soil Science Department, Institute of Arable Crops Research, Rothemsted Experimental Station, Harpenden, Hertfordshire, Britain; *David Tucker*, Professor-Emeritus, Citrus Research and Education Center, University of Florida, Lake Alfred, Florida, USA; *George Gobran*, Department of Ecology, Swedish Agricultural University, Uppsala, Sweden; *Gustavo Madonni*, University of Buenos Aires, Buenos Aires, Argentina; *Harold Browning*, Director, Citrus Research and Education Center, Lake Alfred, Florida, USA; *Ilse Ackermann*, Soil and Water Science, Cornell University, New York, USA; *Jeffrey Burley*, Oxford Forestry Institute, University of Oxford, Oxford, England; *John Houghton*, Woodshole Research Center, Massachusetts, USA; *John Syverstien*, Physiologist, Citrus Research and Education Center, University of Florida, Lake Alfred, Florida, USA; *Joseph Albano*, USDA-Citrus Research Station, Fort Pierce, Florida, USA; *Kate Sebastian*, International Food Policy Research Institute, Washington, D.C.; *Kieth Kelling*, Professor, Extension Service—University of Wisconsin, Madison, Wisconsin, USA; *Maria da Graco Serrao*, Ministry of Agriculture, Portugal; *Micheal Keller*, International Institute for Tropical Forestry, Puerto Rico, USA; *Onema Oene*, Green World Research, Department of Water and Environment, Wageningen, The Netherlands; *Raijo Laiho*, Department of Forest Ecology, University of Helsinki, Finland; *Paul Rasmussen*, Crops Science Department, Oregon State University, Oregon, USA; *Rebecca Hood*, International Atomic Energy Agency, Vienna, Austria; *Reinhardt Howeler*, Centro International Agriculture Tropicale, Thailand; *Ryan, J.,* ICARDA, Aleppo, Syria; *Steven Jarvis*, Institute of Grassland and Environmental Research, North Wyke, Okehampton, Devon, X20 25B, United Kingdom; *Venkateshwarulu, B.*, Central Research Institute for Dryland Research, Santoshnagar, Hyderabad, India; *Victor Timmer*, Professor, Faculty of Forestry, University of Toronto, Ontario, Canada; *Colleagues* and *friends* at International Crop Research Institute for Semi-Arid Tropics, Hyderabad, India; *Compatriots* from Indian Agricultural Research Institute, New Delhi; *Librarian*, British Council Library, Bangalore, India; *Librarian* and *colleagues* at the University of Agricultural Sciences, Bangalore, India.

Photo credits: Dr. K.R. Krishna, ICRISAT, Hyderabad, India; Soil and Water Science division, University of Florida, Gainesville, Florida, USA. Dr. Roger Atkins, Rothamsted Experimental Station, Harpenden, England. CGIAR—Web Photos, cgiar.org. The Citrus Experimental Station, Lake Alfred, Florida, USA http//www.ufl.edu/crec—web photos.

Contents

	Preface	*vii*
	Acknowledgements	*ix*
1.	The Agrosphere	1
2.	The Temperate Wheat Agroecosystem in Great Plains of North America: Nutrient Dynamics, Ecology and Productivity	28
3.	The Temperate Wheat Zones of European Plains, Australia and Pampas in Argentina: Nutrient Dynamics and Productivity	58
4.	The Intensive Rice Culture in South and Southeast Asia: Nutrients in Rice Land Ecosystem	105
5.	The Rice-Wheat Agroecosystem of The Indo-Gangetic Plains: Nutrients and Productivity	141
6.	The Dryland Agroecosystem of West Asia, North Africa and South Asia: Nutrients, Water and Productivity	179
7.	The Semi-arid Tropical Agroecosystem: Nutrient Dynamics	208
8.	The Humid Tropical Agroecosystem: Nutrient Dynamics	241
9.	The Citrus Agroecosystem of Florida: Nutrient Dynamics and Productivity	257
10.	Tropical Forest Ecosystems: Nutrient Dynamics	280
11.	The Temperate and Boreal Forests: Nutrients and Productivity	303
12.	Epilogue	330
	Index	337

CHAPTER 1

The Agrosphere

> *"Agrosphere is an ecosphere concerned exclusively with the agricultural activity on earth, and has been carved out of biosphere. It is predominantly man-made, attributable to his ingenuity, perseverance, and tireless toil aimed at satiating his food requirements. Today Agrosphere serves a dense global population of 6 billion individuals and larger number of domestic animals—achieved never before. Our future, to a good extent, rests on its nurture, upkeep and productivity"*
> Let us be alert on this fact!

CONTENTS

1. Agrosphere: Concept and Expanse
2. Agrosphere versus Biosphere
3. Agrosphere in relation to other Ecospheres
4. Ingredients of Agrosphere
5. Nutrients in Agrosphere
6. Summary

1. The Agrosphere

Concept: Agrosphere could be explained as a man-made ecosphere on earth, developed through constant mending of soils, crops and related factors since 8 to 10 thousand years (Krishna, 2002). The agrosphere confines to terrestrial zones on earth where agricultural activity flourishes. In other words, agrosphere is that terrestrial portion carved out of naturally evolved biosphere, where agricultural activity such as crop production is possible. Biologically, agrosphere deals with crops, and their ecosystematic interactions with other flora, fauna and abiotic factors. Geographically, agrosphere constitutes the cropping zones that occur between the equator (0° latitude) to sub-Arctic/Antarctic circle until 63 to 65° N or S. It practically intrudes into any part of the globe that is congenial, and supports cropping systems. The basic ingredients that allow the development of agrosphere in these latitudes are fertile soils, well adopted crop genetic stocks, water and suitable environmental parameters. On a broader horizon, agrosphere may be defined as a man-made conglomerate of agroecosystems that occurs world-wide and is integrated through constant interaction with other ecospheres namely lithosphere (soils), hydrosphere, atmosphere and biosphere.

The evolution of agrosphere is synonymous with the natural history of agriculture, spread of cultivable land, agrarian revolutions and improved farming practices. Indeed, as the seeds for primordial agriculture were sown in the 'Fertile Crescent' and agrosphere took roots, perhaps there was no inkling of the vast influence that mankind had initiated on the naturally evolved biosphere of this earth. At present, the agrosphere as defined above comprises a series of cropping ecosystems (Pearson, 1992). For example, the rice agroecosystem in Asia, temperate wheat zone in North America and Europe, the corn belt, citrus agroecosystem in Florida, pearl millet in Sahel, coffee in South America etc. These agroecosystems are unique, and well adopted to the geographic region. There may be several different ways of defining such an agroecosystem. To mention a recent example, Wood *et al.* (2000) explain agroecosystem as an agricultural and natural resource system managed by *humans* for the primary purpose of producing food as well as other socially valuable non-food goods.

Clearly, these man-made agroecosystems are dissimilar with natural vegetation, or forests and other biomass classified under biosphere. In order to avoid confounded inferences, ecologists often try to identify this distinction by using prefixes such as 'disturbed' which is different from 'undisturbed' natural ecosystem. Some scholars may still prefer to identify agrosphere as a component of biosphere with predominantly agrarian activity. It may look reasonable. However, consider its origins, excessive human intervention, the vastness, intensity and variety of influences on global environment and human welfare, the self-sustaining nature, and the rapidity of its spread in recent decades. In view of these above characteristics, we may have to resist the idea of studying agrosphere (all agroecosystems), as just disturbed ecosystems under biosphere.

Expanse: Presently the agrosphere extends into 13,500 mha, and on vegetation basis, prairies constituted by cereals, legumes and oil-seed crops occupy the greatest portion of the agrosphere. Cereals spread into 704 mha equivalent to 60% of the agrosphere. The wet-land rice-based prairies which are highly concentrated within Asia contribute 90% of global rice grain harvest (Wood *et al.* 2000). Whereas, wheat-based cropping ecosystems are more widely distributed into temperate and subtropical zones. Legume-based ecosystems occupy 70 mha, oil-seed crops cover 208 mha and horticultural fruit crops are grown in 47 mha (FAO/IIASA, 1999). Overall, 80 to 90% of the agrosphere which comprises prairie vegetation supplies carbohydrates, protein and fat required by human kind. In other words *"Man depends immeusely on prairies of the agrosphere for his nourishment and sustenance"*.

2. Agrosphere versus Biosphere

A few different definitions for agrosphere have already been listed above. Similarly, we may define biosphere in different ways depending on the context. Biosphere includes the flora and fauna encompassing a wide range of living organisms, that reside and perpetuate on, at and in the earth's surface. On a wider horizon, biosphere is defined as the integrated living and life system comprising the pedosphere, atmosphere and hydrosphere (Gerard, 2000; Odum, 1984). Within such a biosphere, man is considered just another biotic factor. Similarly the vast expanse of the agricultural world that includes several agroecosystems is considered yet another activity. Sometimes termed just an aberration leading to intensive growth of a single crop species. Agroecosystems are called disturbed ecosystems/habitats because of excessive human intervention. Perhaps, this is not a feasible explanation, which can account for the influence that cropping systems and man impose on biosphere. In recent times, natural biospheric zones or the so-called undisturbed ecosystems are being progressively eaten into by different agroecosystems. Currently agroecozones are conspicuous in terms of expanse and their influence on global environment. They occupy a third of total landmass. Today, agrosphere has grown into 13,500 mha and in the process, has interacted and influenced other ecospheres such as pedosphere, hydrosphere, atmosphere and biosphere.

Protagonists of biosphere may wish to continue identifying various agroecosystems and the agrosphere as part of biosphere. However, reasons and suggestions to identify agrosphere as an independent ecosphere and a distinct topic within ecology seem stronger.

This situation is parallel to crop specialists asking to identify their subject matter prominently, and in exclusion from general plant science or botany. In this present case, an agroecologist carves out and erects a new ecosphere called 'Agrosphere', which is different from a generalized term biosphere commonly professed by a general ecologist.

The two ecological concepts namely biosphere and agrosphere perhaps can be understood, discerned distinctly, and a clearer perspective attained by comparing them on a range of aspects. Such comparisons may also sustain the claim that agrosphere is predominantly a man-made ecosphere. Now, let us compare and contrast agrosphere with biosphere for different aspects.

The sole purpose in developing the agroecosystems that constitute agrosphere is human welfare—to serve man's food, fiber and fuel requirements. Whereas biosphere is a natural manifestation on earth. Its sustenance and the direction of evolution is influenced by a variety of natural factors, where man is perceived as just another factor. As the ecological intricacies of biosphere get expressed, human welfare, and his needs for nourishment/sustenance are not at all the primary aims of biosphere.

In terms of organismic evolution, biosphere began with the chance formation of the first primordial life forms on earth several billion years ago. It includes all the diverse forms of living organisms, microbes, plants and animals. Whereas agrosphere began with deliberate seeding, domestication of crop species and standardization of terrestrial farming techniques. It is a well directed effort by human beings, that focuses on crops. It is not yet replicated by another species of primates!

In terms of expanse, biosphere extends from atleast below the floor of the ocean (benthic flora), into the atmosphere wherever biological activity is possible. Life has been detected upto 6.5 km above the earth's surface. If life on space stations is included, then it extends until 300 km above earth's surface. These days life in space stations are gaining in permanence! In contrast, the agrosphere confines to well vegetated and cropped land on the earth's surface, wherever agricultural activity is possible. Mainly, between 0° latitude (equator) to 63 to 65° N or S, and from altitudes ranging from sea level (coastal agriculture) till 3000 m above mean sea level in montane forest clearings. We have not yet used space stations to cultivate crops.

Biosphere extends into zones that experience below freezing temperatures approximately –5°C to 45°C, and into surroundings of hot spring and geysers that support microflora capable of growth at temperatures 47 to 50°C. Certain organisms overcome unfavorable temperatures by sporulating or hybernating. Agrosphere does not extend into these extreme climates. Most congenial temperatures known for cropping systems to flourish, range from 10 to 25°C in temperate agroecosystems, and between 17 and 35°C in tropical and sub-tropical cropping zones. However, for shorter durations crops withstand frost in temperate zones, or higher temperatures in humid or arid tropics. Seeds of several crop species do withstand extremes of temperature. Moisture is equally crucial to sustenance of both undisturbed natural habitats and agroecosystems. Still, crops' ability to adapt to extremes of moisture stress or excess is limited compared with other life forms in biosphere, including a few non-agricultural plant species.

Soil types encountered within agrosphere are those which support cropping, pasture and related agricultural activity. Not all soil types are amenable for cultivation. Presently, nearly 30% of land surface falls within the realm of agrosphere (FAO STAT, 1999; Wood *et al.* 2000). Agrosphere is definitely large enough to be an independent ecological entity on the globe. Whereas the diverse life forms, flora and fauna encountered in biosphere extend into any type of soil depending on the species' preferences.

Agrosphere and biosphere vary enormously in their anthropogenic effects. Influence of human ingenuity is greatest with reference to soils and their management. This is in addition to effects of natural factors. Anthropogenic processes such as soil preparation, intercultural operations, amendments, nutrient inputs and incessant cropping affect soil immensely. Whereas, in natural undisturbed ecosystems human influence on soils is minimal, if any. Soils are not mended deliberately, hence alteration will be least. Soil erosion is conspicuous with agricultural soils, because soils are loosened during preparation for cultivation. In a well stabilized natural habitat soil erosion is minimal.

Nutrient management in any of the agroecosystems that constitute the agrosphere is heavily influenced by human preferences and agricultural practices. Nutrient inputs into agrosphere occur both through natural weathering, atmospheric deposition, and via chemical fertilizers, recycling of farm wastes and organic manures. Fertilizer-based nutrient inputs into agrosphere are an example of *human ingenuity*, not seen in natural habitats of biosphere. It enormously alters the nutrient dynamics in agrosphere. Internal nutrient fluxes, rates of nutrient transformation, extent of nutrients loss through erosion, emission etc. too differ between agro-ecosystems and natural vegetation.

Soil organic matter (SOM), particularly its transformations, loss and accumulation rates differ enormously between agricultural soils, and those in natural undisturbed ecosystems. In undisturbed ecosystems, SOM attains a steady state influenced only by soil formation processes, climatic parameters, vegetative flora and fauna. Whereas, in disturbed agroecosystems SOM declines without an equilibrium (Batjes, 1996; Jarvis *et al.* 1996; Schlesinger, 1999; Wood *et al.* 2000; Bayer, 2001).

Vegetation within agrosphere is predominantly field crops such as annuals, biannuals or perennials, and horticultural or forest plantations. Plant genetic diversity is restricted to crops and weeds. An undisturbed portion of biosphere may support more diverse plant species. Generally, a clear pattern of natural succession is discernible in biosphere, depending on biotic and abiotic factors, for example, the pioneers, secondary flora, dominant species, extinctions etc. Such a natural succession is totally absent in agricultural fields. Preference to crops and cropping pattern, and succession if any is heavily influenced by human decisions. Interestingly, when a new crop or its genotype is introduced into an agroecosystem its growth, survival, perpetuation, productivity and storage of that germplasm are carefully managed by human beings. Whereas, any plant species or wild germplasm of domesticated crops found in undisturbed natural ecosystem are left to fend for themselves, without human intervention.

Clearly, human ingenuity as a factor is felt immensely with reference to crop genetic diversity in the agrosphere. In fact, agrosphere began with genetic selection and domestication of major food crops, like wheat in West Asia, rice in the Southeast Asia, and several other crops. Today crop breeders world-wide, create vast genetic diversity and churn out several agronomically elite genotypes suitable to different agro-environments. This aspect is absent in natural habitats of biosphere. The plant genetic variation traced in natural ecosystems are solely the gifts of natural crossing and evolutionary process. At the same time, one must realize that introduction and withdrawal of crop species, and their genotypes in the outfield (agrosphere) is governed largely by human preference and judgment. These aspects again are absent in any natural ecosystem.

Within the agrosphere, seed dispersal, preservation of the crop species, its land races and elite varieties are thoroughly mediated through human ingenuity (eg., crop germplasm centers, seed production and distribution agencies). Influence of natural factors is either inconsequential or negligible, and is carefully avoided if any. Whereas, in a natural biosphere reserve, seed-dispersal, seed-survival and storage are natural processes. Only, a few endangered and identified species are handled by plant ecologists/germplasm specialists.

Agrosphere differs from natural, undisturbed ecosystem with respect to seasonability, flushes in vegetation, appearance of disease/pestilence and productivity. Irrigated summer crops are distinct examples of human ingenuity. Wherein, irrigated crops flourish in summer but a natural ecosystem suffers due to scanty precipitation and/or drought. Post-season/harvest events too are enormously different between agrosphere and biosphere. Processing of harvested material for consumption or storage is unique to agrosphere.

3. Agrosphere in Relation to other Ecospheres

While enunciating several differences between agrosphere and biosphere, the purpose was to convince that agrosphere is to be studied independently and prominently as an ecosphere with massive human influence. Although, agrosphere is self sustaining like other ecospheres, it is in constant state of interaction with other ecospheres, namely pedosphere (soils), hydrosphere, atmosphere and biosphere (Figure 1.1). Such interactions could be either detrimental or beneficial to ecosystematic functions within agrosphere. At this point, elaborate discussions on other ecospheres are not within purview of this treatise. Hence, smaller but

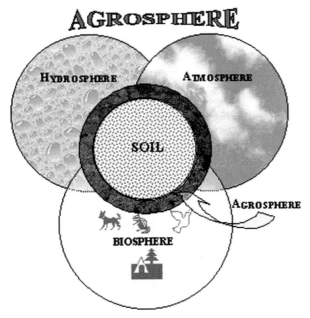

Fig. 1.1 Agrosphere: Its diagrammatic representation in relation to other ecospheres

useful information related to other ecospheres, and their interactions or confluences with agrosphere only are highlighted. Discussions confine to nutrients in different ecospheres.

Hydrosphere: The hydrosphere is a very large repository of water. Water exists in solid, liquid and gaseous state. Water is intimately related with all agroecosystems. Water is a basic ingredient for crop growth and productivity, therefore its interactions are crucial to viability of agrosphere. Water is easily reachable in many places on earth. It occurs as ground water, in rivers or lakes. However, water that resides in soil micropores (soil moisture) is the only fraction which is tapped by crops. Soil moisture is actually a very small fraction of hydrosphere, and any variation, particularly deficiency affects agrosphere activities severely. Water mediates nutrient transfer at soil-root interface, and facilitates several chemical transformations in soil, which are useful for sustaining crop ecosystem. However, variations in soil moisture are common, both dearth and excess could be perceived at the confluences between agrosphere and hydrosphere (e.g. drought or floods). At this point, we may recognize that human ingenuity has played a great role in moderating water availability through irrigation and water conservation procedures, damming excessive water etc., which help in the development of congenial hydrosphere-agrosphere interactions (see Wood *et al*. 2000). Water quality, measured in terms of pollutants, salts nutrients, pH etc. are important indicators of the health and congeniality of agrosphere-hydrosphere interactions. For such confluences between these ecospheres to be harmonious, drastic changes in water quality, or crop management procedures that may be detrimental or disturbing to ecosystematic equilibrium should be avoided.

Lithosphere (Pedosphere): Lithosphere is the rocky, solid outer shell of earth which is different from core. It is the uppermost layer of the earth's crust, and is the essential parent material from which soils develop. Pedosphere could be defined as that layer of the earth's crust, or a portion of lithosphere in which soil-forming processes predominate (Bates and Jackson, 1980; Huggat, 1995). Agrosphere is anchored on pedosphere. Most pedologists agree that pedosphere is that part of the lithosphere which supports plant life and cropping ecosystem. Clearly, at this point pedosphere and agrosphere are intricately and constantly linked to each other. Pedosphere (soils) is the major repository of nutrients that sustains agrosphere. The physico-chemical and biological interactions that develop within soil, and between plant and soil are crucial to agrosphere viability. Agrosphere, through plant roots may penetrate upto 2 to 10 meters or even more into pedosphere. This confluence partially regulates nutrient dynamics, and productivity of any agroecosystem. Intensive cropping activity, excessive chemical inputs, pollutants etc. can disturb, firstly the physico-chemical and biological aspects of pedosphere, consequently the agrosphere functions.

Atmosphere: Events and changes in atmospheric composition can immensely influence sustenance and productivity of agrosphere. Vice versa is also true. Events in agrosphere, such as gaseous emissions (CO_2, CH_4, NO_2, NO etc.) can alter atmospheric composition. Atmosphere predominantly consists of nitrogen (N_2), oxygen (O_2) and argon (Ar) totaling 99% of it. The rest 0.04% is contributed by hydrogen (H_2), ozone (O_3), carbon dioxide (CO_2), methane (CH_4) nitrous oxide (N_2O) and aerosol.

Fixation of N and C from atmosphere into pedosphere through soil microbes (N fixers) and plants (photosynthesis) respectively, involves interactions at confluences formed by atmosphere, pedosphere and agrosphere. Conversely, nitrogen is released into atmosphere via NH_3 volatilization, or gaseous emissions (N_2O, N_2) caused by nitrification-denitrification reactions. Oxygen is released into atmosphere whenever organic matter decomposes or oxidative processes occur in the pedosphere. Photosynthesis releases O_2 into atmosphere. Whereas, respiration by plants, soil microbes and animals constitutes movement of O_2 out of atmosphere into agrosphere and biosphere. Carbon dioxide is fixed into biomass during photosynthesis by agroecosystems. Conversely, it is released into atmosphere through respiration by crops, related flora and fauna. Combustion of stover or other organic portions too release CO_2. Burning fossil fuel is known to have increased CO_2 in the atmosphere by 18% since 1980, increasing it from 296 ppm to 364 ppm in 1999 (see Houghton, 1997). We must realize that such increases in CO_2, or more accurately the CO_2 depletion/accumulation ratio has enormous consequences on global warming and glass house effects. In addition to gaseous components, atmosphere may contain between 1 and 4% water vapour, depending on whether it is a dry zone or humid belt. Equally important is the aerosol present in the atmosphere, which is caused by both natural and anthropogenic processes. Sulphates, carbonates and hydrocarbon particles in aerosols can affect cropping ecosystem. Sulphur aerosol deposition is an important source of S in many agricultural belts. However, at higher concentration, such SO_2 or NO_x containing aerosols can cause acid rains affecting the agroecosystem detrimentally.

Biosphere: Biosphere as defined earlier deals with the biotic component on earth. In nature, the biosphere and its specific components are intricately linked with agrosphere on various ecological aspects, especially energy and nutrient cycling. Several species of plants, animals and microbes are directly/indirectly involved in nutrient dynamics. Their effects on nutrient turnover in an ecosystem can be conspicuous. For example, tropical and temperate trees are major sinks for carbon. Similarly, soil microbes play a vital role in nutrient transformations and cycling. Nitrogen fixation by symbiotic and asymbiotic microbes; nitrification/denitrifications by *Nitrosomonas/Nitrobacter*, immobilization of C in soil or loss of CO_2 through respiration are major functions of soil microbes. Indeed, a large number of treatises are available that deal with biosphere and its ecological manifestations. However, they are not within the purview of this work.

4. The Components of Agrosphere

The components discussed in the following paragraphs are concerned with cropping activity within an agroecosystem. They are derived from the different ecospheres, namely hydrosphere, pedosphere, atmosphere, biosphere and certain others exclusively through human initiative (e.g. fertilizers). Crops, soils, water, nutrients and specific gases are the primary components required for immediate viability, perpetuation and productivity of any agroecosystem. Each of the components has its special effects on the agrosphere. Large number of treatises and reports are available on each of these ingredients. However, within this chapter only cursory remarks needed to develop the theme of this book are provided. Greater details on components and their influence on a few specific agroecosystems are available in chapters 2 to 12.

Crops: Crops are the center piece of agrosphere. A very diverse group of crop species and their genotypes have been utilized by man to develop different agroecosystems. According to Wood *et al.* (2000) nearly 7000 crop species have been utilized for cropping within the agrosphere, although between 10 and 50 thousand plant species are edible. Interestingly, at present only 30 odd crop species are regularly grown within the agrosphere by modern man. These few crop species suffice to provide him with 90% of his calorie and protein requirements (Shand, 1997; FAO, 1998). In addition, a few tree crops that supply nutritious fruits, and nuts are also farmed and utilized. However, Kindt *et al.* (1997) report that seeds of nearly 2632 tree species are currently available for commercial growing.

Crop genetic stocks, their ability to adapt to the given environment, tolerance to abiotic and biotic stress factors, if any, and response to nutrient inputs with higher productivity are the crucial aspects to consider. Genetic selection of crops for useful traits was initiated by man. There is no doubt that *human ingenuity* has played the greatest role in domesticating and learning the art of plant genetic selection and breeding. Crops being part of the agrosphere need to withstand variations/stresses generated by abiotic and biotic fractions of the agro-environment. In modern agriculture, genetic resources of individual crops, their preservation and utilization in breeding programs are important. Both *ex-situ* and *in situ* gene banks are utilized to conserve this primary ingredient of the agrosphere. Major crop gene banks exist with national agricultural centers of China, USA, USSR, India and CGIAR institutes (FAO, 1998; CGIAR, 2000). Wood *et al.* (2000) opine that lacuna with germ plasm banks is that they only preserve germ plasm of crop species relevant to the agrosphere. Focus on other plant species is minimal. More recently, transgenic techniques have been utilized to create greater biodiversity and/or to introduce specific genes. Transgenic crops as part of the agrosphere occupied 1.7 mha in 1996, which increased to 39.9 mha in 1999 in OECD[1] countries (Persely, 2000; Wood *et al.* 2000).

Soil: Soil is a crucial part of the agrosphere. It provides anchorage, supports crop development, supplies nutrients, and determines to a good extent the productivity of individual agroecosystems that form the agrosphere. Soil is an enormously variable ingredient, in terms of its composition,

---
[1] OECD refers to Organization for Economic Cooperation and Development, Paris, France.

surface chemistry, structure, texture, depth, horizons, water regimes, aeration, and biotic component. Most importantly, soil nutrients available to plant roots are highly variable (Webster, 1997). Such variations in soil fertility can be discerned at local scale, say at intervals of few hundred meters in most parts of the world. Vast literature that documents and describes soil resources at different scales, say at field, farm/estate, physiographic region (100 to 10,000 km), national, agroecosystem or world level is available (FAO, 1991; FAO, 1995; FAO, 1999, USGS EDC, 1999; Bayer, 2001). Within this book discussions on soil types, their classification and physico-chemical nature are available separately for each agroecosystem in chapters 2 to 11.

Soil resources utilized for the development of different agroecosystems are not just variable with reference to fertility/mineral content, they come with several constraints/maladies that retard productivity. Again, there are innumerable reports on this aspect (Sanchez, *et al.* 1982; Smith *et al.* 1997; Young, 1998; FAO STAT, 2000; Rosenzweig and Hillel, 2000). Degraded soils constrain development and productivity of agroecosystems. A recent report by Wood *et al.* (2000) states that only 16.2% of soils utilized for global agriculture are free from constraints. At least 13 different types of constraints such as poor drainage, low CEC, Al toxicity, acidity, P-fixation, low K, alkalinity, salinity, sodicity, shallowness, low moisture holding capacity etc, have been identified. Among these, acidity (24.6%), low K (18%), Al toxicity (17%), poor drainage (14%) and shallowness (10%) are the major constraints world-wide. Soil quality which is indicated by soil organic matter (SOM) also affects agrosphere productivity. Agricultural activity generally reduces SOM, hence reduces the quality of this component.

Water Resources: Like crop genetic stocks, soils and nutrients, water is a primary ingredient of agrosphere. Water directly influences physiological growth and yield formation by crops. Wood *et al.* (2000) have remarked that as time lapses, water is becoming an increasingly critical component affecting expansion of various agroecosystems. To a fair extent, water as an essential ingredient has dictated feasibility, type of agroecosystem possible and its viability in a geographic area. For example, low precipitation levels, and scarce water resources in sub-Saharan Africa permits only dry or semi-dryland cropping system. Whereas, a sumptuous supply of water in Southeast Asia can easily support intensive wetland rice culture. Water needs of individual crops vary. Water, as an ingredient for cropping can create sharp differences in cropping pattern, its sequence and productivity. For example in WANA a mere 50 to 100 mm higher precipitation totaling 350 mm allows wheat belt to thrive, but < 250 mm total rain supports only barley cultivation.

Dryland crops (e.g. pearl millet, finger millet) may require between 300 and 600 kg, whereas an arable crop of wheat or maize needs 900 kg, and a wetland rice crop needs >2000 kg water per season. Around 95% of all agricultural land (i.e., agrosphere) and at any instance 83% of cropping zones depend on rainfall as the sole source of water. Often, it is not just the precipitation levels, but a resultant situation between precipitation and evapotranspiration rates that decides cropping pattern and productivity (FAO/IIASA, 1999; Wood et al. 2000).

Globally, 9000 to 12,500 km^3 of water is estimated to be available, out of which 2700 km^3 could be extracted for irrigating crops (Postel et al. 1996; UN, 1997; WMO, 1997). Nearly 30 to 40% of global crop production arises from such irrigated farms which constitutes 17% of the agrosphere in area (FAO STAT, 1999). Irrigated farms support a major share of global rice and wheat harvest equivalent to 60% output. During the past decades irrigation has been expanding at 1.5 to 2.0% per year (FAO STAT, 1999). Globally, 243 mha are irrigated, out of which the major portion, nearly 155 mha occurs in Asia. Information maps pertaining to water resources and irrigated zones are available (see Doll and Siebert, 1999; Alcamo et al. 2000).

In certain parts of the agrosphere, excessive extraction of ground water has resulted in alarming depletion of this resource. For example, Postel (1993) stated that water table in Northern China plains is receding due to excessive removal of ground water. The situation is similar in many locations in the Indo-Gangetic plains. Again, Wood et al. (2000) report that at present global irrigation use efficiency is 43%, but varies with crops, agro-environment and location. For example, Seckler et al. (1998) estimated that dry agroecosystems are more efficient (54 to 58%), compared with 30% achieved in zones without water related constraints. Water quality particularly dissolved salts, nutrients and other ions, salinity, solids, organo-chlorine pollutants, and pesticide contents are important.

To avoid repetitions, nutrients as component of the agrosphere are discussed separately in the following paragraphs. Similarly environmental para-meters and their influences have been dealt distinctly under each chapter dealing with different agroenvironments (chapters 2 to 12).

5. Nutrients in the Agrosphere

Carbon: Carbon is a foremost component, as well an essential nutrient required for the agrosphere and its sustenance. It is directly concerned with biomass accumulation and yield formation in different agroecosystems, and in improving soil fertility and quality through enhanced soil organic matter. On a different scale, cropping and related activities within the agrosphere emit gases such as CO_2 and CH_4, which can induce global warming. Therefore, knowledge on C stocks in nature, its dynamics and relevance to agroecology, as well as human welfare is of utmost importance. At present terrestrial ecosystems, mainly forests are considered major sinks for carbon. For example, Houghton *et al.* (1998) state that during the 1990s average annual emissions at 7.1 Pg C (5.5 from combustion and 1.6 from changes in land use), were higher than annual accumulation of C in atmosphere (3.3 Pg C yr^{-1}), and that by oceans (2.0 Pg C yr^{-1}). Hence, a 'missing sink' (1.8 Pb C yr^{-1}) is to be accounted to balance the global carbon budget. Further, Schimel *et al.* (1995) suggest that accurate estimate of C sequestration into terrestrial ecosystem, that accounts for the missing sink is necessary for forecasting CO_2 concentrations in atmosphere.

According to Houghton *et al.* (1998) there are four major terrestrial processes that determine carbon storage in the short term. They are:
(a) Carbon fertilization (Amthor, 1995; Koah and Mooney, 1996);
(b) Increased anthropogenic mobilization of nitrogen from fertilization (Townsend *et al.* 1996; Holland *et al.* 1997);
(c) Warming enhanced mineralization of nitrogen (Melillo *et al.* 1993); and
(d) Recovery of forests from previous disturbance (Kurz *et al.* 1995; Houghton, 1996).

Possibly multiple mechanisms are operative with reference to CO_2 fertilization and N deposition as factors leading to increased C storage. Insights gained through simulation models, such as Terrestrial Ecosystem Model (TEM) indicate that N mineralized through warming is utilized by plants leading to increased biomass and C storage in the ecosystem (McGuire *et al.* 1995; Melillo *et al.* 1996). Whereas, warming enhanced mineralization of C, respiration and most importantly, conversion of forests to agricultural fields do not help in explaining terrestrial C storage ('missing sink'), because these mechanisms support loss of C to atmosphere.

Generally, terrestrial C storage is dependent on variations in forest cover. If forested area declines, it leads firstly to lowered C storage and then enhances C emissions into the atmosphere. In terms of C sequestration ability, tropical dry agricultural soils possess least C at 0.1 to 2%, temperate agricultural fields may contain higher levels at >10% organic matter. Whereas, forests and peat lands often average >20% organic matter. Clearly, missing 'C sink' could be traced more to terrestrial forest ecosystems, rather than agroecosystems which often reduce C sequestered in soils and vegetation.

Therefore, Wood *et al.* (2000) quoting UN convention on climate change, state that the agricultural community, beyond reducing emissions from fossil fuel should undertake to mitigate build up of green house gases. Mainly, by adopting land use patterns that enhance C sequestration into terrestrial agroecosystem. In fact, C storage in vegetation and soils underneath are primary indicators of the influence of agricultural activity on climate change (Kumar and Goh, 2000; Izzarulde *et al*. 2001). However, value of agroecosystem based vegetation in terms of C storage is limited. The biomass densities of crop land, or pastures are much less compared to forests. The rates of C accumulation too varies (Fig. 1.2). In addition, biomass is regularly harvested and used in ways that release stored C. Only deep-rooted, woody crops could contribute perceivably to C storage in agricultural zones.

Globally, 750 to 800 Gt C* present in atmosphere, and 2000 to 2500 Gt C in biomass and soils of terrestrial ecosystems are said to be relevant to cropping activity within agrosphere (IPCC, 1995). Batjes (1996) and Eswaran (1993) have reported that in the first 100 cm of stone-free soil, where influences of cropping is greatest, the global C storage within this layer could be around 1460 to 1550 Gt C.

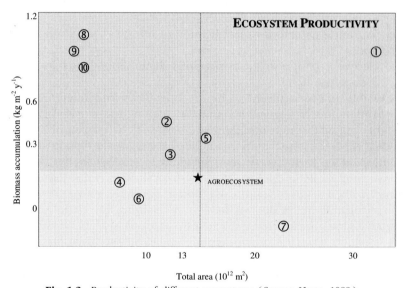

Fig. 1.2 Productivity of different ecosystems (*Source*: Harte, 1988.)
Note: 1. Tropical forest, 2. Temperate forest, 3. Boreal forest, 4. Shrub land, 5. Savanna, 6. Grass land, 7. Desert zones, 8. Swamp/Marsh, 9. Algal beds, 10. Estuary ★ Agroecosystems/Cropping zones)

Carbon Emissions: During the 1990s, net carbon increment into the atmosphere was 3.3 Gt C per year. It is believed that land use changes and practices were responsible for 20% of C emissions equivalent to 1.6 Gt C as CO_2 (IPCC, 2000). Agrosphere activities such as (a) clearing forests and woody savanna for cropping; (b) burning crop stubble and

---
* Gt = Giga tons

pastures for promoting soil fertility; and (c) soil degradation are the primary reasons of CO_2 emission (Lal, 2000). Houghton and Hackler, (1995) report that CO_2 emission from agricultural zones in developing countries alone increased from 110 Gt C in 1850 to 1400 Gt C in 1980, and then to 1600 Gt C in 2000 A.D. During this same period, CO_2 emissions in developed countries increased from 400 Gt C in 1850 to 1600 Gt C in 2000 A.D. Again, during this same period CH_4 emissions from agricultural cropping zones increased form 135 Tg yr^{-1} in 1860 to 325 Tg yr^- in 1997 (Stern and Kaufman, 1998). Wood et al. (2000) report that on a rough estimate, agricultural operations account for 20% of greenhouse gas and global warming effects, wherein CH_4 emissions could be conspicuous. Hence, they caution that even if the predicted decrease in CO_2 emission occurs, small increases in CH_4 emission, which is 20 times more potent in causing global warming can be detrimental. Methane emissions are derived mainly through rice paddy cultivation, biomass burning and livestock farming. Incidentally, agrosphere is the largest anthropogenic source of CH_4. It accounts for over 40% of global CH_4 emission. Methane concentration in the atmosphere has risen from 700 ppb to 1721 ppb in 1994, at an average increase of 4 to 8 ppb per year (IPCC, 1995; IPCC, 2000). Overall, loss of carbon from global agricultural soils is estimated to be around 800 Mt C yr^-. Wherein tropical soils alone loose 500Mt C yr^{-1} (Schlesinger, 1990; Lal, 1995). Within the agrosphere, any cropping pattern or farming systems that involves least soil disturbance, causes less erosion, avoids burning of crop residue, increases litter formation, utilizes deep-rooted crops or trees as components conserves, soil organic matter and reduces CO_2 emission. Lal et al. (1998) report that a mere good control over soil erosion can reduce CO_2 emission by 12 to 22 Gt C yr^{-1}

Nitrogen: Nitrogen is a major nutrient required for normal physiological activity of all the crops grown in different agroecosystems. Optimum N availability during a crop season is crucial to sustenance and productivity of crops. Further, De Willigen (1986) reported that maintaining appropriate levels of different N forms, particularly NH_4^+–N and NO_3–N in soil solution is essential to obtain optimum crop productivity. It is generally accepted that NO_3 is the predominant N form suitable for recovery by crops, except rice (Wiesler, 1997). Therefore, both in natural vegetation and agrosphere the ratio of NO_3–N to NH_4–N is crucial. At present, soils in most agroecosystems receive N inputs through inorganic N fertilizers; from atmosphere and deposits along with precipitation; as N credits derived from biological nitrogen fixation, and through recycling plants and animal residues. Overall, global agroecosystems are said to contain 2.4×10^{12} t of N (Jarvis et al. 1996). Clearly, this estimate does not reflect

---

Tg = Tera gram; ppb = parts per billion

the variation, especially deficiencies of N encountered in major cropping belts. However, the fraction of N relevant to cropping activity is mostly determined by internal fluxes, transformations, storage, N balance and equilibrium. In fact, processes such as mineralization and immobilization predominantly regulate N flow within any agroecosystem. More crucial in terms of practical farming activity, is the N-use efficiency attained by crops.

Nitrogen inputs into the agroecosystem have increased enormously since mid 20th century. It has risen from 9.6 Tg N in 1960 to 80 Tg N in 2000 (FAO STAT, 2000). The magnitude of increased N usage has generally depended on the ecosystem and yield goals. When compared with a global average input of 53 kg N ha^{-1}, greatest N inputs have been occurring in the intensive wetland rice belt of East and South east Asia which averaged 131 kg N ha^{-1} in 1999. Least N inputs are encountered in Sahelian belt of West Africa, where mere 6 kg N ha^{-1} are utilized (see Fig. 1.3). Clearly, intensification of agroecosystem through increased inputs of N and nutrients is dependent on water resources and environmental parameters. Since several decades, major amounts of N inputs into agrosphere have been garnered by cereals such as rice, wheat, and maize. Presently, N inputs to these three cereals average between 90 and 130 kg N ha^{-1}. Despite N fixation ability, legumes too receive N inputs ranging between 0 and 60 kg N ha^{-1}. However, within the same zone rice would have received 75 to 125 kg N ha^{-1} (IFA, 1992).

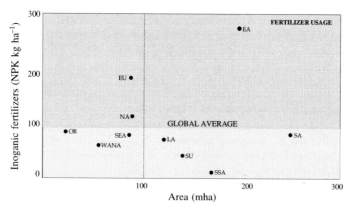

Fig. 1.3 Nutrient input rates (inorganic fertilizers) in different agricultural zones of the world (figure drawn based on data for FAO STAT: 1999

Note: EA = East Asia, SA = South Asia, SEA = Southeast Asia, OR = Orient, SSA = Sub-Saharan Africa, WANA = West Asia & North Africa, EU = Europe, SU = Soviet Union, NA = North America, LA = Latin America). Darker zone indicates above global average inputs and lighter zones below the average.

Legumes and N Credits to Agrosphere: Nitrogen credits to soil received through nitrogen fixing legumes, shrubs and trees occurring naturally, and those via leguminous agricultural crops can be enormous to the ecosystem. Leguminous crops are cultivated in a variety of agroenvironments, under different cropping systems, sequences and rotations. During 1990s about 11 to 12% of arable land on this globe was utlilzed for cropping legumes. In addition, vast expanses of leguminous pastures too contribute N (Peoples *et al.* 1995). Globally, biological nitrogen fixation (BNF), including asymbiotic fixation is known to provide 44 to 200 Tg N yr^{-1}. Bockman (1997) considers that agrosphere derives 140 Tg N yr^{-1} through BNF. However, N derived through BNF varies depending on the agroecosystem, and various other local factors related to crops, soils, microbes and environment.

Major avenues of N loss from agroenvironments are:

a) Leaching of nitrate through percolation to lower horizons or alongwith irrigation water can be severe. Within any agroecosystem, leaching losses of N fertilizer is under the influence of several factors. N leaching is minimized if crops grown are vigorous, and rapidly established. Whereas leaching loss is accentuated if crop stand is poor or land is allowed to lay fallow. Generally, leaching losses of N is less in heavy soils, or those with poor drainage. Whereas N loss is higher in light soils with good drainage. Nitrogen applied at the beginning of the growing season leaches less compared to that introduced at later stages of the crop. Obviously, N loss through leaching is higher if rate of N input is high. Rainfall affects N loss through leaching. High rates of precipitation, and erratic distribution of rainfall pattern also enhances N leaching loss.

b) Volatilization of N as ammonia is common in many agroecosystems. Ammonia emissions from arable cereal cropping zones are modest, and are usually 1 to 2 kg N ha^{-1} yr^{-1}. There are indications that legumes too loose N as NH_3. If complex fertilizers or N and P are applied together, then, NH_3 volatilization becomes feeble. However, it may reach 15 to 20% if urea is applied (Bockman, 1997; Hagin and Tucker, 1988). The crop and prevailing agroenvironment too influence NH_3 volatilization. Loss of N through NH_3 volatilization from rice fields are generally higher, reaching upto 46 to 54% within 8 days immediately after N inputs. There are several strategies that reduce NH_3 volatilization. Obviously some of them are specific to the agroecosystem. In general, dilution increases infiltration into soil; other measures such as shifting NH_4^+/NH_3 equilibrium using acid so that pH is lowered, mulching the field, banding N fertilizers at depth, all tend to reduce NH_3 volatilization.

c) Emissions due to nitrification-denitrification can be severe. Denitrification is a microbially mediated process, wherein NO_3 is reduced to N_2O, NO_x and N_2. In a given agroenvironment, denitrification is favored by lower oxygen tension, high soil nitrate, higher temperature and easily available carbon substrate (Watson *et al.* 1998).

In addition to microbial processes, chemo-denitrification caused through strictly chemical transformations also contribute N_2O, NO or N_2 (Firestone and Davidson, 1989; Gabally, 1989). The proportion of individual gases is dependent on N flow into the system and soil oxygen tension.

Watson and Poland (1999) report that such nitric oxide emissions from agrosphere and other sources is radiatively active accounting for 2 to 4% of total greenhouse warming potential. Global estimates of NO_x emission range from 20 to 100 Tg N yr^{-1}. The soil biogenic emission of NO_x to trophosphoric N budget is said to be 4 to 20 Tg N yr^{-1} (Veldkamp and Keller, 1997; Delmas et al. 1997). Allen et al. (1996) infer that in United Kingdom, NO_x fluxes into atmosphere is increasing by 3% yr^{-1} due to imbalances in biological processes. Natural grasslands and pastures may account for 10% of total N_2O flux, while arable cropping systems account for 2 to 3% of such fluxes. It is obvious that N_2O emissions vary depending on agroecosystems, the cropping pattern and soil nutrient management procedures (Watson and Mills, 1998; Jarvis and Barraclough, 1991).

Phosphorus: Phosphorus is a major essential nutrient crucial to sustenance of crops in agrosphere. It occurs both in inorganic and organic fractions of soils, in soluble form in water (soil moisture), and in crop biomass. Phosphorus status, particularly the plant available fraction (HPO_4^-) is variable in soils of different agroecosystems. For example, the acidic infertile soils in Latin America, sandy Oxisols in Sahelian West Africa, or Alfisols and Laterites in Southern Indian dry lands are low in available P pools. Phosphorus participates in different physico-chemical and biological transformation within the soil environment. Many of them directly influence crop sustenance. To mention a few, adsorption-desorption, precipitation, chemical fixation, buffering, microbial immobilization-mineralization, all of these transformations influence P availability to different extents.

Phosphorus replenishments into agroecosystems were generally through residue recycling until 20th century. This situation changed with the advent of P fertilizers. At present, P fertilizer inputs into agrosphere are substantial. Unlike C, N or S, phosphorus deposits from atmosphere is negligible, often not considered in calculating P budgets. Mainly because, P cycle in nature confines to soils, water and crops, whereas atmospheric aspect is absent. Again, unlike N which is inexhaustible (80% of atmosphere), P deposits in lithosphere are finite. Major phosphate rock ores are distributed in USA, USSR, Morocco, Jordan and Israel. Together these deposits contribute 65% of global P fertilizer consumed within different agroecosystems. Guano and Insular P sources are common in South America and Pacific Islands. Globally, around 3.2 to 3.5 mMT of phosphate rock are mined, processed and utilized each year to augment productivity of Agrosphere (FAO STAT, 1999). The global average input is 21 kg P ha^{-1}, but intensive wetland rice belts in south and southeast Asia are impinged with 30 to 50 kg P ha^{-1}. Whereas, low productivity based pearl millet ecosystem in Sahalien West Africa utilizes <3.1 kg P ha^{-1}. Obviously, water resources, productivity and yield goals, as well as human preferences have governed P inputs in a given agroecosystem (Wood et al. 2000).

In any agroecosystem, loss of P either inherent or that applied as fertilizer occurs. There could be several causes, such as soil erosion, chemical fixation or precipitation as Fe and Al phosphates, immobilization into organic fraction or removal through harvested portion. Phosphorus held in sparingly soluble form or exchangeable surface soils, is generally termed as residual P. This residual P which is released slowly can be utilized by crops in due course. Average recovery of P fertilizer through cropping is very low, and ranges between 15 and 20%. Hence, methods that enhance agronomic use efficiency of P fertilizers are continuously sought. To quote examples, over 16,000 field trials conducted on this aspect revealed that efficiency of P inputs on rice ranged between 18 and 28 kg grains per kg P (De Datta et al. 1990). Greater details on P transformations, P inputs and concomitant productivity of individual agroecosystems are available within chapters 2 to 11.

Sulphur: Atmosphere is the major natural source of SO_2 for all the cropping ecosystems. Atmospheric S is generally present in gaseous state as SO_2 or H_2S, and as particulate SO_4. These S sources are natural in origin. However, in industrial or polluted zones, anthropogenic contribution of S in the atmosphere may reach 220×10^6 ton S per year globally (Fox and Pasricha, 1993). Sea water contains SO_4, hence air surrounding coastal agroecosystems usually have 10 times higher SO_4 than other terrestrial locations. Sulphur deposition from atmosphere/aerosol into crop land diminishes exponentially with distance away from sea coast. Sulphur received through precipitation is variable. Sulphur concentration in rain water in humid tropics is usually low (1.58 to 3.8 kg S ha^{-1}). Whereas in certain semi-arid locations, low S in soil could be attributed to meager S replenishments received through rain-water (<1.0 kg S ha^{-1}). Now compare it with locations near industrial belts in Southern Great Plains in USA. Here, the rain fall, air deposition and particulate sources of SO_4 ranged between 1.8 and 10.7 kg S h^{-1} yr^{-1} (Suarez and Jones, 1982). However, higher levels of SO_2 in the atmosphere causes acid rain and disturbs normal physico-chemical and ecological functions of soils. This leads to retarded crop productivity.

Irrigation and groundwater resources too contribute S to cropping zones. Sulphur in the streams or surface water depends on contribution of S from geologic weathering. For example, runoff from mountains in South America are known to be deficient in S (Fox and Pasricha, 1993).

Sulphur levels in irrigation water are crucial. If S is low in irrigation source, then crops respond to application of fertilizer S. Minimum levels of S required in irrigation vary with location.

In temperate North America, irrigation with river water containing 1.4 mg S L^{-1} was more than sufficient for crop production. In the intensive rice belt, irrigation water containing more than 2.7 mg S L^{-1} meets the crops requirement of S, and upper limit is supposed to be 6 mg S L^{-1}. In several other locations within the rice belt, irrigation water containing 2.8 mg S L^{-1} removed S deficiency if any. In general, irrigation water with 4 to 6 mg S L^{-1} is known to satisfy requirements of crops cultivated in most agroecosystems. There is no doubt that sulphur transactions also occur at the confluences between atmosphere and agrosphere through emissions and depositions. Similarly, interactions between hydrosphere and agrosphere (actually pedosphere), that occur mainly through S inputs from irrigation sources are crucial. These interactions are crucial because they contribute to S equilibrium and productivity of agroecosystems. Sulphur resources in agricultural soils generally occur as plant available SO_4-S, ester–S, organic–S and unidentified organic–S. Sulphur imbalances may ensue intensive cropping using only N, P, and K replenishments. Beringer and Mutert (1997) suggest that due to unpredictable leaching losses, microbiological immobilization and lack of sufficient knowledge on S dynamics at ecosystem level necessitates study of S balance at more local/regional basis. Physico-chemical transformations of S specific to different agroecosystems are discussed under individual chapters hence not discussed here. Overall, major factors that contribute to sulphur deficiency are consistent use of S-free fertilizers, decreased S emission from industrial sources, use of S-free insecticides and fungicides, reduction in organic matter recycling, and higher crop harvests (Hagin and Tucker, 1988).

Other essential elements particularly K, Ca, Mg, including the microelements Zn, Fe, Mn, Co, and B are important, because they too influence crop growth and yield formation in any agroecosystem. However, to maintain brevity of this chapter, discussions are confined to C, N, P and S. Literature on other elements are sited under each of the several agroecosystems in chapters 2 to 12 wherever appropriate.

Nutrient Budgets: During the past century and a half before it, farming communities residing in different agroecosystems have been consistently adopting methods that replenish/recycle nutrients removed through crops. These procedures have mainly involved organic manures, stover/residue recycling and inorganic fertilizer inputs. In fact, choice of cropping systems/sequences have depended on their ability to suit the crop to the soil nutrient status. Therefore, the concept of nutrient budgeting in farms is not new. By definition, an approximate, or if possible accurate idea regarding nutrient inflow, transformations, fluxes and outputs versus crops' demand in a given environment is referred as nutrient budget. While preparing such nutrient budgets for a field plot or large area or even an agro-ecozone, it should be noted that the flow of nutrients into and out, as well as intervening physico-chemical and biological transformation concerns only a fraction of the total nutrient wealth of the ecosystem. However, data based on such inflow/outflow is more crucial while drawing inferences. Onema and Heinen (1998) point out that despite forming fertilizer schedules on as much thorough knowledge about nutrient dynamics in soils, and matching them with crop demands, seldom does the equation balance. Within intensive farming belts (eg. wheat in Europe and North American Great Plains; rice-wheat in Indo-Gangetic belts, rice in Southeast Asia) nutrient inputs often exceed outputs through crop harvests. Whereas, in more extensively cropped zones (eg. wheat or barley in West Asia, pearl millet in Sahelian West Africa) inputs, if any, may insufficiently match crop removals. These situations induce nutrient imbalances in the agroecosystem. Plus, there are no known avenues whereby, such deficiencies/excesses are buffered, or even out locally and quickly to rectify the situation. Such nutrient imbalances may persist for long durations, leading to declining productivity. Let us consider an example related to persisting nutrient deficiencies. Within the Indo-Gangetic belt, during the 1950s, nitrogen deficiencies were common which was overcome through fertilizer N inputs. Next, in the 60's N, P and S deficiencies appeared, and since the 70's, all three major and several micronutrient deficiencies have occurred. It necessitates yearly replenishments of many nutrients (Sundaram, 2001). Such situations are attributed to mining of soil nutrients with a total disregard to local field or agroecosystem level nutrient budgeting. In general, accuracy of nutrient budgets are basically dependent on ecological objectives and economic significance.

The complexity of the agroecosystem, and our understanding of the nutrient cycle, actually determine the accuracy levels possible within nutrient budgets (Onema and Heinen, 1998). A wide range of biases and errors may creep into nutrient budgets. Sampling errors/biases can alter nutrient recommendations, estimation on nutrient recovery and removals. For example, errors related to fertilizers may range between 1 and 3%, that for organic manure ranges from 10 to 20%. Atmospheric depositions could be estimated with an error of 10 to 30% imbedded into it. Among outputs, leaching, runoff and volatilization loss figures may come with 50 to 200% error (Onema and Heinen, 1998). Nutrient input related errors range between 5 and 15% and those for output ranges from 10 to 20%. Overall, nutrients and their inputs, transformations, fluxes, removal into crop, losses etc. are key aspects that influence sustenance and pro-ductivity of all agroecosystems to a certain extent.

6. Summary

Agrosphere is a term used for explaining global cropping activity, and excludes in its purview the natural vegetation within the biosphere. Presently, the agrosphere occupies 35% of global land resources. We are not sure of the limits to its spread. Perhaps, it could be debated, tested, simulated, forecasted as accurately as possible, and acted upon with great care. Demarcation of agrosphere firstly may allow both, naturalists and agricultural specialists to study their topics with a better perspective, and inferences could be more accurate. In the context of nutrient dynamics, the physico-chemical transformations, rates of nutrient turnovers, nutrient threshold levels stipulated and productivity attained may not apply equally to both agrosphere and natural vegetation. Agrosphere is a useful concept in many other ways. It allows policy-makers to judge cropping systems better and channel infrastructure, fiscal and material inputs (e.g., nutrients, water, pesticides etc.) appropriately.

Agrosphere exists in constant interaction with other ecospheres. Its confluences with hydrosphere, atmosphere and pedosphere are frequent, intricate, and crucial for its own sustenance and productivity. Hence, disturbance to such ecosystematic equilibria can affect cropping activity and deteriorate ecosystematic functions. For example, excessive NO_3 inputs to enhance citrus production in Florida, may contaminate groundwater/aquifers. Modern techniques, such as fertigation (drip system) localizes nutrients into upper horizons and avoids groundwater contamination. In this case confluences between pedosphere, hydrosphere and agrosphere were disturbed, which could be rectified to a certain extent. We may quote innumerable examples on this aspect. Essentially agronomic measures that reduce deterioration of soils, irrigation sources and atmosphere are to be upgraded periodically. For example, an improvement in N-use efficiency firstly increases productivity, then reduces chemical fertilizer N usage, delays groundwater contamination, and reduces harmful emissions. Similarly, methods that improve carbon sequestration can improve soil quality. On the other hand, agrosphere contributes 23% of glass house/global warming effects through gaseous emissions, and 40% of the methane emission. In this regard, all agroecosystems are not equally culpable. Cereal belts in Southeast Asia, Europe and North America are impinged with high levels of nutrients, but other agrozones utilize moderate or low levels of chemical fertilizers.

Productivity of individual agroecosystems differ. Among them tropical forests at 1.3 kg m^{-2} yr^{-1} are most efficient biomass converters. Cereal/legume-based agroecosystems average only 0.3 kg m^{-2} yr^{-1}. Intensification of these agroecosystems through higher nutrient inputs, water and better genotypes may proceed further. Limits to such intensification should be deciphered. Again, crops which are the central pieces of agrosphere are being genetically modified, and new genotypes regularly introduced. Such introductions may affect nutrient dynamics significantly. Prior knowledge about it, developed through experimentation and simulations can be useful.

Overall, agrosphere plays a crucial part in sustaining and nourishing human beings and domestic species. Potentially higher productivity is a clear possibility in most agroecosystems. Projections indicate that demand for food grains, and other products will be higher in the next quarter century, and this needs to be achieved either through expansion and/or intensification.

It has been clearly stated in the Preface, that agrosphere includes several aspects other than just nutrient dynamics, ecology and productivity dealt in this book. For example, understanding the dynamics of pests, diseases or economics is equally important. However, in the next ten chapters of this book, emphasis is greatest on nutrient dynamics within prairie-based cereal and legume agroecosystems. Then, a chapter on citrus in Florida and two on forestry have been included to represent plantation ecosystems.

REFERENCES

Allen, A.G., Jarvis, S.C. and Headon, D.M. (1996). Nitrous oxide emissions from soils due to inputs of nitrogenous excrete return by livestock on grazed grassland in the U.K. *Soil Biol. Biochem.*, **28**: 597-607.

Alcamo, J., Henrichs, T. and Rosch, T. (2000). World Water in 2025—Global Modeling and Scenario Analysis for the World Commission on Water for 21st century. Report A0002. University of Kassel, Kassel, Germany.

Amthor, J.S. (1995). Terrestrial higher-plant response to increasing atmospheric CO_2 in relation to the global carbon cycle. *Global Changes Biol.*, **1**: 273-279.

Bates, R.L. and Jackson, J.A. (1980). Glossary of geology. American Geological Institute, Falls Church, Virginia, USA.

Batjes, H.H. (1996). Total carbon and nitrogen in soils of the world. *European Journal of Soil Science,* **47**: 151-163.

Bayer, L. (2001). Soil geography and sustainability of agriculture. *In:* Soil fertility and crop production. K.R. Krishna (ed.), Science Publishers, Enfield, New Hampshire, USA, pp. 33-64.

Beringer, H. and Mutert, E. (1997). Sulphur nutrition of crops with special reference to K_2SO_4. International Rice Research Institute, Manila, Philippines, 352 pp.

Bockman, O.C. (1997). Fertilizers and biological nitrogen fixation as sources of plant nutrients: Perspective for future agriculture. *Plant and Soil*, **194**: 11-14.

CGIAR (2000). Consultative Group on International Agricultural Research, Rome, Italy, http://www.cgiar.org.

Delmas, R., Serca, D. and Jambert, C. (1997). Global inventory of NO_x sources. *Nutrient Cycling in Agroecosystems,* **48**: 51-60.

De Datta, S.K., Biswas, T.D. and Charoenchamratcheep (1990). Phosphorus requirements and management in low land rice. *In:* Phosphorus Requirements for Sustainable Agriculture in Asia and Oceania. International Rice Research Institute, Manila, Philippines, pp. 307-323.

De Willigen, D. (1986). Supply of soil nitrogen to the plant during the growing season. *In*: Fundamental ecological and agricultural aspects of nitrogen metabolism in higher plants. H. Lambers, J.J, Nicolson and I. Stulen (eds.), Martinus Nijhoff Publishers, Dordrecht, Boston, pp. 417-432.

Doll, P. and Siebert, S. (1999). A digital global map of irrigated areas. Report No. A9901. Center for Environmental Systems Research, University of Kassel, Kassel, Germany.

Eswaran, H., Van den Bergh, E. and Reich, P. (1993). Organic carbon in the soils of the world. *Soil Science Society America Journal,* **57**: 192-194.

FAO (1991). How good the Earth? Quantifying land resources in developing countries: FAO's agroecology series. Food and Agricultural Organization, Rome, Italy, 32 pp.

FAO (1995). Food and Agricultural Organization of the United Nations. Digital Soil Map of the World (DSMW) and derived soil properties. Version 3-5 CD-ROM.

FAO (1998). The state of the world's plant genetic resources for food and agriculture. Food and Agricultural Organization of the United Nations. Rome, Italy.

FAO (1999). Food and Agricultural Organization of the United Nations land and water digital media series. The soil and terrain database for Latin America and the Caribbean. CD-ROM, FAO, Rome, Italy.

FAO/IIASA (1999). Food and Agricultural Organization of the United Nations and International Institute of Applied Systems Analysis, Global Agro-ecological zoning. FAO Land and Water Digital Media. Series No. 11.

FAO STAT (1999). Food and Agricultural Organization of the United Nations. Statistical databases. Online at: *http://apps.fao.org*.

FAO STAT (2000). Food and Agricultural Organization of the United Nations. Statistical databases. Online at: *http://www.fao.org/waicent/agricult/landuse-e.html*.

Firestone, M.K. and Davidson, E.A. (1989). Microbiological basis of NO and N_2O production and consumption in soil. *In:* Exchange of trace gases between terrestrial ecosystems and the atmosphere. M.O. Andrae and D.S. Schimel (eds.), John Wiley and Sons, Chichester, England, pp. 1-21.

Fox, R.L. and Pasricha, N. (1993). Plant nutrient sulphur. *Advances in Soil Science,* **13**: 209-267.

Galbally, I.E. (1989). Factors controlling NO_x emission from soils. *In*: Exchange of trace gases between terrestrial ecosystems and atmosphere. M.O. Andrae and D.S. Schimel, (eds.), John Wiley and Sons, Chichester, England, pp. 23-37.

Gerard, S. (2000). Fundamentals of Soil Science. Routledge Publishers, New York/London.

Hagin and Tucker (1988). Fertilizers for dryland and irrigated soils. Springer Verlag, Heidelberg, 385 pp.

Harte, J. (1998). Atmospheric Chemistry. University Science Books, Bill Valley, California, 283 pp.

Holland, E.A., Braswell, B.H., Lamarque, J.F., Townsend, A., Sulzman, J., Muller, J.F., Deuter, F., Brasseur, G., Levy, H., Penner, J.E. and Roelofs, G.J. (1997). Variations in the predicted spatial distribution of atmospheric nitrogen deposition and their impact on carbon uptake by terrestrial ecosystems. *J. Geophys. Res.,* **102**: 15849-15866.

Houghton, R.A. (1996). Terrestrial sources and sinks of carbon inferred from terrestrial data. *Talleas Ser B.,* **48**: 420-432.

Houghton, J. (1997). Global warming: The complete briefing. Cambridge University Press, Cambridge, U.K.

Houghton, R.A., Davidson, E.A. and Woodwell, G.M. (1998). Missing sinks, feedbacks, and understanding the role of terrestrial ecosystems in the global carbon balance. *Global Biogeochemical Cycles,* **12**: 25-34.

Houghton, J.T. and Hackler, J.L. (1995). Continental scale estimates of the biotic carbon flux from land cover change: 1850 to 1980. online at *http://cdiac.esd.ornl.gov/ndps/ndp050.html*.

Huggat, R.J. (1995). Geo-ecology. Routledge, London/New York, pp. 3-27.

IFA (1992). World fertilizer use manual. International Fertilizer Industries Association. Paris, France.

IPCC (1995). Intergovernmental panel on climate change—Radiative forcing of climate change and an evaluation of the IPCC 1592 Ericsson scenarios. J.T. Houghton *et al.* (Eds.) Cambridge University Press, Cambridge, U.K., 337 pp.

IPCC (2000). Intergovernmental panel on climate change. *In*: Land use, land use change and forestry. R.J. Watson, I.R. Noble, B. Bolin, N.H. Ravindranath, D.J. Verarda and D.J. Dokken (eds.), Cambridge University Press, Cambridge.

Izzarulde, C.R., Rosenberg, N.J. and Lal, R. (2001). Mitigation of climatic change: By soil carbon sequestration: Issues of sience, monitoring, and degraded lands. *Advances in Agronomy,* **70:** 2-62.

Jarvis, J.C. and Barraclough, D. (1991). Variation in mineral nitrogen under grazed grassland swards. *Plant and Soil,* **138:** 177-188.

Jarvis, J.C., Stockdale, E.A., Shepherd, M.A. and Powlson, D.S. (1996). Nitrogen mineralization in temperate agricultural soils: Process and measurement. *Advances in Agronomy,* **57:** 188-237.

Kindt, R., Muasya, S., Kimotho, J. and Waruhice, A. (1997). Tree seed suppliers—N sources of seeds and microsymbionts. International Center for Agro-forestry, Nairobi, Kenya.

Koah, G.W. and Mooney, H.A. (1986). Carbon di-oxide and Terrestrial Ecosystem. Academic Press, San Diego, CA, USA.

Krishna, K.R. (2002). Nutrient dynamics in Agroecosystems *In*: Soil fertility and crop production. K.R. Krishna (ed.), Science Publishers, Enfield, New Hampshire, USA. pp. 387-410.

Kumar, K and Goh, K.M. (2000). Crop residues and management practices: effects on soil quality, soil nitrogen dynamics, crop yield and nitrogen recovery. *Advances in Agronomy,* **68:** 197-320.

Kurz, W.A., Apps, M.J., Benkewa, S.J., and Lakstrum T. (1995). The 20th Century carbon budget of Canadian Forests. *Tallus Ser,* **47:** 170-177.

Lal, R. (1995). Erosion-Crop productivity relationship for soils of Africa. *Soil Science,* **165:** 57-72.

Lal, R. (2000). Soil Management in the Developing Countries. *Soil Science,* **165:** 57-72.

Lal, R., Kimble, J.M., Follett, R.F. and Cole, C.V. (1998). The potential of US cropland to sequester carbon and mitigate the green house effect. Ann Abor Press. Chelsea, MI, USA.

McGuire, A.D., Melillo, J.M., Kicklighter, D.W., and Joyce, L.A. (1995). Equilibrium responses of soil carbon to climate change: Empirical and process based estimates. *J. Biogeography,* **22:** 785-796.

Melillo, J.M., McGuire, A.D., Kicklighter, D.W., Moore, B., Vorosmarluy, C.J. and Schoss, A.L. 993. Global climate change and terrestrial net primary production. *Nature,* **363:** 234-240.

Melillo, J.M., Prentice, I.C., Ferquehas, G.D., Schulze, E.D., and Sala, O.E. (1996). Terrestrial biotic responses to environmental change and feedbacks to climate. *In*: Climate change 1995. Houghton *et al.* (eds.), Cambridge University Press, pp. 445-481.

Odum, E.P. (1984). Properties of agroecosystems. *In*: Agricultural ecosystems: unifying concepts. R. Lawrence, B.R. Stinner, and G.J. House (eds.), John Wiley and Sons-Interscience, New York, pp. 3-12.

Onema, O. and Heinen, M. (1998). Uncertainties in nutrient budgets due to biases and errors. *In:* Nutrient disequilibria in Agroecosystem. E.M.A. Smaling, O. Oenema, and L.O. Fresco (eds.), CAB International, Oxford, pp. 75-98.

Pearson, C.J. (1992). Field Crop Agroecosystems. Elsevier, Amsterdam, 576 pp.

Peoples, M.B., Herridge, D.F. and Ladha, J.K. (1995). Biological nitrogen fixation: An efficient source of nitrogen for sustainable agricultural production. *Plant and Soil,* **174:** 3-28.

Persely, G.J. (2000). Agricultural biotechnology and the poor: Promethean Science. *In*: Agricultural Biotechnology and the Poor—proceedings. Consultative Group on International Agricultural Research, Washington, D.C., pp. 3-24.

Postel, S. (1993). Water and Agriculture. *In*: Water in crisis: A guide to the world's fresh water resources. P.H. Glick (ed.). Oxford University Press, Oxford, England, pp. 56-66.

Postel, S., Daily, G.C. and Elrich, P.R. (1996). Human appropriation of renewable freshwater. *Science,* **271:** 785-788.

Rosenzweig, C. and Hillel, D. (2000). Soils and global climatic change: challenges and opportunities. *Soil science,* **165:** 47-56.

Sanchez, P.A., Conto, W. and Bhol, S.W. (1982). The fertility capability soil classification system: Interpretation, application and modification. *Geoderma,* **27:** 283-309.

Schimel, D.S., Entiny, I.G., Heimann, T.M.L., Wighy, D., Raynaud, D., Alves, D. and Siegenthaler, U. (1995). CO_2 and the carbon cycle. *In*: Climate change—1994. D.S. Houghton, L.G.M. Filho, J. Bruce, H. Lee, B.A. Callander, E. Haites, N. Harris, and A. Maskcell (eds.), Cambridge University Press, New York, pp. 76-86.

Schlesinger, W.H. (1990). Evidence from chronosequence studies for a low carbon storage potential of soils. *Nature,* **348:** 232-234.

Schlesinger, V.H. (1999). Carbon sequestration in soils. *Science,* **284:** 2095.

Seckler, D., Amarsinghe, U., Molden, D., de Silva, R. and Barker, R. (1998). World water demand and supply, 1990 to 2025: Scenarios and issues. Research Report No. 19. International Water Management Institute, Colombo, Sri Lanka.

Shand, H. (1997). Human nature: Agricultural biodiversity and farm-based food security. Rural Advancement Foundation International, Ottawa, Ontario.

Smith, C.W., Buol, S.W., Sanchez, P.A. and Yost, R.S. (1997). Soil fertility capability classification system (FCC). 3rd approximation: the link between pedologist and soil fertility manager. Cited in PAGE—Agroecosystems. S. Wood, K. Sebastian and S. Scherr (eds.), International Food Policy Research Institute, Washington, D.C. p. 93.

Stern, D.I. and Kaufmann, R.K. (1998). Annual estimates of global anthropogenic methane emissions: 1860 to 1994. US Department of Energy, Oak Ridge National Laboratories, online: *http://cidac.esd.ornl.gov/trends/meth/ch4.html.*

Sundaram, K.P. (2001). *Fertilizer News,* New Delhi, **46**(4): 1-120.

Suarez, E.L. and Jones, U.S. (1982). Atmospheric sulphur as related to acid precipitation and soil fertility. *Soil Sci. Soc. Am. J.,* **45:** 87-90.

Townsend, A.R., Brasswell, B.H., Holland, E.A., and Penner, J.E. (1996). Spatial and temporal patterns in terrestrial carbon storage due to deposition of fossil fuel nitrogen. *Ecol. Appl.,* **6:** 806-814.

UN (1997). United Nations, comprehensive assessment of the freshwater resources of the world. Stockholm: Commission for Sustainable Development, Stockholm Environment Institute. Section 37.

USGS EDC (1999). United States Geological surveys earth resources observation systems (EROS) data center. 1 km land cover characterization database revision for Latin America., Sioux falls, SD: USGS EDC/(cited in Wood *et al.* 2000).

Veldkamp, E. and Keller, M. (1997). Fertilizer-induced nitric oxide emissions from agricultural soils. *Nutrient Cycling in Agroecosystems,* **48:** 69-77.

Watson, C.J. and Mills, C.L. (1998). Gross nitrogen transformations in grassland soils as affected by previous management intensity. *Soil Biology and Biochemistry,* **30:** 742-753.

Watson, C.J. and Poland P. (1999). Change in the balance of ammonium—N and nitrate—N content in soil under grazed grass swards over 7 years. *Grass and Forage Science,* **54:** 248-254.

Watson, C.J., Stevens, R.J., Steen, R.W.J., Jordan, C. and Lennox, S.D. (1998). Minimizing nitrogen losses from grassland. *In*: 71st Annual report—1997/98. Agricultural Research Institute of Northern Ireland, pp. 49-60.

Webster, R. (1997). Soil resources and their assessment. *Philosophical Transactions of Royal Society of London B.* **353:** 963-973.

Wiesler, F. (1997). Agronomical and physiological aspects of ammonium and nitrate nutrition of plants. *Z. Pflanzenernater./*Bodenk. **160:** 227-238.

WMO (1997). World Meteorological Association, Comprehensive assessment of the fresh water resources of the world. Geneva: World Meteorological Association..

Wood, S., Sebastian, K. and Schess, S.J. (2000). Pilot analysis of global ecosystems—Agroecosystems. International Food Policy Research Institute, Washington, D.C., 110 pp.

Young, A. (1998). Land resources: Now and for the future. Cambridge University Press, Cambridge, U.K., pp. 57-62.

CHAPTER 2

The Temperate Wheat Agroecosystem in Great Plains of North America
Nutrient Dynamics, Ecology and Productivity

> *The Great Plains of North America are a living example of more recently developed agricultural practices and expansion of agrosphere, attained predominantly through human ingenuity. Agriculturally, the greatness of 'Great Plains of North America' is perhaps immeasurable. Particularly, its effects through introduction of new crops from Europe, Asia and the Orient into the New World. During the twentieth century, progress in productivity of this agroecosystem has been remarkably rapid. High nutrient inputs, large farming units, improved crop genetic stock, high mechanization and proportionately higher harvests have become characteristic of this part of the agrosphere. These great plains supply food grains, to not only population residing within its expanse, but also to those in other continents of the globe. Long-term upkeep of this agroecosystem and its equilibrium is a necessity. Equally so, a better understanding and good control over nutrient dynamics is crucial to sustain the high-productivity of this agro-belt in future years.*

CONTENTS

1. Introduction
   A. Soils and cropping sequence
   B. The wheat agro-climate
2. Nutrient Dynamics
   A. Nitrogen dynamics
   B. Crop residues, soil carbon and other nutrients
   C. Phosphorus dynamics during wheat culture
   D. Physiological genetics and nutrient dynamics
3. Concluding Remarks

1. Introduction

The Great Plains of North America have hosted a wide range of both native food crops and those introduced by settlers. However, the Great Plains are famous world-wide for wheat production, and their fertile Mollisols which dominate the landscape from Canada in the North to Texas in the South, then Colorado in the West to Tennessee/Ohio in the East. During the past century, wheat-based cropping schemes and agronomic procedures adopted by the farming community have been dynamic. For example, tillage procedures have changed from maximum tillage in 1920s to conventional in the 60s, then to the present reduced till on no-till systems. Essentially, these changes have aimed at deriving better water and nutrient efficiency. Nutrient inputs into this agroecosystem have increased steadily, mainly to support farmers' wheat yield aspirations. Overall, water and nutrients have regulated intensification and productivity of this agroecosystem. In this chapter, an introduction on soils, cropping sequence and agroclimate is followed by detailed discussions on nutrient dynamics. The intention is to provide information on recent developments, and greater insights regarding this temperate wheat agroecosystem (Fig. 2.1).

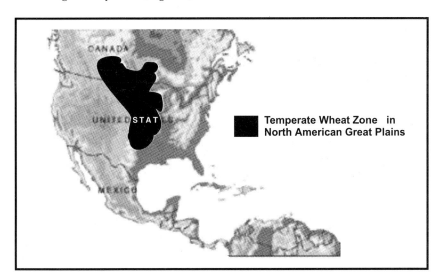

Fig. 2.1 The Temperate Wheat Agroecosystem of North American Great Plains (Shaded area represents approximate location and expanse of the cropping belt. Demarcation is not to scale)

A. Soils and Cropping Sequence

The major soil types encountered in the wheat belt of the Great Plains of North America, including Pacific Northwest and Canadian Prairies are:

- Mollisols (Borolls, Udolls, Ustolls, Ustorthents, Fluvents) which are spread across as large tracts in Central and Northern plains, occupying nearly 25% of the total area;
- Alfisols (Ustalfs, Ustolls) are predominant in Central and Southern plains, and cover nearly 18% of the plains zone;
- Aridisols (Argids), Entisols (Psamments, Aqults), and Ultisols are frequented in Western, Northeastern and Southern plains respectively.

Soils in the Central plains are neutral to alkaline in reaction, owing to calcareous parent material at shallow depths, and low precipitation levels. In comparison, soils in the Pacific Northwest, Midwest and Eastern plains are acidic due to leaching (Bowman and Halvorson, 1998). Although, the plain soils are characterized as Mollisols, their fertility as measured by soil organic matter (SOM) content, cation exchange capacity (CEC), and nutrients is generally low. Soil specialists attribute this to soil degradative processes, traditional clean-till procedures and wheat-summer fallow system (Campbell and Paul, 1978). Under native prarie or cultivated wheat, the Northern plain soils have high base exchange capacity, possess Ca and Mg in sufficient quantities, and are fairly rich in organic matter (4 to 8%). Continuous cultivation has reduced soil C to 40 to 70% of the original levels (O'Hallaran *et al.* 1987). Soil bulk density (D_b) changes due to continuous wheat farming have been conspicuous. For example, at soil depths of 0 to 7.5 cm increasing N inputs resulted in decreased bulk density. Such effects are accentuated whenever fractions of wheat biomass returned to soil increased and N applications were incessant, say 90 kg N ha^{-1} each season (Halvorson *et al.* 1999a). Soil D_b at depths below 7.5 cm were unaffected despite 11 years of wheat cropping and N inputs. In general, such changes in D_b through soil organic matter allow better discing, easier water infiltration and nutrient movement in the plow layer, and induces higher rooting. Essentially, soils become more arable and better in quality (Herrick and Wander, 1998). However, Clean—till accentuates loss of surface soil. Subsequent tilling cycles mix soils in lower layers that possesses higher pH and low SOM. Therefore, it results in a buffering effect. In certain areas, farmers in Great Plains do face soil acidity problems, such as P fixation, Al toxicity, reduction in root growth, and herbicide ineffectiveness etc. Mineral nutrient deficiencies are frequently discernible in annual cropping systems, than under wheat-fallow system. It is attributable to shorter time lapse with annual cropping, whereas wheat-fallow systems allow sufficient period for mineralization process, and for nutrients to become available. Long-term evaluations of tillage effects indicate that among nutrient deficiencies, P is less affected (Rasmussen and Douglas, 1992).

Tillage and Soil Quality: Crop-fallow system with conventional tillage practices have been leading to deterioration of soil quality in Great Plains (Wienhold and Halvorson, 1998). During fallow, soil deteriorates, looses quality, becomes vulnerable to wind and water erosion. Also, the soil is held in an oxidative state. These processes result in loss of organic C and mineral nutrients. Generally, as incidence of fallow decreases, loss of soil and organic C decreases (Biederbeck *et al.* 1984). Therefore, conventional systems are being increasingly discarded as not sustainable. More intensive management strategies of wheat production in Great plains include, reduced or no-tillage, fertilization to meet a yield goal, and annual cropping systems (Norwood and Dhuyvetter, 1993; Halvorson and Black, 1996). Such conservation tillage combined with annual cropping improves several soil quality attributes. Firstly the water-use efficiency improves (Peterson *et al.* 1996). This modern practice then enhances organic C, conserves N, therefore induces higher microbial numbers and activity (Linn and Doran, 1984; Karlen *et al.* 1994; Campbell *et al.* 1999). Wienhold and Halvorson (1998) have stated that 'soil quality is becoming as large a public concern as air and water quality in Great Plains'. Traditional farming practices, such as crop-fallow systems with conventional tillage actually are leading to a deterioration in soil quality. Whereas, increases in C or N are attributable to enhanced, recycling of crop residue during annual cropping (Table 2.1; Havlin *et al.* 1990). Higher fertilizer doses applied to meet the yield goal, also induce higher biomass (stover) which gets recycled into soils. To quote an instance, the no-till and reduced-till systems caused 25% and 14% increases in organic C respectively over conventional—till (CT) wheat fallow system. Such increases in organic C represent 1000 kg organic $C/ha^{-1}/3yr^{-1}$. In the CT plots, bulk of the C lost was in top 2 cms layers, attributably to erosion and decomposition (Halvorson *et al.* 1994a, b)

Table 2.1: Soil chemical quality attributes as a function of tillage in the annual cropping system

Attribute	Conventional tillage	Minimum tillage	No-tillage
Total N (Mg ha^{-1})	1.5 ± 0.1	1.6 ± 0.1	1.8 ± 0.08
Organic C (Mg ha^{-1})	21.9 ± 1.5	22.5 ± 1.6	25.8 ± 1.20
Inorganic N (kg ha^{-1})	8.4 ± 0.7	8.3 ± 0.8	10.4 ± 0.70

Source: Wienhold and Halvorson, 1998; Mg = mega gram

Wheat-based Cropping Sequence: Historically, cropping sequences and the ensuing nutrient dynamics in the Great Plains have fluctuated, depending on a range of factors related to crop introduction by new settlers, soil and environmental conditions, economics and human preferences. Nutrient dynamics, particularly nutrient inputs in the Great

Plains changed enormously during early mid 20th century, owing mainly to use of chemical fertilizers. Within the limitations imposed by soil and agro-climate, fertilizers allowed changes and/or intensification of cropping systems. Consequently, productivity levels were raised and corresponding nutrient equilibria were achieved.

The conventional wheat-fallow (WF) system with stubble mulch has been in vogue in the Great Plains since several decades. It continues to be practiced in several locations. More recently, minimum or no-till intensive wheat culture, which may include corn, sorghum, sunflower, proso-millet or forages in sequence have been gaining in area. Typical intensive cropping sequences replacing the traditional WF system in the central plains are wheat—corn—fallow (WCF), wheat-sorghum-fallow (WSF) and wheat-prosomillet-fallow-(WPMF) (Westfall *et al.* 1996). In the Great Plains region, moisture and nutrients (N and P) are the major constraints. With reference to water, minimum-till (MT) or no-till (NT) crop production systems enhances precipitation storage during non-crop periods. Therefore, it enables intensive cropping, say 2 crops in 3 years, or 3 crops in 4 years or annual cropping (Halvorson *et al.* 1997). Intensification of cropping sequence and adoption of no-till system, normally depends on availability of nutrients and moisture at reasonable levels to attain optimum productivity. For example, intensification through inclusion of sunflower in the sequence is highly feasible because, soil-water and nutrients localized at depths below that scavenged by wheat is utilized by sunflower efficiently. Evaluations in Central Plains by Halvorson *et al.* (1994a; 1997, 1999a) suggested that sunflower scavenged soil NO_3–N efficiently, leaving only very low levels of soil NO_3–N at lower depths. Sunflower also added diversity with acceptable productivity levels both under minimum-till and no-till systems.

B. The Wheat Agro-climate

Weather data from over 80 years, as summarized by wheat experts indicate that in Great Plains, the average growing season precipitation (GSP) varies between 95 and 290 mm depending on location. The annual flux through precipitation ranges between 150 and 600 mm (Halvorson and Black, 1985b; Nielson and Anderson, 1993; Rasmussen *et al.* 1997; Halvorson *et al.* 1999; Weinhold and Halvorson, 1999).

In the North American Great Plains, water and its efficient use is the driving variable that influences productivity and the underlying nutrient dynamics in soil, as well as the agroecosystem in general (Peterson *et al.* 1996). The corollary, that management of crop residue, soil nutrients and their availability, in turn influences efficiency of water-use by crops is also true (Nielson and Anderson, 1993). Considering it chronologically, the summer fallows became a common practice to conserve moisture received through precipitation around 1920's. That is when improved ploughing through powered tractors and good control over weeds, started supporting higher fallow efficiency. Since then, summer fallows have been widely used to maintain a higher water-use efficiency (WUE) in this intensified agroecosystem. Variations in tillage practices during fallows, such as conventional-shallow discs, maximum tillage, stubble mulching, and minimum tillage have differential effects in terms of fallow efficiency, precipitation use efficiency, nutrient usage and productivity of wheat. However, most modern fallows adopt no-tillage, least disturbance to soil, and enhanced residue cover. Since 1970's, efficiency of such fallows have reached 40%, but variations in soil water storage are to be expected depending on other variables such as location, soil type, precipitation levels, farming systems and fallow tillage.

The agro-climatic parameters at or around wheat harvest time influences fallow efficiency. Mainly because, during that period soil profiles are only partially charged with water. Subsequent precipitation during the actual fallow period may be less efficiently stored (Peterson *et al*. 1996). Knowledge regarding the underlying soil and agro-climatological factors that influence fallow efficiency and precipitation-use efficiency is crucial. Peterson *et al*. (1996) lists several such factors, that include:

(a) better interception of rain and steady infiltration;
(b) decreased evaporation caused by cooler temperatures under better mulches;
(c) reduced wind speed at surface because of residue cover;
(d) improved fertilizer and nutrient status;
(e) reduction in opportunistic weed control; and
(f) improved wheat cultivars.

Timing of fallow operations are equally crucial in enhancing soil water storage capacity. Actually, not one or couple of them, but complex interactions among different factors, acting at different proportions both in time and space govern the fallow efficiency and precipitation-use efficiency during wheat culture in Great Plains (Peterson *et al*. 1996; Halvorson *et al*. 1997). The precipitation-use efficiency is best under annual cropping systems (80%), followed by spring wheat-winter fallow rotation (55%) and least with spring wheat-fallow system (40%).

Tillage is an important aspect of wheat-fallow system because it directly affects water storage. The no-till system in vogue during modern times is superior to reduced till or a plowing exercise during fallow. The percentage precipitation storage efficiency generally increases as tillage intensity decreases. On a comparative scale, precipitation storage efficiency was highest at 49% with no-till system > reduced-till (40%) > stubble mulch (32%) > rod weeder (27%) > dust mulch (24%) in that order in Central

Plains (Nielson and Anderson, 1993). The WUE of wheat fallow system, with no-till practiced in the North American Great Plains ranged between 78 and 203 lb acre in^{-1}. Locations in Montana averaged 180 to 200 lb acre in^{-1}, while in Texas it was lowest at 78 to 128 lb acre in^{-1}. Peterson et al. (1996) observed that there has been no significant increase in WUE since 1970s, but believe that modifications such as 2 wheat crops in 3 years with 10 to 11 months fallow, rotation with sunflower will improve the situation. To quote an example, the three-year rotations increased WUE in every subclimate regime from Texas to Kansas to Colorado. The WUE consistently averaged at 196 lb acre in^{-1} (Peterson et al. 1996). A comparison of WUE of various cropping systems revealed that wheat-fallow system averaged 143 lb acre in^{-1}, but wheat-corn-fallow averaged best at 226 lb acre in^{-1}. Continuous cropping at Akron in Colorado, which included winter wheat, spring barley and corn improved WUE further to 273 lb acre in^{-1}.

As a bottom line, we may realize that individual crop (or genotype), and its ability for adaptation to an agro-zone too affects WUE. Depending on each location within Great Plains, the WUE of wheat varies when compared with other crops such as barley, sunflower, corn, millet or sorghum. For example, spring wheat (216 lb acre in^{-1}) in North Dakota, or corn (at 215 lb acre in^{-1}) possessed better WUE than wheat at 138 lb acre in^{-1}, when grown in Central Plains (Colorado). Similarly, sorghum grown in Southern Plains (Texas) had better WUE at 188 lbs acre in^{-1} than wheat at 87 lb acre in^{-1}. As summarized by Peterson et al. (1996), the average WUE of wheat in various locations of North American Great Plains including Canada ranged from 78 to 236 lb acre inch^{-1}. In general, this intensified wheat-based cropping system has improved farmers' ability to harness precipitation better and achieve higher productivity.

2. Nutrient Dynamics

Discussions on nutrient dynamics, and factors that influence it form the core of this chapter on wheat culture in Great Plains. It is agreed that knowledge about dynamics of all essential nutrients is important. However, considering the importance bestowed by the farming community and trends with wheat scientists, discussions are confined to salient aspects. They are nitrogen dynamics, carbon recycling and soil quality, phosphorus and other nutrients. In addition, effects of changes in crop physiological aspects on nutrient dynamics have been included.

A. Nitrogen Dynamics

Soil Nitrogen: Both, NH_4-N and NO_3-N are available to wheat roots. NO_3-N is mobile and translocates freely with water in dissolved state. Therefore, it is easily absorbed by roots. At the same time, NO_3-N percolates fast into soil layers below the root zone. Although NH_4-N is less subject to leaching, loss through volatilization and denitrification can be severe, if soil temperatures reach above 10°C in the Great Plains (Grant *et al.* 1996). A 12-year evaluation in Northern Plains by Halvorson *et al.* (1999b) revealed that soil NO_3-N retention varied depending on the type of tillage practices. The no-till treatment consistently retained less soil NO_3-N (9 kg N ha^{-1}), compared with conventional or minimum tillage system (137 and 132 kg N ha^{-1} respectively). Significant tillage versus moisture interactions, higher N removal by crop, soil erosion and precipitation level variations, are commonly listed as causes for lowered soil NO_3-N with modern no-till system. These observations tally with the general opinion that intensive cropping systems are needed for efficient N and water exploitation (Black *et al.* 1981).

Nitrate Leaching: Avoiding nitrogen leaching and undue accumulation at lower horizons has been a major concern with researchers dealing with wheat in Great Plains. Recently, Shock *et al.* (2000) have remarked that transitions form low input, irrigated farming of mid-20th century towards intensive, high input supported wheat production was actually accompanied by irrigation induced soil erosion and nutrient loss through excessive leaching. Further, Westfall *et al.* (1996) have cautioned that traditionally, low fertilizer inputs (40 kg N ha^{-1}) and low rainfall pattern in Central Plains, actually causes us to believe that nitrate leaching towards lower horizons, i.e., the Vadose zone or groundwater could be negligible. However it may not be so. The chemical N forms used by farmers in Great Plains, mainly ammonium salts and urea are oxidizable to nitrate, which leaches easily to lower horizons and into ground water. High inputs common with intensive farming, combined with furrow irrigation too induces excessive N leaching risks. In fact, excessive use of furrow irrigation accentuates N leaching and its percolation. Therefore, Bruch (1986) suggested that it is not surprising to find ground water contaminated with nitrate and other herbicide residues. Isotope based studies indicate that ^{15}N loss from soils in Northern Plains in Canada varied depending on sub-regions and soil types. Black, grey luvisolic and sodic soils allowed higher ^{15}N leaching, than the black and dark brown soils of southern Alberta. ^{15}N mass balance indicated that 10 to 90% loses from fall applied fertilizer N, could be partly attributable to N leaching. Long-term trials of over 7 years, clearly showed residual nitrate-N accumulation at lower horizons. Analyses indicate that W-F system causes higher leaching, hence N accumulations at Vadose zone were also high

(120 kg N ha^{-1}). Intensification of cropping through WCF or WCMF system is a method to reduce leaching losses of N. Peterson and Westfall (1994) observed that under such intensified cropping in Western Plains, the nitrate-N leaching to lower horizons were much low at 61 to 88 kg N ha^{-1}. Clearly, intensification reduces fallow period, and enhances roots' ability to scavenge soil-N, thereby avoiding its movement to, and accumulation at lower horizons.

Leached NO$_3$-N may reach and accumulate in ground water. If it happens, then ground water quality deteriorates. Chances of such leached nitrate-N accumulation in ground water increases, whenever farmers shift from conventional tillage (WF) to no-till wherein precipitation storage ratio is higher. However, variations in leaching depending on locations are to be excepted. Generally, adoption of no-till intensive cropping reduces ill effects of N-leaching even in higher fertilizer input situations (Westfall *et al.* 1996). Straw mulching or incorporation, as it increases soluble C in soils, results in N immobilization. It therefore, reduces N leaching to lower horizons (Kochler *et al.* 1987). Shock *et al.* (2000) are of the view that reduction of irrigation-induced soil erosion, runoff and nutrient leaching have mainly come through adoption of furrow mulching, sediment ponds, filter strips, laser leveling furrow-irrigated fields or drip irrigation methods. Obviously, these modern techniques, firstly reduce N-leaching, therefore modify the N dynamics favorably.

Nitrogen Mineralization: In the Great Plains, change from crop-fallow to intensive SW-WW-SF cropping systems needs appropriate modifications in N management procedures. During this effort detailed knowledge regarding nitrogen mineralization rates and factors affecting it, helps in improving the efficiency of N fertilizer schedules. In addition, to set up suitable yield goals for intensive cropping systems, quantification and appropriate mathematical adjustments pertaining to previous N inputs made, its timing, and extent of N mineralized from crop residues etc. are required. However, accurate quantification of some of these variables may not be easy. Nitrogen mineralization rates may vary depending on location within Great Plains, the specific environmental parameters, soil, cropping sequence practiced etc. As an example, let us consider N mineralization rates and factors affecting them in Northern Plains. Wienhold and Halvorson (1999) reported that the average N mineralization rates were superior with intensive cropping systems SW-WW-SF than the conventional SW-F, and this was irrespective of tillage systems followed or nitrates applied (Table 2.2). Obviously, in this case cropping system imposed a predominant effect on N-mineralization. We may search for causes for it. Commonly, the quantity of residue recycled, its quality (C : N ratio), soil moisture, temperature, fallow period, microbial diversity and a few other factors are quoted as reasons. However, crop sequences

Table 2.2: Nitrogen mineralization rates (kg ha^{-1} week^{-1}) in soil from 0 to 0.5 m depth, supporting spring wheat-fallow (SW-F) and Spring wheat-winter wheat-fallow (SW-WW-F) systems in Northern Great Plains under three tillage systems.

Cropping systems	Conventional Tillage	Minimum Tillage	No-Tillage
SW-F	5.0 ± 1.5	6.3 ± 1.3	5.6 ± 1.0
SW-WW-F	7.6 ± 0.9	9.1 ± 1.1	13.1 ± 2.1

Source: Excerpted from Weinhold and Halvorson, 1999.

were not a major factor influencing N-mineralization rates in Southern Great Plains. It was actually attributed to high intensity of cropping and raised soil temperatures (Weinhold and Halvorson, 1999). Within SW-WW-SF systems, N-mineralization rates at 0 – 0.5 m soil depth were higher under no-till system (NT), compared with either minimum tillage (MT) or conventional tillage (CT) system. Influence of nitrate fertilization on N-mineralization seems variable. Instances, wherein N-mineralization increases are easily discernible due to N inputs have been reported for crop-fallow systems (Campbell *et al.* 1991). Whereas, in Southern Plains, Franzleubbers *et al.* (1994; 1995) found no such effect of N fertilization on N-mineralization rates.

The quantity of crop residue, its C : N ratio and succulence seem crucial to N-mineralization rates possible. Nitrogen is freely translocated within plant parts. Much of it converging into storage sinks-seeds. This, N stored in grains is removed at harvest and residue low in N contents are recycled into soil. Such residues with low C : N ratio, upon incorporation will induce N immobilization. Hence, low N-mineralization rates may be encountered after cereal residue incorporation. The cropping system induced changes in N-min rates spread upto 0-0.15 m depth in the soil profile, compared with no effect due to N application. Whereas, the tillage effects confine to 0.05 m depth. Therefore, Weinhold and Halvorson (1999) suggest that eliminating fallow may be a better way of managing long term soil N fertility in this wheat belt. Under WW-F system in the Central Plains (Nebraska), the net N-min that occurred in 14 months from August to next year September, that is during fallow up to wheat planting, amounted to 40 kg N ha^{-1}. It was 25 kg N ha^{-1} for WCF system during the 10-month fallow period. There occurs a break in N-min in both the cropping systems, due to extreme cold and freeze during the winter (Rouppet, 1994). However, percentage of total N mineralized during the summer fallow period, just before wheat planting seems same, in both the no-till systems—i.e., WF and WCF (Rouppet, 1994).

Intensification of wheat production in the Great Plains and adjoining agrozones necessitates high N fertilizer input to satiate crop demand. It, in turn, stipulates precise estimates of N supplied by native soil. So that, we may prevent over or under fertilization, and their effects on plant nutrition and soil quality. Under such situations, information from long-term assessments of N mineralization may be useful in devising appropriate agricultural practices. With such goals in view, Rasmussen *et al.* (1998) made long-term evaluation of N mineralization as influenced by N inputs, tillage, cropping systems etc. They found that soil mineralization rates were 32, 42 and 51% for wheat-fallow, wheat-pea and wheat-wheat sequences respectively. Then, stubble mulches supported 10 to 20% more N mineralization compared to board-plowed soils. Nitrogen mineralization rates were actually directly proportionate to N inputs in soil and/or tillage intensity. Previous N inputs and accumulations in soil (Fig. 2.2) and increasing soil organic contents also influenced N mineralization favorably. Therefore, it implies that long-term soil and crop management strategies affect soil N mineralization potential, and they should be considered appropriately when estimating fertilizer requirement (Rasmussen *et. al.* 1998; Rasmussen and Rohde, 1998).

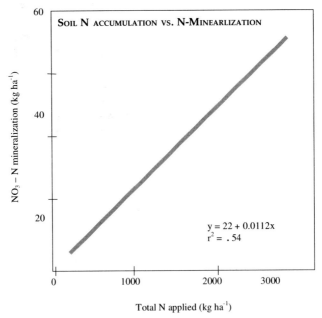

Fig. 2.2 Previous N accumulation in field soil (0–20 cm depth) under wheat-fallow sequence 1942–1990, and its influence on N mineralization in North Western USA
Source: Redrawn from Rasmussen *et al.* 1998

Nitrogen Requirements/Inputs: In the Central Plains, need for N fertilization was minimal during 1950's and 1960's, mainly because of inherently higher levels of available mineralizable N generated during summer fallows (Greb *et al.* 1967, 1974). However, with intensive cropping and lower N mineralization capacity, additional N inputs became mandatory. While summarizing an 8-year evaluation in Colorado, Halvorson and Reule (1994) detected that to obtain acceptable yield (3000 kg^{-1} grain ha^{-1}), 67 kg N ha^{-1} were required, and nitrogen applied beyond 90 kg N ha^{-1} lead to increase in residual N. Normally, maximum residual N accumulations could be detected at 30 to 90 cm dept in the soil profile. On less fertile Ardic Argic Ultisols in Colorado, wheat productivity at 5.5 bu acre^{-1} required 100 kg N ha^{-1}, irrespective of cropping sequence. Whereas, for a similar 100 kg N ha^{-1} input a more productive Kieth clay loam supported 6.4 bu acre^{-1} wheat under W-F system, and 12.8 bu acre^{-1} under WCF system. Further, a five-year evaluation of W-F and intensive WCF systems revealed that removal of N into grains reached 75% or greater under WCF. In yet another study, in Kansas (Central Plains), wheat crop required between 45 and 90 kg N ha^{-1} for maximum productivity. High need for N coincided with higher rainfall year or zones (Westfall *et al.* 1996). Long-term analyses indicate that optimum N rates depended much on the grain levels. During the years 1963 to 1990, when wheat productivity ranged from 3.5 to 5.0 t ha^{-1}, optimum N rates fluctuated between 30 and 90 kg ha^{-1}, where as optimum rates were 90 to 135 kg N ha^{-1} when grain productivity was higher around 5 to 7 t ha^{-1} (Rasmussen, 1996). Primarily, nitrogen inputs for winter wheat depends on yield goals. For example, in Northern Plains N inputs have ranged between 22.5 and 45 kg ha^{-1} for an yield goal 20 to 40 bu acre. Whereas, 67 kg-95 kg N ha^{-1} is suggested if expected yield is 60-90 bu acre^{-1} (Oplinger *et al.* 1998; Kelling *et al.* 1998). Such standardizations are derived after careful considerations on SOC, soil fertility (residual N), crop genotype and physiological aspects, environment etc.

Soil-Plant N Buffering—A Concept: Inorganic N buffering in soil profile is a concept suggested in recent years (Raun and Johnson, 1995). It differs from the traditionally accepted version, that if N inputs exceed plants' demand, the remaining N is either volatilized, leached or is available as residual NO$_3$-N. Instead, in the new concept, a soil-plant inorganic buffer system is proposed. Based on a 20 year soil-N trial, Raun and Johnson (1995) infer that inorganic (NO$_3$ + NH$_4$) N does not begin to accumulate until N input exceeds N requirement for maximum yield by 22 to 55 kg N ha^{-1} yr^{-1}. In one of the examples, nearly 62 kg ha^{-1} yr^{-1} was required to achieve maximum yield, but to discern a change in residual N, an input of 85 kg N ha^{-1} yr^{-1} was needed (see Fig. 2.3). Thus, an inorganic N buffer of 23 kg N ha^{-1} yr^{-1} develops each year. Further, Raun and Johnson (1995) enlist a few mechanisms by which an inorganic N buffer seems to establish in soil. These relate to excessive N removal by plants, denitrification, ammonia volatilization etc. Although, data to support this concept—i.e., 'inorganic N buffer' in soil are available (Halvorson and Reule, 1994), there is a feeling that it needs further evaluations. In a jist, Westfall *et al.* (1996)

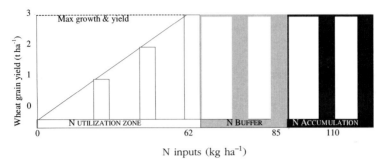

Fig. 2.3 Soil-plant N-buffering during wheat culture in Great Plains of United States of America. *Source:* Raun and Johnson, 1995.
Note: For the example quoted here N max is 62 kg ha^{-1} yr^{-1}, N application is 85 kg N ha^{-1} yr^{-1}. This will have an yearly N-buffer (N buffer = N max—N acm) of 23 kg N ha^{-1} yr^{-1}

pointed out that, under this plant inorganic-N buffer concept, 'excessive N' means N amounts larger than the sum of the crop demand plus buffer N. Obviously, the buffer size, the level of N at which maximum yield is attained, and N input rates at which accumulations are possible should be expected to vary with the crop genotype, soil and environmental parameters in the Great Plains.

Fertilizer N Recovery: During the recent cropping history in the Great Plains, fertilizer-based N inputs have been greater than crop N removal. This practice of higher N input is based on the assurance of maximum yield, and not precise or economic rates. Further, fertilizers were inexpensive, hence nutrients were lavishly replenished and enriched in soil. However, in more recent times, the intensive wheat culture has changed from liberal nutrient inputs and water supply, towards precise nutrients and their amounts, accurate placement that closely match requirements, and net nutrient recovery rates from fertilizer (Shock and Steiber, 1991). In general, the more intensive, no-till wheat based cropping systems in Great Plains require N inputs in quantities excess compared with conventional crop fallow systems. Let us now consider a few previous studies on fertilizer N recovery by wheat crop. Field trials on spring wheat, conducted in early 1980's utilized ^{15}N labeled $(NH_4)_2 SO_4$ to understand the N dynamics in no-till SW-F system. Nearly 36 to 39% of the fertilizer N was garnered by plants. Whereas, 36 to 42% was retained at 0 to 60 cm soil profile. When a winter crop followed in the same field, another 3 to 6% ^{15}N was traced in plants (Frederickson *et al.* 1982). Whereas, if labeled ^{15}N crop residue from spring wheat was incorporated into an unlabeled field and winter wheat cultivated in it, then 8 to 11% ^{15}N was recovered. Variations in the extent of ^{15}N recovery are obvious, and they depend on factors related to soil, crop, environment and fertilizer formulation. For example, ^{15}N recovery in plants as reported by Olson and Swallow (1984) was 26 to 33%, while, 45 to 55% remained in soil, and 16 to 23% was lost from the soil system—perhaps through volatilization and/or leaching. In yet another study by Campbell and Paul (1978), soil retained 34% of fertilizer N, grain and straw accumulated 49%, roots retained 3% and 13% was either lost or accounted for by error. In a different trial, Power *et al.* (1986) surface applied ^{15}N fertilizer formulation and traced 16 to 18% in wheat grains, and an additional 3-6% in second wheat crop under a no-till SW-WW-F system. In a more recent 4-year study, Porter *et al.* (1996) accounted for nearly 90% of ^{15}N fertilizer inputs. Nitrogen losses perceived were attributed to denitrification and/or NH_3 volatilization. Of the accounted N, 24 to 28% remained in soil within 0 to 60 cm profile. Nitrogen uptake by crops was mainly from the soil NO_3 pool. Recovery of N from labeled wheat residue was 6 to 7%, which is indicative of N mineralization. Follett *et al.* (1995) observed that, ^{15}N incorporated into soil as residues from previous crops was slowly assimilated into microbial biomass and localized itself within 0 to 10 cm soil profile. After due consideration to previous data, and other recent investigations, Porter *et al.* (1996) summarized that, under no-till intensive cropping sequences, the applied fertilizer N is conserved in organic forms through the soil biological process and assimilation of NO_3 in plants. It is followed with residue recycling and slow decomposition at surface, which then gets assimilated in soil microbial biomass. In summary, ^{15}N studies in no-till SW-F sequence in Great Plains suggest that between 24 and 49% fertilizer N could be recovered by wheat crop, 24 and 55% may be retained in soil, and between 3 and 12% could be lost due to error in experimentation, leaching, NH_3 volatilization or denitrification.

Fertilizer Placement, Soil Temperature and N Dynamics: Soil temperatures obtained at the time of nutrient placements are crucial to soil N dynamics and its recovery by roots. With NT, soil temperatures during spring are lower by 5° to 10°F compared to those obtained with CT. If N and/or P fertilizers are banded closely to plant roots, it enhances nutrient use efficiency of no-till system. Most no-till seed drills have built in capacity to place fertilizers close to seeds. For example, at 55°F placement of N and P at 2, 5 or 8 inches away from seed influenced nutrient recovery, but at 77°F placement effects were nullified, because of rapid root proliferation. There are numerous reports regarding comparisons of banding versus broadcasting fertilizer nutrients during wheat culture in North American Great Plains. Banding nutrients close to seeds or even below seeds have often proved better than broadcasting. Banding results in higher nutrient scavenging and recovery, reduced nutrient loss from the root zone, better productivity and nutrient use efficiency. To confirm it, a long term (10-year) evaluation reported by Kochler (1987) indicates that banding, on an average produced 10 bu acre^{-1} higher than broadcasting with spring wheat. Also, such productivity increase through banding increased yield by 4 bu acre^{-1} for winter wheat. Yearly variations in nutrient recovery and consequent productivity advantages possible through banding, are to be excepted because of soil, crop and environmental interactions. Sometimes shift from CT to NT may also require a change in fertilizer disbursement methods. Fertilizer placement options may even get limited, hence physical location of crop residues and fertilizer based nutrients will then need alterations.

Moisture Versus N Recovery: Whether in short or longer durations, N recovery by wheat depended on the level of growing season precipitation (GSP). If GSP are grouped, the mean N recovery during above normal precipitation years was always higher at 81 kg ha^{-1}, compared with 74 kg N ha^{-1} during near normal years, and 70 kg N la^{-1} during below normal years. Interestingly, precipitation effects superceded those due to other introduced factors such as tillage, soil depth etc. Obviously, N versus soil moisture interactions are crucial to total N recovery and yield formation by wheat, or any other cereal grown in sequence in western USA (Rasmussen and Rohde, 1991; Halvorson et al. 1999a, b, c; Table 2.3).

Table 2.3: Influence of precipitation levels on average N recovery and yield formation by wheat in Northwestern United States of America during the period 1963-85

Precipitation category (GSP)	N-recovery (kg ha^{-1})	Wheat grain yield (t ha^{-1})
Above normal (147 mm)	85	5.5
Near normal (88mm)	68	4.1
Below normal (61mm)	62	3.5

Source: Rasmussen and Rohde, 1991
Note: GSP = Growing season precipitation. Average precipitation received in above normal years (Fallow + Winter crop) were 458 mm and 339 mm respectively; that for near normal years were 363 mm and 283 mm; and that for below normal years were 317 mm and 291 mm. The 55-year average GSP was 105 mm.

Nitrogen × Phosphorus Interactions: Long-term analysis of phosphorus and nitrogen interactions has revealed a positive significant effect on wheat grain productivity, under both W-F and annual cropping systems in the Northern Great Plains (Halvorson and Black, 1985b). Under annual cropping systems, P is often a less limiting nutrient than N, hence, no discernible P effects will be possible without first ensuring adequacy of N in soils. For example, a mandatory 45 to 90 kg N ha^{-1} input ensures productivity gains as a consequence of P application under annual cropping systems. Long term analysis of N × P interactions revealed a positive significant effect on wheat grain productivity, under both wheat-fallow and annual cropping system in the Northern Great Plains (Halvorson and Black, 1985b).

N-use Efficiency: Around different locations in Colorado, the fertilizer N-use efficiency averaged 27% for the traditional W-F and 43% for intensive WCF on highly productive Kieth soils. The WCF system utilized 0.5 kg N per 27 kg grain (wheat or corn), whereas W-F utilized 0.64 kg N per 27 kg grain (wheat). Clearly, in this case, intensification of cropping system enhanced fertilizer N-use efficiency (Kolberg *et al.* 1993). Such studies on N-use efficiency in Great Plains are indeed numerous. Several of them deal with factors that influence N-use efficiency and ways to improve it (see Raun and Johnson, 1999).

Legume N Credits to Wheat: In Great Plains, areas with comparatively high humidity, where plant available water is sufficient for annual cropping, forage legumes are inserted into crop sequence. It may improve soil productivity, particularly soil N fertility through symbiotic N fixation (Halvorson *et al.* 1997; Norwood, 2000). Effects of such legumes grown in sequence on nutrient dynamics are mainly associated with increased N supply. Therefore it lowers C : N ratio and improves soil quality. The extent of N credits derived from legumes vary widely between 40 and 170 kg N ha^{-1} (Kelling *et al.* 2000). Such N credits are dependent on soil fertility management practices in different locations, legume species, nutrient and water status of legume crop at harvest and incorporation, extent of rooting, nodulation and N fixation potential, as well as residual N levels attainable in soil. Most commonly utilized legume cover crops in the Great Plains are *Medicago, Trifolium, Mellitus, Vicia, Lens, Phaseolus* and *Pisum*. Obviously, legume N credits in soil benefits the succeeding cereal wheat. Such N credits will partially satisfy the N needs of wheat. However, some researchers believe that N credit advantages may be offset by increased use of soil moisture for green manure production. In summary, insertion of legume into the wheat-based cropping sequence does influence N budgets, and dynamics in general.

Residual Nitrogen: Soil testing, a basic component in deciphering fertilizer N requirements actually assesses and quantifies residual soil NO_3-N. Both, surface and subsoil nitrate is accounted. In addition, most tests ascertain soil organic matter (SOM) component, because N-mineralization during the growing season contributes to soil NO_3 pool. Thus, residual available soil NO_3-N is to be measured when N mineralization is complete during the fallow, and just before planting to arrive at a correct status of residual-N in wheat fields. Westfall *et al.* (1996) pointed out that a mistiming of residual-N sampling/measurement to June 1st can overlook nearly 28-36 kg N ha^{-1}, which goes undetected because of N release through mineralization, that occurs after June, but before planting in September (Lamb, et al. 1985; Roupett, 1994). Halvorson and Reule (1994) reported that residual N which percolates into the soil profile accumulates greatest at 60 to 90 cm depth, but such accumulations vary as a function of annual N input rate. For example, annual N input at 135 kg N ha^{-1} for 6 consecutive years (1984 to 1990) induced an accumulation of residual N equivalent to 160 kg ha^{-1} at soil depths between 60 and 90 cm.

Long-term N Effects on Soil Chemical Constitution: A shift from conventional tillage (CT) and SW-F, to intensive, no–till has necessitated use of enhanced levels of ammoniacal (NH_4^+) fertilizer. This affects the chemistry of surface layers of soil differently. For example, under no-till practices, surface soils with low pH are not mixed with lower horizons of comparatively higher pH. Therefore, neutralization of acidity caused by NH_4^+ or urea based N inputs is minimal. Volatilization as NH_3-N is much less under no-tillage system. It is both, interesting and useful to know the longterm effects of incessant nitrogen inputs under intensive cropping systems. Also, basing on it, alternative N management strategies can be devised. Soil acidification is a natural process. It accelerates with farming activity and as C, N, or S cycles set in. On a long run, excessive NO_3-N input acidifies the soil, mostly in the range between pH 4.5 to 5.5. For example, long-term (9-year) evaluation in Central Plains revealed that for an increase in N input rate from 0 to 112 kg N ha^{-1}, the soil NO_3-N levels increased from 10 to 65 kg N ha^{-1}, soil organic matter raised from 378 mg kg^{-1} to 657 mg kg^{-1}, and pH dropped from 6.8 to 5.5. Concurrently, Ca contents decreased from 8.2 to 5.3 c mol kg^{-1} (Bowman and Halvorson, 1998).

It is believed that acidification of soils in Great Plains under no-till system is mainly achieved by two factors: (a) lack of soil mixing and oxidation of ammonium ions; (b) hydrogen ions are produced when organic-N and NH_4–N are transformed to NO_3. Decomposition and mineralization may also release organic acids in the surface soils. Under such acid conditions, Ca and Mo availability is affected. In nine years, that is 3 crop cycles, NH_4^+–N and NO_3 additions increased organic residue re-incorporation leading to enhanced SOC. Incessant N inputs has also increased Al and Mn concentration. Increased SOC may chelate and reduce Al and Mn ill effects only to a certain extent. Increase in residue recycling enhances CEC. In summary, Bowman and Halvorson (1998) and Halvorson et al. (2001) conclude and caution, that shift to no-till will mean 50 to 100 kg N extra input each year. Therefore, appropriate soil management strategies that minimize disturbance to soil chemical equilibrium need to be thought over and used.

Weeds and N Management: The modern no-till W-C-F system allows weed growth and nutrient diversion. Hence, competition for nutrients applied into soil ensues. Broadcasted N is equally accessed by weeds or wheat roots. Nitrogen dynamics is then decided by the weed infestation versus rate of wheat root proliferation. However, appropriate fertilizer placement can reduce such N loss to weeds. For example, N placement below the wheat seeds or band placement avoided N diversion to the main weed-out plants, better than broadcasted N. Nitrogen recovered by weeds reduced from 50 kg ha^{-1} to 44 kg ha^{-1} due to banding. Simultaneously, the main crop-wheat recovered 64 kg N ha^{-1} when broadcasted (Kochler *et al.* 1987). Gain in wheat grain productivity due to placement/banding of N, over broadcasting equaled 8 kg N ha^{-1}.

B. Crop Residues, Soil Carbon and Other Nutrients

A certain level of accumulation of soil organic carbon (SOC) is essential for long-term sustainability of agricultural crop production. Within mid-Western Great Plains, greater amount of manuring and C inputs have occurred on maize compared with wheat. Although C inputs via residue recycling is on the increase, it has not been dramatic because of the changes in the wheat plant stature. That is, semi-dwarfing and higher harvest indices have reduced stover portions. A decline in soil C can retard crop productivity. Hence, during the recent years, there has been greater interest in studying the soils as a major component sequestering C in the wheat belt of United States of America. Essentially, soil then acts as a sink to atmospheric CO_2 (Rasmussen and Albrecht, 1998). Major factors affecting carbon sequestration are (a) crop rotation, (b) manure inputs, (c) crop residue recycling, (d) tillage, (e) land conversion from pastures to crop land (Hao *et al.* 2001). Residue incorporation enhances soil C levels. Repeated analyses at different locations in Great Plains suggest that such increases in Soil C are linearly related to extent of residue recycling or fresh inputs. (Rasmussen *et al.* 1980; Havlin *et al.* 1990; Paustian *et al.* 1992).

Certain summarized equations derived by Rasmussen *et al.* (1980) suggest that nearly 5 t crop residue ha^{-1} yr^{-1} is required to maintain SOC at a steady level under wheat-fallow system. Sometimes high levels of residue inputs may hamper normal nutrient dynamics in soil, and become determental to wheat productivity. Such effects are commonly mediated through greater N immobilization, weed related competition for nutrients, low early season soil temperatures (2 to 6°C) that limits rooting etc. All of these aspects decrease nutrient scavenging ability of the wheat crop (Rasmussen *et al.* 1997).

Wheat residue management also affects physiologically, the crop development, its tillering, root-proliferation. All of which ultimately determine productivity. Under no-till (NT) system, flattening the stover rather than leaving it standing in the field improves plant development, nutrient absorption and utilization. Further, burning the straw under NT system increased tillering and productivity. In certain situations, the lowered productivity response to N inputs under NT system has been attributed to poorer tillering due to standing stover and stubbles (Rasmussen, 1993).

The influence of wheat residue burning on soil nutrients, its quality and crop growth has been termed controversial. According to Rasmussen and Rohde (1988), the short-term benefits related to nutrient availability, reduction in N immobilization, weeds and propagules of pathogen, may at times be offset by long-term detriment to soil quality. Stubble burning induces N and S volatilization. On soils already deficient for N, this effect could be pronounced. Burning removes carbonaceous residue, therefore reduces microbial immobilization of soil and fertilizer N. Clearly, burning alters C and N mineralization patterns. Higher stover yield, if burnt leads to higher nutrient availability—therefore better vegetative growth. Conversely, in non-burn conditions there is greater N-immobilization early in the season, with eventual release later at anthesis, through mineralization during the crop season. In that case non-burn situation is beneficial (Wagger *et al.* 1985). In summary, factors affecting SOC in the semi-arid temperate Great Plains and Canadian prairies are the level of C input through residues and/or manure, frequency of fallow and crop season. Currently, adoption of no-tillage, seems the only good way to conserve soil C.

Carbon and Nitrogen Interrelations: Stover from wheat or other cereals have high C : N ratio. Upon incorporation into soil, such straw reduces amount of N available to crops via immobilization. Such reduction of N is accentuated whenever C-based energy source is sufficient for enhanced microbial activity (Kochler *et al.* 1987). As a corollary, when N was added, more of soluble C gets utilized by microbes. Hence, as a consequence, less of soluble C is leached to lower horizons. Actually, a complex interaction ensues *in situ* among factors such as manures, tillage practices and organic C component in soil. Fresh, succulent green manures are rich in water and relatively low in nutrients. This necessitates high rates of manure application. Composting, which lowers moisture content actually concentrates nutrients. Still, their availability to plant roots may be low.

Schlegal (1992) reports 13% N-use efficiency for compost, but 36% for chemical fertilizers. Fresh manure or compost enhances soil organic matter content, but the extent may vary. Linear increases in SOC depending on manure rates up to 29 t ha^{-1} have been reported (Sommerfeldt *et al.* 1988). In some cases SOC increases have been marginal even for manure inputs at 22 to 45 t ha^{-1}. Whereas, Schlegal (1992) reported that SOM increased by 14% through application of only 16 t ha^{-1} manure for three years. Intensive tillage works in opposite direction on SOM contents in soils. Nearly 50% of SOM could be lost through intensive cropping for a few years. Havlin *et al*. (1992) state that intensive tillage induces loss of top soil, and with it SOM depreciates. Productivity decreases were upto 19.3 kg C ha^{-1} m^{-1}. Halvorson *et al*. (1997) opine that best option to conserve SOM is to suppress excessive SOM loss, and simultaneously enhance C input via recycling residues. In wheat-fallow systems, soil C declined by 2% after 12 years under no-till practices, but it depreciated by 15% if maintained under conventional tillage for same duration. Clearly, conventional tillage reduces SOM, whereas no-till preserves it.

Carbon Sequestration: In the Great Plains, change to no-till, intensive cropping system needs appropriately higher nutrient inputs. The need for N fertilization is accentuated due to low mineralization capacity of these soils (Kolberg *et al*. 1996). Further, higher inputs result in proportionately greater quantities of crop residues that can be recycled through incorporation. Nitrogen fertilization also influences SOC content and total soil N in the NT-annual cropping system (Jones *et al*. 1997; Rasmussen and Smiley, 1997; Campbell *et al*. 1997; Halvorson *et al*. 1999c). During a 11 year study, the total N uptake, as well as biomass and crop residue returned to soil increased with increasing N rates. On an average, 28% of the total N removed by crop was returned to soil as residue. Such increase in crop residue based N, on being returned to soil localized itself and accumulated at 0 to 15 cm depth. Nitrogen fertilization often induces higher levels of SOC within cropping systems (Rasmussen and Rohde, 1988). To quote an example, annual accumulation of SOC at 45 kg N ha^{-1} input was 94 kg C ha^{-1} yr^{-1}, and it was 182 kg C ha^{-1} yr^{-1} at 134 kg N ha^{-1} input into the system. Greatest accumulation of SOC was confined to surface layers between 0 and 7.5 cm depth (Follet *et al*. 1997; Huggins *et al*. 1998; Halvorson *et al*. 1999c).

Carbon sequestration in soil is an important aspect of cropping in Great Plains, because of its effects on farming practices, and on greenhouse gas emissions such as CO_2 (Lal *et al*. 1998). Clearly, farming methods that enhance sequestration of atmospheric CO_2 are beneficial, hence they need to be encouraged. The no-till annual cropping system increases SOC (Halvorson *et al*. 1999b). Further, it has been very well demonstrated that N input levels have positive effects on SOC accumulation. However, C-sequestration efficiency will vary depending on soil microbial processes, root growth and general productivity of the wheat cropping system. To summarize, Halvorson *et al*. (1999a) opine as well profess that managing no-till cropping system for optimum productivity with sumptuous N supply to soil will have positive environmental effect. Mainly, a good N fertility program helps in sequestering atmospheric CO_2 into SOC, via increased productivity. The crop residue, upon returning to soil allows C storage. Such positive effects of N-fertilization on SOC have been demonstrated through a long-term, annual cropping system under no-till conditions (Halvorson and Reule, 1999; Jauzen *et al*. 1998).

Carbon Sequestration and Microbial Dynamics: Understanding microbial dynamics, in relation to changes in soil nutrient status has great relevance, especially while deciding on the management strategies that restrict soil fertility deterioration (Collins et al. 1992). In general, the fallow season in Great Plains supported lower levels of microbial activity and population, when compared with wheat, or wheat-pea, or wheat-wheat rotations, or grass pastures (Table 2.4). It seems obvious, that wheat roots, or those of peas or grass provided the necessary soil conditions, for better proliferation of microbes compared with unplanted fallows. Wheat root exudates rich in carbohydrates, proteins/amino acids, growth promoting factors, plus organic acid that alter root zone pH or chelators, aid higher microbial activity. From their analysis, Collins et al. (1992) suggest that dynamics of microbial populations was well reflected in microbial biomass, N, C:N ratio, C% or N% changes monitored during W-F or annual cropping systems (Table 2.5). Also, nitrifying microbes multiplied to higher population levels under annual cropping system, compared to those with wheat-fallow system.

Table 2.4: Differences in soil microbial populations induced by long-term wheat-based cropping sequence practiced from 1931 to 1991.

Microbial flora	Wheat-fallow rotation		Annual cropping	
	in fallow	in wheat	Wp + WW	Grass pasture
Total Bacteria ($\times 10^6$)	34.7	78.3	87.5	79.7
Fungi ($\times 10^4$)	11.0	11.4	33.3	8.6
Actinomycetes ($\times 10^4$)	2.4	2.0	5.9	6.0
Nitrosomonas ($\times 10^4$)	7.4	8.3	30.0	96.0
Nitrobacter ($\times 10^4$)	1.8	3.5	4.5	16.3

Source: Collins et al. 1992
Note: Microbial counts are average of three observations made during November to May under each treatment. Microbial counts are expressed as organisms g^{-1} dry soil. WP = wheat-pea rotation; WW = wheat-wheat rotation

Table 2.5: Soil microbial aspects (0.20 cm depth) as influenced by wheat-fallow, annual crop rotations, and continuous grass pastures in the adjoining areas of Western Great Plains—a 58-year average from long-term trials

Microbial component	Wheat-fallow	Annual cropping	Grass pasture
Biomass C (mg C kg^{-1})	321	424	1731
Biomass N_2 (mg N kg^{-1})	52	120	192
Biomass C/N	20	16	18
Biomass C (% of soil C)	4	6	9
Biomass N (% of soil N)	6	10	11

Source: Excerpted from Collins et al. 1992

C. Phosphorus Dynamics during Wheat Culture

Phosphorus is one of the limiting plant nutrients in the glacial till soils of Northern Plains. It is equally so, in Mollisols and Alfisols of Central, Southern and Western Plains. Obviously, knowledge regarding dynamics of P in soil, and within this wheat ecosystem is crucial. Phosphorus is an important nutrient input in Great Plains, because it ensures adequate productivity of wheat (Ozanne, 1980). However, P sufficiency is not common, instead excessive P fixation by calcareous soils, the top soil erosion through wind and water, P depletion through incessant cropping create deficiency of this element. Phosphorus deficiency can be rampant in localized zones. For example, soil tests reveal that 73% of soils in the Central plains of Colorado, 51% in Kansas and 60% in Nebraska exhibit dearth for P. Further, in view of intensification of wheat cropping system, there is need to optimize P inputs, assess short-term and long-term effects of P fertilizers on soil P status, the extent of residual P accumulated, P fertilizer efficiency and then relate it to wheat productivity. O'Halloran *et al.* (1987), remarked that excessive inputs of crop residue and litter during no-till, intensive cropping system, makes it less possible for appropriate prediction of available P levels, inorganic or organic P pools. However, more recent evaluations by Bowman and Halvorson (1997) suggests that under more intensive cropping systems, P recycling through residue and litter is an important mechanism resulting in additional pools of plant—available P. Additionally, both total and organic P in soil increased with cropping intensity. Actually higher levels of organic residue recycling supports better accumulation of organic P in soil (Selles *et al.* 1999). Obviously, fallows with meager cropping/vegetation recycled negligible amounts of organic residue, hence organic P pools were least improved.

Phosphorus Recovery: Irrespective of levels of P, for a single dose of P input, wheat crop grown in Central Plains recovered between 4 and 13 kg P ha^{-1} in grains. When averaged for 10 consecutive cropping seasons from 1968 to 1983, 8.0 kg P ha^{-1} was recovered into grains by each crop, and approximately 70 to 80 kg P ha^{-1} was removed by 10 wheat crop cycles (Halvorson and Black, 1985a, b). On an average, 4.0 g P were recovered into one kg wheat grain. Next, during the 17 years of evaluation in the Central Plains, fertilizer P recovery was < 50% and rarely exceeded this limit (Table 2.6). Further, it was concluded that a one time broadcast application of P fertilizer at rates as high as 90 kg P ha^{-1} seems efficient, and meets the crop needs for this element.

Table 2.6. Long-term—total P uptake, and fertilizer P recovery by wheat crop (10 crops) in Central Plains

P input (kg P ha^{-1})	Total P uptake (10 crops) kg P ha^{-1}	Recovery (%)
0	62	—
22	76	62
45	86	53
90	102	44
180	117	30
Mean	89	47

Source: Halvorson and Black, 1985
1. Nitrogen supplied at 45 kg N ha^{-1}

Long-term wheat-fallow or annual cropping (i.e., wheat-millet-corn or wheat-millet) have generally depreciated available P levels. Such decline seems to occur irrespective of levels of P replenishments. In an experiment involving 12 crops in sequence, plant available P depreciated from 44 mg P kg^{-1} soil to 12 mg P kg^{-1} soil during the first 7/8 cropping seasons (Halvorson and Black, 1985a). The rate of P depreciation then platued between 7 and 12th cropping season. In terms of P dynamics or residual P effects, it then becomes clear that P replenishments are needed after a few cropping seasons, despite a single large P fertilizer input. Residual P levels stabilized around 6 to 10 mg P kg^{-1} soil as wheat cropping proceeded. After giving due weightage to local soil and environmental variations, P recovery and depletions rates in this wheat ecosystem may be expected to follow the above trend. Even economically, the P management strategies adopted by wheat farmers in Great Plains indicate that a one time high rate of 90 kg P ha^{-1}, either broadcasted or incorporated, resulted in long-term profitability. Also, as the initial P fertilizer input increased, the long-term net profits too increased.

D. Physiological Genetics and Nutrient Dynamics

Genetic manipulation of crops is among the major factors that has been influencing the nutrient dynamics in this extensive wheat-based cropping ecosystem. Changes in nutrient inputs, recovery, loss and commensurate biomass/grain productivity have actually been accentuated, since the introduction of semi-dwarf and high yielding wheat genotypes. Modifications in crop duration, plant stature, tillering, total biomass, harvest indices, grain weight and nutrient partitioning, rooting pattern etc., have enormously influenced the net nutrient dynamics in this wheat based ecosystem. With the use of semi-dwarfs, the extent of crop residues recycled, and alongwith it C, N, P and other nutrients have proportionately dwindled. Equally important is the net nutrient loss to the ecosystem through excessive grain harvests. Alterations, in crop duration, mostly shortening of maturity periods, has meant less period for mineralization and release of nutrients. Change in growth physiology has resulted in enhanced nutrient demand for biomass conversion. Proportionate rescheduling of higher quantities of nutrient inputs, particularly N and P has been necessitated. To quote an example, on 'Zadocks' decimal scale', wherein wheat growth stages are expressed from 00 to 100, the N split dosages are provided at GS 10 to 20, which coincides with seedling growth, then at GS 20 to 29 that is tillering, and later during GS 25 to 32 which is the stem elongation and flag leaf formation stage (Oplinger et al. 1998). Quantity of N (or other nutrients) inputs will also vary depending on components of yield, and aimed yield levels.

In accordance with nutrient input schedules, nutrient recovery too commensurately differs depending on physiological stages. Both the aspects of fertilizer N-use efficiency, that is uptake and later utilization in the tissue are largely dependent on genetic nature of wheat cultivars. In turn, it decides the extent of biomass produced, residue recycled and net grain N removed out of the ecosystem. During the past several decades, wheat breeding programs have improved the harvest indices of various genotypes, from 0.3 to 0.56. Concurrently, genetic traits related to grain yield, total N recovery (N_t), N-use efficiency (NUE) and utilization efficiency (UTE) have also been improved (Table 2.7). Ortiz-Monastero et al. (1997) have pointed that at certain level of nutrient application, say 150 kg N ha^{-1}, the genetic progress in NUE, was attributable to an extent of 30% to uptake efficiency (UPE), while UTE accounted for the rest 70%. Whereas, at an input level of 300 kg N ha^{-1}, only 37% of progress in NUE was accounted by uptake efficiency, and 63% UTE. Further, they concluded that UTE had greater relevance to N recovery by grains, whenever N inputs were in higher range (around 300 kg N ha^{-1}). However, at lower levels of N inputs (or availability), N recovery, NUE and productivity were governed better by UPE. Together with harvest index, UPE and UTE and related genetic traits can significantly influence N dynamics in this wheat belt.

Table 2.7: Genetic advance in wheat for grain yield, nitrogen use efficiency, total N recovery, N utilization and harvest index

Genotype	Grain yield (kg ha^{-1})	NUE (kg grain kg N^{-1})	N_t (kg N ha^{-1})	UTE (kg kg N^{-1})	HI
Tall variety (Yagni, 1950)	2985	26	105	31	25.7
Semi-dwarf (Opata, 1985)	5120	42	124	39	40.0
Net gain in 35 yrs	2135	26	19	8	14.3

Source: Ortiz-Monasterio et al. 1997
NUE = Grain dry wt (GDW)/Fertilizer N supplied
UTE = GDW/N_t; HI = Harvest index; N_t = Total Nitrogen recovery

3. Concluding Remarks

Productivity of wheat in the Great Plains of North America has improved steadily during the past century, owing, mainly to improved soil/crop management techniques, and high yielding genetic stocks. However, steady intensification of cropping has been affecting soil fertility and productivity. Rampant loss of nutrients due to soil erosion, depreciation in soil organic carbon, and nutrient imbalances have been perceived in many locations in Great Plains. Lavish nutrient replenishments to regain lost fertility is common. Progressive changes from maximum tillage in early 1900s to conventional, then to reduced or no-tillage systems, have all actually aimed at conserving nutrients and moisture. When practiced in conjunction with efficient residue/stubble recycling, the reduced/no-till system preserves soil organic carbon (soil quality) and recycles essential nutrients effectively. More importantly, it enhances precipitation-use efficiency and storage.

Mainly soil N, then P and other elements are key to wheat productivity in Great Plains. A clearer understanding about N dynamics in individual fields is useful for making appropriate fertilizer recommendations. Agricultural scientists suggest that subtle changes in N transformations, particularly N-mineralization rates, emissions and other losses, plants' demand, soil-N buffering level etc. need to be monitored accurately. For example, small error in soil sampling data can under estimate upto 20 kg residual N. At present, nutrient inputs are managed at comparatively higher levels in Great Plains. Depending on precipitation use efficiency/irrigation, further intensification of this wheat based cropping system through enhanced nutrient inputs and better yielding genotypes seems the predictable course (Peterson et al. 1996). This may affect firstly, N and C dynamics, and in general nutrient turnover rates. One set of suggestions made include adoption of integrated nutrient management procedures. Since these procedures encompass a range of different nutrient sources and agronomic methods, they may sustain soil fertility and productivity, but with feeble effects on environmental parameters. For example, devising wheat-based cropping sequences that include nitrogen-fixing green manures can enhance soil N fertility and preserve it. Still, the general trend is to utilize greater chemical fertilizers and reap higher harvests. At this point, we should realize that wheat productivity in Great Plains is important to people in North America, plus those residing elsewhere in other continents. In this regard there is a prudent suggestion, that carefully managed intensification, for example use of no-till, high nutrient input methods may outweigh environmental consequences if any. Perhaps, higher production trends need primacy here.

Several other factors, such as pestilence, disease, economics, and human preferences may partly influence nutrient dynamics. Ascertaining and quantifying these effects could be useful. Overall, the North American Great Plains is a major food bowl of this world. Maintenance of soil fertility, regulating nutrient dynamics and preserving congenial agroenvironment are crucial to sustain its productivity.

REFERENCES

Biederbeck, V.O. and Bouman, O.T. (1994). Water use by annual green manure legumes in dryland cropping systems. *Agronomy Journal* **86**: 543-549.

Biederbeck, V.O., Campbell, C.A and Zentner, R.P. (1984). Effect of crop rotation and fertilization on some biological properties of a loam in southwestern Saskatchewan. *Can. J. Soil Sci.*, **64**: 355-367.

Black, A.L., Brown, P.L., Halvorson, A.D and Siddoway, F.H (1981). Dryland cropping systems for efficient water-use to control saline seeps in northern Great Plains. *Agric. Water Manag.*, **4**: 295-311.

Bowmen, R.A. and Halvorson, A.D. (1997). Crop rotation and tillage effects on phosphorus distribution in the Central Great Plains. *Soil Sci. Soc. Am. J.*, **61**: 1418-1422.

Bowman, R.A. and Halvorson, A.D (1998). Soil chemical changes after nine years of differential N fertilization in a no-till dryland wheat-corn-fallow rotation. *Soil Science*, **163**: 241-247.

Bruch, G. (1986). Pesticide and nitrate contamination of ground water near Ontario, Oragon. Impacts on Groundwater Conference, Proc. Am. Water Well Assn. Omaha, Nebraska, USA.

Campbell, C.A., Biederbeck, V.O., Schnitzer, M., Selles, F. and Zentner, R.P. (1999). Effect of 6-year zero tillage and N fertilizer management on changes in soil quality of Orthic brown Chernozem in southwestern Saskatchewan. *Soil and Tillage Research*, **14**: 39-52.

Campbell, C.A., Laford, G.P., Mouten, A.P., Townley-Smith, L. and Zentner, R.P. (1997). Crop production and soil organic matter in long-term rotations in sub-humid northern Great Plains of Canada. *In:* Soil organic matter on temperate agroecosystem: Long-term experiments in termperate agro-ecosystem in North America. CRC Press, Boca Raton Fl, USA, pp. 317-334.

Campbell, C.A. and Paul, E.A. (1978). Effects of fertilizer N and soil moisture on mineralization, N recovery and A values, under spring wheat grown in small lysimeters. *Can. J. Soil Sci.*, **58**: 39-51.

Collins, H.P., Rasmussen, P.E and Douglass, C.L. (1992). Crop rotation and residue management effects on soil carbon and microbial dynamics. *Soil Sci. Soc. Am. J.*, **56**: 783-788.

Follet, R.F., Paul, E.A., Leavitt, S.W., Halvorson, A.D., Lyon, D. and Peterson, G.A. (1997). Carbon isotope ratios of Great Plains soils wheat-fallow systems. *Soil Sci. Soc. Am. J*, **61**: 1068-1077.

Follett, R.F., Porter, L.K. and Halvorson, A.D. (1995). Nitrogen-15 labelled fertilizer dynamics in a soil, in a 4-year, no-till cropping sequence. *In:* Proceedings Int. Symp. on nuclear and related technique in soil/plant studies on sustainable agriculture and environmental preservation. Vienna, Austria, pp. 43-47.

Frenzeleubers, A.J., Hons, F.M and Zuberer, D.A. (1994). Long-term changes in soil carbon and nitrogen pools in wheat management systems. *Soil Sci. Soc. Am. J.*, **58**: 1639-1645.

Franzileuber, A.J., Hons, F.M and Zuberer, D.A. (1995). Tillage and crop effect on seasonal soil carbon and nitrogen dynamics. *Soil Sci. Soc. Am. J.*, **59**: 1618-1624.

Fredrickson, J.K., Kochler, F.E., and Cheng, H.H. (1982). Availability of ^{15}N labeled nitrogen in fertilizer and under reduced till and no-till. *Soil Sci. Soc. Am. J.*, **46**: 1218-1222.

Grant, C.A., Ferguson, R., Lamond, R., Schlegel, A. and Thomas, W. (1996). Uses of urease inhibitors in Great plains. *In:* Proceedings of 1996 Great plains soil fertility conference. Kansas State University, Manhattan, Kansas, **6**: 318-329.

Greb, B.W., Smika, D.E. and Black, A.L. (1967). Effects of straw mulch rates on soil water storage during summer fallow in the Great plains. *Soil Sci. Soc. Am. Proc.* **31**: 556-559.

Greb, B.W., Smika, D.E., Woodruff, N.P. and Whitefield, C.J. (1974). Summer fallow in Central Great plains. *In:* Summer fallows in the western United States. United States Department of Agriculture Conservation Research Report 17.

Halvorson, A.D and Black, A.L. (1985a). Fertilizer phosphorus recovery after seventeen years of dryland cropping. *Soil Sci. Soc. Am. J.* **49**: 953-937.

Halvorson, A.D and Black, A.L. (1985b). Long-term dryland crop responses to residual phosphorus fertilizer. *Soil Sci. Soc. Am. J.* **49**: 923-933..

Halvorson, A.D and Black, A.L. (1996). Dryland cropping systems and economics. *In:* Proceedings of 18th Annual Monitoba—North Dakota Zero Tillage Farmers Association Workshop. North Dakota, USA, pp. 157-165.

Halvorson, A.D. and Reule, C.A (1994). Nitrogen fertilizer requirement in annual dryland cropping system. *Agronomy Journal*, **86**: 315-318.

Halvorson, A.D. and Reule, C.A (1999). Long-term nitrogen fertilization benefit soil carbon sequestration. *In:* Better crops with plant food. Potash and Phosphate Institute, Georgia, USA.

Halvorson, A.D., Havlin, J.L and Schlegal, A.J. (1997). Nutrient management for sustainable dryland farming systems. *Annals of Arid Zone*, **36**: 233-254.

Halvorson, A.D., Reule, C.A and Follet, R.F (1999b). Nitrogen fertilization effects in soil carbon and nitrogen in a dryland cropping system. *Soil Sci. Soc. Am. J.*, **63**: 912-917.

Halvorson, A.D., Wienhold, B.J. and Black, A.L. (2001). Tillage and nitrogen fertilization influences on grain and soil nitrogen in a spring wheat-fallow system. *Agronomy Journal*, **93**: 1130-1135.

Halvorson, A.D., Anderson, R.L., Henkle, S.E., Nieben, D.C., Bowman, R.A and Vigil, M.F. (1994a). Alternative crop rotations to winter wheat. *In:* Proceedings of Great Plains soil fertility conference, Havlin, J.C.,(ed.) Kansas State University, Manhattan, Kansas, pp. 6-11.

Halvorson, A.D., Anderson, R.L., Tomain, N.E. and Welsh, J.R. (1994b). Economic compression of three winter wheat-fallow tillage systems. *J. Prod. Agric.*, **7**: 381-385.

Halvorson, A.D., Black, A.L., Krupinsky, M. and Murill, S.D. (1999c). Dryland winter wheat response to tillage and nitrogen within an annual cropping system. *Agro. J.* **91**: 702-707.

Halvorson, A.D., Black, A.L., Krupinsky, S.M., Murill, S.D. and Tanaka, D.L. (1999a). Sunflower response to tillage and nitrogen fertilization under intensive cropping in a wheat rotation. *Agro. J.*, **91**: 637-642.

Hao, Y., Lal. R., Izzaurralde, C., Ritchie, J.C., Owens, L.B. and Hothem, D.L. (2001). Historic assessment of agricultural impacts on soil and soil organic erosion in an Ohio watershed. *Soil Science*, **166**: 116-126.

Havlin, J.L., Kissel, D.E., Maddrek, L.D., Claosen, M.M. and Long, J.H. (1990). Crop rotation and tillage effects on soil organic carbon and nitrogen. *Soil Sci. Soc. Am. J.*, **54**: 448-452.

Havlin, J.L., Kok, H. and Wehmueller, W. (1992). Soil erosion-productivity relationships for dryland winter wheat. *In:* Proceedings of the Great plains soil fertility conference. Denver, Colorado Havlin, J.L. (ed.) Potash and Phosphate Institute, Georgia, USA, pp. 60-65.

Herrick, J.L and Wander, M.M. (1998). Relationship between soil organic carbon and soil quality in cropped and ragi lands: The importance of distribution, composition and soil biological activity. *In:* Soil process and carbon cycle. Advances in Soil Science, CRC Press, Boca Raton, Fl., pp. 405-425.

Huggins, D.R., Allan, D.L., Gerdener, J.C., Karlen, D.L., Bezdicek, D.F., Rosak, M.J., Flock M., Miler, B.S. and Staben, M.L. (1998). Enhancing carbon sequestration in CRP-managed land. Management of carbon sequestration in soil. Advances in Soil Science, Lal, R. et al. (eds.), CRC Press, Boca Raton, pp. 323-334.

Janzen, H.H., Campbell, C.A., Izzaurralde, R.C., Ellert, B.H., Juma, McGill, W.B. and Zentner, R.P. (1998). Management effects on soil C storage on the Canadian Prairies. *Soil and Tillage Research*, **47**: 181-195.

Jones, O.R., Stewart, B.A. and Unger, P.W. (1997). Management of dry-farmed Southern Great plains soils for sustained productivity. *In:* Soil organic matter in temperate agroecosystems. Long-term experiments in North America. Paul E.A. (ed.) CRC press, Boca Raton, FC USA, pp. 357-401.

Karlen, D.L., Wollenhasept, N.C., Erback, D.C., Berry, E.C., Swen, J.B., Nash, N. S and Jordehal, J.L. (1994). Crop residue effects on soil quality following 10 years of no-till corn. *Soil and Tillage Research,* **31**: 149-167.

Kelling, K.A., Bundy, L.G., Combs, S.M. and Peters, J.B. (1998). Soil test recommendations for field, vegetable, and fruit crops. University of Wisconsin, Co-operative Extension publication, Madison, Wisconsin, USA, pp. 1-45.

Kelling, K.A., Speth, P.E. and Wood, T. (2000). Effect of tillage on alfalfa N credit to winter wheat. Wisconsin Fertilizer, Aglime and Pest Management Conference **39**: 405-410.

Kolberg, R.L., Westfall, D.G., Peterson, G.A., Kitcher, N.R. and Sherrad, L. (1993). Nitrogen fertilization of dryland cropping system. Colorado State University Agric. Exptl. Stn. Tech Bull. 93-96.

Kolberg, R.L., Kitcher, N.R., Westfall, D.G. and Peterson, G.A. (1996). Cropping intensity and nitrogen management impact of dryland no-till rotations in the semi-arid Western Great Plains. *J. Prod. Agric., * **9**: 517-522.

Kochler, F.E., Cachran, V.L. and Rasmussen, P.E. (1987). Fertilizer placement, nutrient flow, and crop response in conservation tillage. *In:* STEF—Conservation concepts and accomplishments. L.F. Ellert (ed.), Washington State University Publication, Pullman, WA, pp. 57-65.

Lal, R., Kimble, J., Folet, R.F. and Cole, C.V. (1998). The potential of U.S. cropland to sequester carbon and mitigate the greenhouse effect. Ann. Arbor Press, Chelsea, MI, USA.

Lamb, J.A., Peterson, G.A and Fenster, C.R. (1985). Wheat-fallow tillage systems effect on a newly cultivated grassland soils nitrogen budget. *Soil Sci. Soc. Am. J.* **49**: 352-356.

Lin, D.M. and Doran, J.W. (1984). Aerobic and anaerobic microbial populations in no-till and plowed soils. *Soil Sci. Soc. Am. J.*, **48**: 794-799.

Nielson, D.C. and Anderson, R.L. (1993). Managing residue and storing precipitation. Colorado State Soil Conservation Board, Denver, CO. Conservation Tillage fact sheet 2-93, 2.

Norwood, C.A. (2000). Dryland winter wheat as affected by previous crops. *Agronomy Journal,* **92**: 121-127.

Norwood, C.A. and Dhuyvetter, K.C. (1993). An economic comparison of the wheat-fallow and wheat sorghum-fallow cropping systems. *J. Prod. Agric.,* **6**: 261-266.

O'Halloren, I.P., Stewart, J.W.B and Jong, E. (1987). Changes in P forms and availability as influenced by management practices. *Plant and Soil,* **100**: 113-126.

Olson, R.V and Swallow, C.W. (1984). Fate of labeled nitrogen fertilizer applied to winter wheat for five years. *Soil Sci. Soc. Am. J.,* **48**: 583-586.

Oplinger, E.S., Wiersma, D.W., Gran, D.R. and Kolling, K.A. (1998). Intensive wheat management. University of Wisconsin—Extension, Madison, WI. USA pp. 1-18.

Ortiz-Manasterio, J.I., Sayre, K.D., Rajaram, S. (1997). Genetic progress in wheat yield and nitrogen-use efficiency under four nitrogen rates. *Crop Science*, 37: 898-904.

Ozanne, P.G. (1980). Phosphate nutrition of plants—A general treatise. *In:* The role of Phosphorus in Agricultural Khaswhale, F.E. (ed.), American Society of Agronomy, Madison, WI, USA, pp. 559-589.

Paustian, K., Paston, W.J. and Persson, J. (1992). Modeling soil organic matter in organic-amended and nitrogen fertilized long-term plots. *Soil Sci. Soc. Am. J.*, **56**: 476-488.

Peterson, G.A., Schlegel, A.J., Tanaka, D.L. and Jones, O.R. (1996). Precipitation use efficiency as affected by cropping and tillage systems. *J. Prod. Agric.*, **9**: 180-186.

Peterson, G.A. and Westfall, D.G. (1994). Economic and environmental impact of intensive cropping systems—semi-arid region. Proc. conf. on Nutrient management of highly productive soils. PPI/FAR Special Publication 1994-1.

Porter, L.K., Fillet, R.F. and Halvorson, A.D. (1996). Fertilizer nitrogen recovery in a no-till wheat-sorghum-fallow-wheat sequence. *Agronomy Journal*, **88**: 750-757.

Power, J.F., Doran, J.W. and Doran, J.W. (1986). Recovery of fertilizer nitrogen by wheat as affected by fallow method. *Soil Sci. Soc. Am. J.*, **50**: 1499-1503.

Rasmussen, P.E. (1993). Surface residue and nitrogen fertilization effects on no-till wheat. *In:* Plant nutrition from genetic engineering to field practice Barrow, N.J. (ed.) Kluwer Academic Publications, Armsterdam, Netherlands., pp. 555-558.

Rasmussen, P.E. (1996). Fertility management in dryland conservation cropping systems of the Pacific Northwest. *American Journal of Alternative Agriculture*, **11**: 108-114.

Rasmussen, P.E. and Albrecht, S.L. (1998). Crop management effects as organic carbon in semi-arid Pacific Northwest soils. *In:* Management of carbon sequestration in soil. R. Lal, J.M., Kimble, R.F., Follett, and B.A. Stewart (eds.), CRC Press, Boca Raton, New York, pp. 209-219.

Rasmussen, P.E., Allmaras, R.R., Rohde, C.R. and Roeger, N.L. (1980). Crop residue influences in soil carbon and nitrogen in a wheat-fallow system. *Soil Sci. Soc. Am. J.*, **114**: 596-600.

Rasmussen, P.E. and Douglass, C.L., Jr. (1992). The influence of tillage and cropping-intensity on cereal response to nitrogen, sulphur and phosphorus. *Fertilizer Research*, **31**: 15-19.

Rasmussen, P.E., Douglas, C.L., Collins, H.P. and Albrecht, S.L. (1998). Long-term cropping system effects on mineralizeable nitrogen in soil. *Soil Biology Biochemistry*, **30**: 1829-1837.

Rasmussen, P.E., Reckman, R.W. and Klapper, B.L. (1997). Residue and fertility effects as yield of no-till wheat. *Agronomy Journal*, **89**: 563-567.

Rasmussen, P.E., and Rohde, C.N. (1991). Tillage, soil depth, and precipitation on wheat response to nitrogen. *Agronomy Journal*, **55**: 121-124.

Rasmussen, P.E. and Rohde, C.N. (1988). Stubble burning effects as winter wheat yield and nitrogen utilization under semi-arid conditions. *Agronomy Journal*, **80**: 740-942.

Rasmussen, P.E. and Rohde, C.N. (1998). Long-term tillage and nitrogen fertilization effects on organic nitrogen and carbon in a semi-arid soil. *Soil Sci. Soc. Am. J.*, **52**: 1114-1117.

Rasmussen, P.E. and Smiley, R.W. (1997). Soil carbon and nitrogen in long-term agricultural experiments at Pendilton, Oregon. *In:* Soil organic matter in temperate agroecosystems. Longterm experiments in North America Pacel, E.A. *et al.* (eds.), CRC Press, Boca Raton, Fl., USA, pp. 356-360.

Raun, W.R. and Johnson, G.V. (1995). Soil-plant buffering of inorganic nitrogen in continuous winter wheat. *Agronomy Journal*, **87**: 827-834.

Raun, W.R. and Johnson, G.V. (1999). Improving nitrogen use efficiency for cereal production. *Agronomy Journal*, **91**: 357-363.

Rouppet, B. (1994). *In situ* soil nitrogen mineralization in no-till dryland agro-ecosystems. Ph.D. Dissertation, Colorado State University, Fort Collins, USA.

Schlegal, A.J. (1992). Effect of composted manure on soil chemical properties and nitrogen use by grain sorghum. *J. Prod. Agric.* **5:** 153-157.

Selles, F., McConkey, B.G. and Campbell, C.A. (1999). Distribution and forms of P under-cultivator and zero tillage for continuous and fallow-wheat cropping systems in semi-arid Canadian Prairies. *Soil and Tillage Research*, **51:** 47-59.

Shock, C.C. and Steiber, T.D. (1991). Nitrogen uptake and removal by selected crops. Oregon State University. Special report, **882:** 182-186.

Shock, C.C., Feibiot, E.B.G., Jensen, L.B., Jones, R.L., Capps, G.W. and Ghun, E. (2000). Changes toward sustainability in the Malheur-Owynu Watershed. Oregon State University, Report.

Sommerfeldt, T.G., Chung, C. and Eutz, T. (1988). Long-term annual mature applications increase soil organic matter and nitrogen, and decrease carbon to nitrogen ratio. *Soil Sci. Soc. Am. J.*, **52:** 1668-1672.

Wagger, M.G., Kissel, D.E. and Smith, S.J. (1985). Mineralization of nitrogen from nitrogen-15 labeled crop residues under field conditions. *Soil Sci. Soc. Am. J.*, **49:** 1220-1226.

Westfall, D.G., Havlin, J.L., Hargest, G.W. and Raun, W.R. (1996). Nitrogen management in dryland cropping systems. *J. Prod. Agric.*, **9:** 192-199.

Wienhold, B.J. and Halvorson, A.D. (1998). Cropping system influences on several soil quality attributes in the Northern Great Plains. *J. Soil and Water Cons.*, **53:** 254-258.

Wienhold, B.J. and Halvorson, A.D. (1999). Nitrogen mineralization responses to cropping, tillage and nitrogen rate in the Northern Great Plains. *Soil Sci. Soc. Am. J.*, **63:** 192-196.

CHAPTER 3

The Temperate Wheat Cropping Zones of European Plains, Australia and Pampas in Argentina
Nutrient Dynamics and Productivity

CONTENTS

1. The Temperate Wheat Cropping Zones in Europe
   A. Expanse, agroclimate, soils and cropping systems
   B. Soil tillage practices and nutrients
   C. Soil organic carbon
   D. Nitrogen dynamics
   E. Phosphorus dynamics
   F. Long-term nutrient balance in wheat fields
2. The Wheat Agroecosystem of Australia
   A. Expanse, agroenvironment, soils cropping system and tillage
   B. Soil organic carbon
   C. Nitrogen dynamics
   D. Phosphorus dynamics
3. The Wheat based agroecosystem of Pampas in Argentina
   A. Expanse, agroclimate and cropping sequence and soils
   B. Nitrogen dynamics and crop productivity
4. Concluding Remarks

1. The Temperate Wheat Cropping Zones in Europe

The seeds for a Temperate European Wheat Agrosystem were sown 5 millenniums ago by migrants from the "Fertile Crescent". Currently it occupies the vast central plains, and adjoining areas in the European countries. It supplies a very large human population with carbohydrates. Modifications in nutrient management technologies and cropping sequences played a crucial role in enhancing grain productivity during 18 and 19th century A.D. Later, in the Early/Mid 20th century this agroecosystem experienced drastic intensification

> through large inputs of chemical fertilizers and change to high yielding (semi-dwarf) wheat genetic stocks. At present, it is a major "bread basket", and one of the highest yielding cropping systems of the world.

A. Expanse, Agroclimate, Soils and Cropping Systems

Expanse: Geographically, the European wheat agroecosystem spreads across vast areas, which can be classified into five major regions. They are: northern Europe, northwestern and western Europe, central Europe, Russia and eastern Europe, southern Europe and Mediterranean Europe. Among these, north and northwestern Europe is a major wheat producing zone of the world. Intensive cultivation of wheat on the central European Plains extends into European Community nations such as United Kingdom, Ireland, Denmark, Netherlands, Belgium, Germany and France (Spiertz et al., 1992). Wheat production in central Europe occurs between 41° 41' and 54° 54' N latitude, and 9° 54' and 28° 37' E longitude. The central European wheat based cropping ecosystem extends into Poland, Bulgaria, Hungary, Czechoslovakia, Romania and parts of eastern Germany. These were formerly called, the Eastern Block countries (Nalborczyk and Czember, 1992). Within the former Soviet Union, wheat cultivation extends into 188 m ha, between 45 and 56° N latitude (Gataulina, 1992). The sub-agroecological zones identified are the non-Chernozem zones of European Russia, the central Chernozem; middle and lower Volga territory; northern Caucasus; Ukraine; northern Kazakhstan and Steppe Siberia. The central Chernozems, Ukraine and parts of Volga territory which are predominantly wheat producing zones constitute the 'Bread basket' of former Soviet Union'. The European wheat also extends well into the mildly temperate and Mediterranean areas in southern France, Spain, Portugal, Greece and Italy (Lopez-Bellido, 1992).

Agroclimate: The wheat agroecosystem in Europe experiences a wide range of climatic variations such as cool temperate, moderately temperate and cool Mediterranean environment. In north and northwestern Europe cool temperate climate predominates. Mean temperatures during the season range from 3°C to 15°C. However mean maximum temperature can reach 20-25°C in July. Mean monthly precipitation fluctuates between 60 mm (January) and 95-100 mm (August), and the total annual rainfall varies between 550 mm (Paris, France) and 950 mm (Schliesing, Germany). The evapotranspiration rates are too low owing to temperate climatic conditions. Wheat productivity is linearly dependent on incident radiation, mean daily temperature and the photothermal equations. Within limits, the climatic pattern in north Europe can become uncertain. Hence, an array of futuristic climatic scenarios developed using General Circulation Model (GCM) for France, Germany and other countries have been assessed by coupling it with wheat growth models such as SIRIUS and

AFRCWHEAT-2 (Barrow and Semenov, 1995; Semenov et al. 1993; Semenov and Brookes, 1999).

Central and eastern European wheat cropping zones extend into temperate climate, with mean temperatures ranging from $-1°C$ to 20 to 24°C. In northeast Holland and Hungarian plains severe winters dip temperatures to $-20°C$. Summer temperature in Bulgaria and Hungary easily reach 25°C. Influx of warm air in September/October generally bring warmth to plains, whereas in May, cool freezing breezes sweep into the zone from the Arctic. Nalborczyk and Czembro (1992) have suggested that central European wheat zones can also be grouped based on total temperatures, into 6 zones, beginning with >4000°C (eg. Bulgaria) to <1600°C (eg. East Germany, Poland). The total radiation impinged into central European wheat belt varies from 60 to 80 K Cal M^{-2} in Poland, to 80 to 100 K Cal M^{-2} in southern Ukraine and Romania. The number of sunshine hours in a year vary between 1100 in Baltic coast and 2100 in Hungarian lowlands and Black Sea coast.

The agroclimate within the Russian wheat belt becomes warmer and dryer as we travel from northwest towards southeast. The Baltic coast and Central Russia receive relatively higher rainfall (650 mm per year), whereas Volga territory receives 380 mm y^{-1} and Kazak area 280 mm y^{-1} year. The evapotranspiration varies between 500 and 1000 mm per year depending on location. The temperature rises to >10°C for nearly 150 days in Russia. Winters here still allow wheat cultivation, which is impossible in the nearby Siberia. The total solar radiation increases from northwest to southeast. The central Chernozem receives 60 to 80 K Cal M^{-2}-y^{-1}, Volga territory and Caucasus 100 to 120 K Cal M^{-2} y^{-1} and Kazakhstan 120 to 140 K Cal M^{-2}. However, photosynthetically active radiation in the Russian plains range between 34.2 K Cal M^{-2} y^{-1} and 55 K CalM^{-2} y^{-1}.

The wheat-cropping system in southern Europe experiences cool Mediterranean climate. The mean summer (June to September) temperatures range from 24 to 27°C, whereas January/February are cooler at 7 to 10°C. The annual precipitation ranges between 500 and 1000 mm in Portugal (Carranca et al. 1999), Spain, southern France, Greece and Italy (Lopez-Bellido,1986). The wettest months are November/December.

Soils: A very wide range of soil types are utilized by European wheat farmers. The most intensively cropped wheat zones in northwestern Europe thrive on Cambisols in France and Britain, on Luvisols in Germany and Fluvisols in Netherlands. These soils are classified as arable, and most suitable with no limitation to wheat production. They extend into 31 M ha, of which 50% occur in France (Spiertz et al. 1992). In central Europe, Loess soils, Podzols, light brown steppe soil, Black earth and sandy soils are frequent. Whereas, in Russia and eastern plains, chernozems are most fertile on which wheat is intensively cultivated. Of the entire plough land of Russia, nearly 48% is Chernozems, 14% Podzols, 10% Solonchaks and rest grey soils. Podzols in western Ukraine, Byelorussia, Lithuania and Estonia also support wheat productivity. In Mediterranean Europe, wheat cultivation proceeds on Xeric and Argillic Cambisols, Alfisols and Entisols. Soils with calcareous sediments are frequent. In general, major soil types encountered in this intensive wheat agroecosystem of Europe can be summarized as follows:

North and northwestern Europe: Eutric Cambisols (Britain); Drystic Cambisols (France); Orthic Luvisols (Germany); Fluvisols and Histosols (Nertherlands).

Central Europe: Loess soils (Eastern Germany, Holland, Hungary); Sandy soils (Romania, Czechoslovakia); Podzols (Poland, Ukraine); Saline/ Alkaline soils in Hungary

Russia and eastern Europe: Chernozems, Podzols, Chestnut soils, Solonetz and Solonchaks.

Southern Europe: Cambisols, Alfisols, Entisols.

Wheat Based Cropping Systems: Cereals, predominately wheat occupies 56% of arable land in northwestern Europe. In this zone, wheat grain and biomass productivity have been steadily increasing since Mid-20th century. It is attributable to inorganic and organic fertilizers, improved varieties and better crop management practices (IACR, 2001). Winter wheat yield may reach 8 to 10 t ha^{-1}. Most rotations include a cereal (wheat or barley), fodder legumes, sugar beet, potato and fallows.

In central Europe, total harvested area for wheat has fluctuated between 7.5 and 8.6 m ha annually. Wheat grain yield has improved from 1.2 t ha^{-1} in 1960's to 3.4 to 4.8 t ha^{-1} in A.D. 2000. Yet, these countries in central Europe experience insufficiency, therefore, import 4 to 5 mt wheat grain annually. Wheat-based rotations include fallow, sugar, potato and legume pastures. In the former Soviet Union, that is Russia and eastern European states, winter wheat occupies over 18 to 19 m ha annually. Winter wheat is sequenced with forage crops such as vetch-oat-pea mixture. In drier zones, bare fallow is a good precursor that retains soil moisture. Winter wheat is intensive and yields between 7 and 8 t ha^{-1}. To achieve it, fertilizers at 80 to 120 kg N, 40 kg P and 25 to 50 kg K plus FYM are supplied to wheat fields. The main wheat-based rotations in southern Europe are wheat-wheat (continuous), wheat-fallow, wheat-grain legumes, wheat-tobacco-sugar beet-grain legumes etc.

B. Soil Tillage Practices and Nutrients

Tillage intensity is a crucial factor, because it affects several soil properties, including quality and nutrients status. Within the German wheat belt comprising mainly, Eutric Cambisols and Fluvisols, the tillage intensity decreases as we shift from conventional plough tillage (CT), to reduced tillage (RT) and no-tillage (NT). This same trend in tillage intensity is encountered in wheat zones of many other European countries, as well as in Great Plains of USA. A recent review of long-term soil tillage practices in German wheat farms indicates that soil physical conditions were moderately affected by RT, which could be considered intermediate between CT and NT (Tebrugge and During, 1999). Overall, NT suited best, wherein the bulk density of upper layers of soil increased, and better aggregate stability enhanced soil fertility. It also reduced surface soil erosion. Accumulation of organic matter and nutrients at surface, and better biological activity were other favorable attributes related to NT. Presently, German wheat farms favor conservation tillage processes because of both, long-term soil quality/ecological and economic considerations. Long-term trials (over 30 years) at Humbolt University, Berlin, have shown that on silty or sandy soils shallow ploughing (reduced till) caused distinct increases in soil organic carbon and phosphorus, and did not affect potassium or soil pH. Shallow ploughing also enhanced grain yield of spring wheat (Ellmer *et al.* 2000). Rasmussen (1999) states that ploughless tillage or reduced tillage was introduced into Scandinavian wheat belt in 1970's. Reduction in soil erosion and concomitant nutrient losses, and avoiding costly or laborious ploughing operations were the major advantages that attracted Scandinavian wheat farmers. Best wheat yield advantages from reduced tillage accrued in heavy clayey soils, which are almost difficult to prepare using conventional tillage systems. Plough-less tillage enhanced bulk density. Mulches led to lower evapotranspiration and better soil water holding capacity. Nutrients and organic matter generally accumulated near the soil surface after a plough-less tillage. Most importantly, omitting tillage had very little impact on wheat yield, and it reached well over 90% of that achievable through constant conventional tillage practice. In the east European plains, again a change from conventional tillage to conservation or reduced tillage system has been guided by considerations such as, better water storage and precipitation use efficiency, reduced soil erosion, decreased energy inputs and optimum ecological condition (Korman and Koller, 1997). Under reduced tillage, N requirements of wheat was comparatively lesser than CT (Knezevic *et al.* 1999). Under conventional tillage, an extra 15 to 45 kg N per ha^{-1} was required to attain yield levels possible under reduced tillage schemes. Reduction in nitrogen diversion to weeds was an additional advantage under RT. According to Lipiec and Stepniewski (1995), tillage and soil compaction affects soil nitrogen balance in Polish cropping zones. Such an effect is attributable to alteration of:

a) Soil aeration status, which directly influences denitrification, gaseous N losses, N mineralization rates and reduces N-fixation by legumes.
b) alteration in soil water status, which enhances N transport and leaching; and
c) alteration in soil aggregates and compaction which affects root growth, its configuration, root-soil contact and ion diffusivity.

Since phosphorus is relatively immobile, the effect of soil compaction is mostly related to wheat root system. Restricted root growth and low P recovery from compact soil results in low yields. Similarly, K^+ uptake by wheat crop reduces with soil compaction. It was attributed to reduced root surface area. In the Mediterranean climate of southern Europe (Andulasia, Spain) too, a change from conventional tillage to no-till has influenced nutrient dynamics and wheat grain yield (Lopez-Bellido et al., 2000). However, long-term assessments of wheat in southern Spain have again indicated that no-till was an appealing alternative to conventional tillage, because it leads to better wheat grain productivity for unit energy/ economic input. Additionally, no-till treatment provided higher residual-N effects. Wheat grain yield and N recovery was also affected by interaction effects of no-till with cropping sequences and weather pattern. Grain yield under no-till system generally decreased in the following sequence of crop rotations: wheat-fallow-bean > wheat-fallow > wheat-sunflower > wheat-chickpea> continuous wheat. Nitrogen inputs enhanced grain yield best, whenever legumes were absent in the rotation. Intractive effects of tillage verses rainfall pattern seems important. Under conventional tillage, wheat fields in southern Spain receive high N inputs, which may be easily lost through runoff, leaching, and denitrification. To reduce N losses, timing of N fertilizer inputs needs to be accurate.

In northern Italy, the standard cropping practices have involved conventional tillage (CT), minimum tillage or restricted tillage (RT) practices or no-tillage (NT). The positive effect of minimum till or no-till can be two-fold: reduction of CO_2 emission, and greater sequestration of C because of reduced mineralization of organic matter. The energy output/ input ratio increases as soil tillage gets reduced. In this particular study by Borin et al. (1997) the output/input ratio was 4.09, 4.3 and 4.6 for CT, RT and NT respectively. Despite advantages, reduced or no-tillage system have been slow to spread in winter wheat zone of northern Italy (Borin et al. 1997). No-tillage is practiced in only 30,000 ha of cereal zone out of the possible 2.5 m ha. Unfavorable rainfall pattern such as spring rainfall, heavy or poorly structured soil, enhanced soil moisture content and related factors seem to be the reasons for lack of spread of no-till system into spring cereal zones.

C. Soil Organic Carbon

Soil organic matter (SOM) content and its composition influence wide range of soil properties that affects wheat cultivation in the European Plains. Firstly, soil physical traits such as aggregates, moisture holding capacity, and impedance to root spread, if any in the profile are affected by SOM. A good fraction of nutrients, especially N, P and S absorbed by wheat crops are actually stored in the organic fraction of soil. There is no doubt the SOM sustains activity of diverse biological population, including the symbiotic N fixers and P solubilizers. In addition to influences on crops, SOM affects carbon cycle to carbon cycle within the cropping ecosystems, especially the extent of CO_2 and CH_4 emission into the atmosphere. It is well known that agriculture operation adopted during wheat cultivation, particularly the soil tillage, interculture, removal of straw without reincorporation, and removal of grains, all lead to decline in SOM. Reversal of this trend towards accumulation of C is not an easy task. To quote an example, located in the "Bread basket zones of Russia", the Kursk area with Chernozems (Pachick Hapludolls) is an intensively cultivated cereal zone. According to land resources division of Russia, this belt occupies 6.6 m ha, and contains 130 to 160 t ha^{-1} of organic matter in the top 20 cms. Hence this zone is both a huge source and sink of CO_2 gas (Orlov and Birukova, 1995). Long-term (50 y) trials indicated that SOC and N concentration decreased both in fallow and cropped fields. Greatest decreases occurred on upper 10 cm of cropped land, wherein SOC dipped by 38 to 43%, and total N by 45 to 53%. Actually, soil management procedures indicated change in SOC and N even beyond the plough layer (Mikhailova et al. 2000). On the other hand, in western Germany, large inputs of SOM and associated nutrients have accumulated during the past three decades. The fertilizer based nutrient inputs have exceeded removal by crops. Organic C and total N balances (1979 to 1998) in Lower Saxony yielded a cumulative increase of 16 t C ha^{-1} and 1 t N ha^{-1} in loess soils (Nieder and Ritcher, 2000).

The extent of change in SOC is obviously dependent on the source and quantity of organic residue inputs into wheat fields. Long-term changes in SOC as consequence of differences in wheat based cropping system and their duration, or due to extent of inorganic and organic replenishments have been extensively investigated in several countries of this temperate wheat agrozone (Table 3.1). Depending on the duration of experimentation, percentage changes in SOC ranged between 9.6 and 50%. In addition to FYM, other common inputs during wheat cultivation are sewage sludge and animal manure. Among these, yearly increases in SOC was often highest (3-12%) with sewage sludge, compared with wheat straw that produced 0.5 to 3% increase in SOC (Smith et al. 1998b). Recycling wheat straw increases SOM content, and the extent of SOM depends on quantity of straw incorporated and duration of rotations (Table 3.2). Cropping systems and soil management practices which are intensive, conventional or those devised to maintain an ecological stability in the agrozone have been employed by European wheat farmers. The guidelines for ecological management practices stipulate smaller dozes of nutrients to avoid soil toxicity. Recently, Beyer et al. (2002) evaluated the pros and cons of such conventional and ecological management practices during wheat

Table 3.1: A long-term comparison of organic verses inorganic amendments on soil organic carbon in the European wheat belt.

Location	Wheat based rotation	No. of years	FYM input (t ha^{-1} y^{-1})	Soil organic carbon (t ha^{-1} in 0 to 30 cm)	Difference from inorganic or zero %
Bad Lauchastatt, Germany	Winter wheat, Spring barley, potato	90	15 0	87.1 71.1	21.5
Meckenheim II, Germany	Wheat and other cereals	22	6.5 0	59.6 49.8	19.7
Weihestephon, Germany	Wheat-based rotations	47	10 0	38.1 34.8	9.8
Praha-Ruzyae, Czeck Republic	Spring wheat/sugar beet	38	11 0	62.6 55.8	12.2
Broadbalk, United Kingdom	Continuous wheat	144	35 0	99.1 49.6	100.0
Woburn, United Kingdom	Continuous wheat and barley	49	18 0	72.0 52.8	36.4
Deherain, France	Wheat/sugar beet	112	10 0	48.7 40.6	20.0
Askov, Denmark	Wheat	100	13.5 0	57.2 52.2	9.6
Essai Permanent, Belgium	Wheat-based rotation	30	2.4 0	41.8 36.5	14.3
Ultana, Sweden	Wheat, arable crops	35	8.7 0	71.2 47.6	49.6

Source: Smith et al. 1997
Note: Under the column FYM inputs zero indicates nil organic manure inputs, however, inorganic fertilizers were supplied

Table 3.2: Influence of wheat straw incorporation on soil organic matter

Experiment	Crop/ rotation	No. of years	Treatment (t ha^{-1} y^{-1})	Rate (t ha^{-1} y^{-1})	SOC (t ha^{-1})	Deference (%) from no straw
Rothamsted, UK	Continuous wheat	7	18 no straw	18 0	78.9 71.6	10.2
Meckenhiem, Germany	Cereals	22	6.5 no straw	6.5 0	59.6 49.8	19.7
Essai Permanent, Belgium	Cereals-based rotation	30	2.4 no straw	2.4 0	41.8 36.5	14.3

Source : Excerpted from a compilation by Smith et al. 1997
Note: SOC measured in upper 30 cm soil layer.

production. According to them, although SOC increases are appreciable under ecological management systems, wheat grain productivity may be low due to insufficient nutrient supply prescribed under this system. On the other hand, intensive management system lead to higher productivity, but retained lower levels of C in soil. Sometimes only half the level of C was retained compared to ecological cropping schemes. These inferences on C storage verses grain productivity between conventional and ecological management schemes hold good for a range of soil types encountered within European Plains, namely Albaaquic Hapludalf, Mollic Endoaquept, Aeric Fluvaquent, or Anthropic Udipsamments. A similar evaluation in Netherlands revealed that organic matter increases were least with conventional practices compared with organic farming (Pulleman *et al.* 2000). Obviously, simulations and testing are required so that a compromise between the yield levels aimed through higher inorganic inputs, concern for soil C dynamics and ecological needs could be arrived at. We must also realize that on a long-term basis, higher N inputs not only provide higher grain yield, but proportionate wheat straw recycling enhances total organic carbon in soil (Beyer *et al.* 2002).

Carbon Sequestration: In general, agricultural soil contains lesser quantities of organic matter compared to pastures or woodlands. Agricultural activity such as ploughing, interculture, and inverting soils enhances mineralization rates leading to C loss, and elevation of CO_2 concentration in the immediate atmosphere. Therefore, sustainable use of soil resources that allow increase in organic C pool need priority. It is generally accepted that organic matter inputs, crop rotation and tillage practices have to change in order to make agricultural soil a sink for C (Powlson *et al.* 1998; Smith *et al.* 1998b). Although, greatest potential for mitigating C is in the tropics, the potential for C sequestration by the vast temperate wheat agroecosystem in Europe needs due consideration. Smith *et al.* (1998b) state that one of the main options for carbon mitigation is to adopt methods that sequester this element in soil. Accordingly, five different scenarios through which carbon status in soils is affected have been studied. They are, a) incorporation of cereal straw; b) amendment with animal manure; c) use of sewage sludge; d) afforestation of surplus arable farming zones and e) extensification of arable wheat/cereal belt. Assessments indicate greater potential of C sequestration through extensification of agricultural land (~40 Tg Cy^{-1}), and through afforestation of surplus cereals zone (~50 Tg Cy^{-1}). In the course of wheat based rotation in Europe, cereal straw is recycled into soil. Such wheat straw incorporation (at 5.1 t $ha^{-1} y^{-1}$) under one of the scenarios tested indicated that in 100 years, 1.4 Pg SOC could be sequestered. It meant an increase of total SOC from 22.95 Pg to 24.35 Pg (Smith *et al.* 1998b). However, we may realize that despite such potential to sequester large amounts of C, the vast temperate wheat belt in Europe can only mitigate a small fraction equivalent to 0.8% of annual global CO_2 emission (Smith *et al.* 1997; 1998b).

The temperate European wheat zone is also an important sink for methane (CH_4). We may note that the net CH_4 fluxes in any agroecosystem is actually determined by the balance between production and oxidation. In soils that support an aerobic (in microsites) and anaerobic process,

---
Pg = Peta gram

both production and oxidation occur. Sometimes, much of the CH_4 generated may be oxidized in soil layers even before it diffuses out and reaches atmosphere. Generally, soils with forest cover possess 3 to 4 times higher rates of CH_4 uptake compared to cropped land. Soil which is almost entirely arable (aerobic) show significant variation in CH_4 absorption, depending on soil management, cropping sequence and fertilizer-based nutrient inputs. For example, within long-term Broadbalk wheat trials, on plots given continuous N inputs at 144 kg N ha^{-1}, the CH_4 uptake (sink) by soil reduced 50% compared with no fertilizers or FYM treated plots (Hutsch, 1993). Powlson *et al.* (1997) explain that such variations in sink strength of soils for CH_4 is determined by relative proportions and activities of methanotrophs and nitrifiers, as well as methanogens which are responsible for CH_4 formation and release. In nature, current CH_4 concentration, soils enzymes and soil physico-chemical environment affect the activity of soil microbes involved in methanogenesis and oxidation of CH_4.

D. Nitrogen Dynamics

Nitrogen Inputs: The European wheat farmers practice intensive agriculture. It may often lead to over-fertilization. Nitrogen inputs into wheat fields have been ever increasing, and are generally higher ranging from 100 to 250 or 300 kg N ha^{-1}. This leads to rapid accumulation of this element in soil. For example, N balance for German and British wheat fields indicate a surplus of about 100 kg N ha^{-1} (Weigel *et al.* 2000; IACR, 1997; Powlson, 1986). In addition, NH_3 and NO_x emitted into atmosphere is re-deposited within a few days. In Germany, long-term quantification of airborne N-input indicated that about 20 to 30 kg N per ha^{-1} y^{-1} could be deposited into wheat fields. Weigel *et al.* (2000) estimated that wheat field at Bad Lauchstadt in Germany may receive between 50 and 58 kg N ha^{-1} y^{-1} via airborne N-deposition. ^{15}N based analysis using the Integral Total Nitrogen Input (ITNI) system indicated that airborne N deposition could be 65 kg N ha^{-1}. Over ten years, average atmospheric N deposition at Broadbalk, England was 45 kg N ha^{-1} (IACR, 1997). There are instances when airborne N deposition could be underestimated. Atmospheric N deposition reported from other European locations vary between 25 and 61 kg N ha^{-1} y^{-1}. For example, in Denmark it was 46 kg N ha^{-1} y^{-1} (Christensen, 1989), in Rothamsted, England 48 kg ha^{-1} y^{-1} (Goulding, 1990), in Prague, Czechoslovakia, 61 kg N ha^{-1} y^{-1} (Klier *et al.* 1995). Therefore, due care is required while computing N fertilizer requirements, so that over-fertilization of N is avoided. Generally, total amount of N returned in the above ground crop residues average 14 kg N ha^{-1}. Assuming that dry residues contain 40% C, and the C : N ratios of recycled material 25 : 1 (Macdonald *et al.* 1997).

Nitrogen Mineralization: Mineralization of organic N which releases inorganic forms of NH_4^+ first then NO_3^-, as well as immobilization of inorganic N (either just released or pre-existing) by soil microbes are simultaneous processes in soil. These transformations can occur in microsites. The mineralization rate measured and reported until recently for wheat zones in Europe actually relates to net differences between the two processes, that is, gross mineralization minus immobilization. However, using ^{15}N dilution techniques it is possible to estimate them separately. During a winter wheat crop actual turnover of N is high. Since, both mineralization and immobilization rates are high and balance out, net NH_4^+ or NO_3^- production estimated could be small (IACR, 1997). In the field, NH_4^+ recycling seems rapid because microbes reutilize the released NH_4^+ rather quickly.

The mineralization:immobilization rates, in general, are influenced by soil C and N status, C : N ratio of FYM inputs, inorganic fertilizer inputs, soil environment and cropping patterns adopted. Under temperate condition, N mineralization is directly related to soil thermal units (Clough et al. 1998). Organic farming is in vogue in many places within the temperate wheat belt in Europe. In such farms where no synthetic (mineral) N fertilizer is supplied, the N availability to non-legumes, for example wheat grown in rotation depends entirely on microbial turnover of soil organic matter (Friedel, 2000; Friedel and Gabel, 2001). Of course, soil microbial biomass and activity are generally on a high, and are site-specific under organic farming. Hence, both N-storage function within microbial biomass and, N-mineralizing activity are pronounced in organically farmed soils (van Lutzow and Ottow, 1994). For example in 8 years, the microbial biomass increased from 230 to 330 $\mu g\ g^{-1}$ soil, and microbial biomass N increased from 47 to 53 $\mu g\ g^{-1}$ soil (Friedel and Gabel, 2001).

Nitrogen Losses: Interrelationships among nitrate leaching, applied fertilizer N, atmospheric N inputs, production of N through mineralization and wheat crop management are complex (Skinner et al. 1997). Nitrate loss from wheat field is also influenced by the type of land preparation procedures. For example, in England, NO_3–N lost from direct drilled land was 24% less than that from regularly ploughed zones (Catt, 1974). Average N lost under conventional ploughing treatment was 39 kg N $ha^{-1}\ y^{-1}$. No doubt, the nitrogen dynamics can be significantly affected by no-till (NT) or minimum tillage practices, which are known to enhance bulk density and proportion of surface connected micropores. There are several studies which point out that greater downword leaching of N takes place in NT than in conventional tillage conditions (Goss, 1988). Such rapid downward displacement of N was attributable to flow in a preserved macrospore network, and to reduction in surface runoff. Despite susceptibility to leaching, N concentration in soil profile was higher in NT compared to CT. Some reasons attributable for this phenomenon are rapid and greater quantity of N removal by crops, greater maximum inflow of N per unit root length, greater denitrification rates, lower mineralization and higher immobilization of N. In most fields nitrate contents in drain-flow fluctuate, but coincide with rainfall events and irrigation over flow (Powlson et al. 1992). Macdonald et al. (1997) point out that much of the nitrate in risk of leaching may actually be derived through mineralization. Goss et al. (1988) summarized that nitrate loss from wheat belt in England are immensely influenced by a) soil mineral N residual status in autumn; b) amount of drain flow; c) fertilizer N rates and rainfall at spring top-dressing; d) crop residue and tillage regimes.

Crop rotations influence the extent of N losses. Actually, development of compact efficient cropping systems, that allow least N losses is a necessity within the European wheat belt. It avoids unnecessary excess N inputs. In England, Smith et al. (1997; 1998a) analyzed N losses using a dynamic simulation model (eg. SUNDIAL), and found a difference equivalent to 132 kg N ha^{-1} between best and worst rotation. Such reduction in N loss achieved through a better crop rotation, also decreased nitrate contents in irrigation/drainage water. Much of nitrate lost through leaching or denitrification is derived from the mineralization of organic N, and mostly during the period when wheat crop is not absorbing it. It is believed that in most wheat-based arable crop rotations, such losses are difficult to minimize. Simple, accurate fertilizer scheduling may not be of great consequence in terms of 'Plugging N leaks'. In wheat fields, nitrate leaching through water flow may occur in a vertical phase (eg. in sandy, light textured soils) or approximately horizontally in poorly drained heavy soil. Nitrate leaching, in whatever direction it be, depends on NO_3 concentration in soil solution. Jarvis et al. (1996) suggest that NO_3 loss patterns from soil could be super-imposed with soil mineral status achieved via fertilizer N inputs. They may follow similar patterns, meaning that NO_3 leaching accentuates with high NO_3 levels. It may still be inaccurate mainly because, mineral N in soil profile is continuously recharged through soil organic N mineralization. Mineralization is also a key factor that enhances NO_3 leaching in wheat field. The crux of nitrate leaching problems is in the 'Untimely nitrate'$^{-1}$, i.e. soil NO_3^- N pool that remains without being synchronously utilized by wheat crop, as and when it becomes available in soil because of mineralization or fertilizer N inputs. Some leaching loss is inevitable, because temperature thresholds for root absorption is higher than microbial aided mineralization and N release. Therefore, whenever fertilizer N recovery is low, a certain quantity of NO_3 accumulates and becomes vulnerable to leaching. In summary, matching N supply, ie. mineral NO_3–N level with crop demand and recovery patterns is crucial, if NO_3–N leaching losses are to be minimized in this wheat zone. Whenever possible autumn application of N fertilizer should be avoided, so that leaching is reduced.

On heavy fertilized (280 kg N ha^{-1}) cereal fields in southern France, the total N loss into atmosphere reaches 30 to 110 kg N ha^{-1}, with <1% as NH_3, 40% as NO, 14% as N_2O and 46% as N_2 (Jambert et al. 1997). On Broadbalk fields, N loss via denitrification was 17% compared with 13% via leaching (IACR, 1997). Volatilization losses could be due to volatilization of ammonia, or due to conversion of NO_3 to nitrous oxide or N gas. In North Europe, livestock farming accounts for 80% of NH_3, N_2O, or N_2 emission. Fertilizer inputs to wheat or pasture in rotation is estimated to result in 5% of N loss as emission equivalent to 9.1 Kt NH_3–N annually (Skinner et al. 1997). Ammonia volatilization from the aerial parts of wheat at maturity, due to relatively warmer temperatures is another source of N loss (Carranca et al. 1999). Nearly 20 kg N ha^{-1} could be lost through such volatilization processes.

Nitrous Oxide Emission: Agricultural soils are a major cause of N_2O emission into atmosphere. In Europe, 80% N_2O emission in arable soils has been attributed to agricultural activity. The annual N-application rates for wheat cultivated on arable soils of Europe is generally high, and ranges from 100 to 350 kg N ha y^{-1}. Hence, N loss as nitrous oxide emission increases proportionately. It affects fertilizer-N use efficiency. Kaiser and Ruser (2000) utilized closed chamber technique to monitor N_2O flux into atmosphere, and found that for an year it ranges from 0.5 to 16.8 kg N_2O ha^{-1}. Time course investigations by Roner et al. (1998) have revealed that microbial processes were responsible for N_2O production in both thawing and freezing soils. Nearly 50% of N_2O emission occurred during winter (Table 3.3). Such N_2O emissions also depended on N inputs, and type of fertilizer-N. Organically enriched soils or those treated with +FYM emitted higher levels of N_2O when compared to wheat fields supplied with mineral-N. Application of 220 kg N ha^{-1} to soil increased significantly high N_2O losses throughout wheat cropping season compared to unfertilized fields.

In many of the unfertilized sites in Germany, Kaiser and Ruser (2000) estimated that a 60 kg N ha^{-1} recovery into crop yield resulted in 2 kg N ha^{-1} emission as N_2O. Such high background emission of N_2O was attributed to rapid turnover of larger quantity of soil organic matter. Further, they suggested that N_2O emission was more site-specific. For a wheat-barley and sugar beet rotation common to many northern European locations, the N_2O emission seems to increase with levels of N-balance. Obviously positive balance can lead to N accumulation, which is susceptible to denitrification and emission losses. According to Addiscott and Powlson (1992), the extent of N loss via denitrification seems greater than that due to leaching.

Table 3.3: Influence of nitrogen fertilization, crop species and geographic location on N_2O emission, N recovery (N in yield) and mean N-balance in German loess zone

Crop/ location	N Input kg N ha^{-1}	N in Yield kg N ha^{-1}	N Balance Kg N ha^{-1}	Total Kg N ha^{-1}	N_2O Emissions Losses in winter %	N_2O/ fertilizer %
Crop						
Wheat	17.1	156	+15	2.81	43	1.64
Barley	160	119	+41	1.68	59	1.05
Legumes	0	243	−243	2.10	55	NA
SB-W-B	71	98	−26	2.39	NA	NA
Location						
Braunschweig	162	117	+45	2.77	56	1.71
Scheyern	151	140	+11	7.58	47	4.99
Gottingen	157	124	+33	1.13	53	0.72
Timmerlan	147	119	+28	2.58	55	1.75

Source: Kaiser and Ruser, 2000

Note: Loss in winter refers to N_2O emission during winter wheat crop; N_2O/Fert = Fertilizer related N_2O emission; SB-W-B = Sugar beet-Wheat-Barley rotation; NA = Date not available.

Nitric Oxide Emission: The major NO sources are combustion of fossil fuel, biomass, burning and soil-microbe mediated emission. NO production in soils is through nitrification, denitrification and other processes, such as chemical reaction (Conrad, 1996). Within the wheat agroecosystem, amount of NO emission is known to coincide with the N inputs to fields. In order to monitor and quantify NO emission on a continuous basis, Gut et al. (1999) utilized membrane tube technique (METT). In their study, largest NO emission occurred following fertilizer application and irrigation/rainfall events. Such emission occurred in pulses of 4 days, and spread across 3 weeks, coinciding with elevated soil NH_4–N concentration. If non-fertilized (control) plots emitted 5g NO–N ha^{-1} 5d^{-1}, those supplied with NH_4NO_3 emitted 95g NO-N ha^{-1} 5d^{-1}, and those given cattle slurry emitted 31 g NO-N ha^{-1} 5 d^{-1}. Irrigation event had distinct influence on NO emissions from wheat fields. In 4 days following irrigation 39 g NO–N ha^{-1} was emitted. The daily pulses of NO–N emission ranged between 0.3 and 0.7 g NO–N ha^{-1} d^{-1}. According to Gut et al. (1999) the NO emission after fertilization or irrigation event is a result of complex interaction between amount of fertilizer N, climatic factors and soil physical properties. Such NO emission patterns in wheat field could be simulated using nitrification algorithms, and by introducing parameteri-zation of NO-production into the model propounded by Galbaly and Johansson (1989). These can improve our ability to predict N loss via NO emission, as well as the extent of its effect on tropospheric ozone.

Nitrogen Removal by Crop: Over a large expanse (>100,000 ha) of agricultural belt in the United Kingdom, winter wheat is rotated with other cereals and legumes. Analysis of nitrogen input verses removal equations has revealed that typical 7.5 t grain yield per ha contains 19 kg N t^{-1} grain. To achieve it, the wheat crop removes 190 kg N ha^{-1} in toto from the soil. Nearly, 140 kg N of it is located in the harvested crop (panicles), and the remaining 50 N ha^{-1} that occurs in crop residue is returned to soil (Jarvis et al. 1996). Such nitrogen removals by wheat are obviously influenced by a range of factors related to crop, soil and environment. Genetic constitution of the crop and its physiological manifestation, particularly harvest index, N uptake and translocation index influence net N removals immensely. During the past five decades, genetic gain in the harvest indices of European wheat genotypes has caused large changes in biomass, stover and grain output, and along with it N removals have changed. Macdonald et al. (1997) at Rothamsted in England, depict one such situation, wherein harvest index increased from 0.32 to 0.48. This has enhanced N allocation into grain by 12 to 45%. Consequently, it improved N-use efficiency in terms of grain productivity. Semi-dwarfing has also reduced the extent of N in crop residue, hence less N is recycled into the wheat ecosystems.

Fertilizer N Recovery: Fertilizer N recovery, within limits may vary depending on soil, crop genotype and environmental parameters. Some times low N recovery could also be attributed to poor crop stand, diseases, pestilence and weeds. In each of these situations, the normally possible levels of N recovery is altered. Soils, in particular, affect fertilizer N-use efficiency due to their physico-chemical composition. Rooting and differential nutrient availability in soil are other factors that affect N absorption.

In England, Powlson et al. (1986) made a 3-year evaluation of wheat crop grown in rotation at Broadbalk. They traced nearly 57% fertilizer N in the aboveground parts of winter wheat, of which, the proportion of fertilizer-N in wheat grains averaged 37%. In an experiment at Prague, Czechoslovakia, the average recovery of ^{15}N labeled fertilizer into wheat crop was 33.8% (Mouchova et al. 1996) Long-term ^{15}N label studies at Rothamsted indicate that around 32% N is allocated to grain formation, while crop residue may retain up to 20% which can be recycled upon incorporation after harvest. Of the remaining N, up to 24% is held in soil, mainly sequestered in organic fraction and very little in available inorganic pool. In some instances soil and fine roots distributed at 0 to 30 cm profile together retained up to 27% of fertilizer N applied. From a later study in 1992, Powlson et al. (1992) reported that N recovery ranged between 46 and 87% for winter wheat grown on three different soil types in southeast England, namely sandy loam, chalky loam and silty loam. Approximately 3% of applied fertilizer N that stays in inorganic form gets localized within 0 to 23 cm soil profile. In general, much of inorganic N ($NH_4^+ NO_3$) measured immediately after harvest of wheat crop was localized in top 0 to 20 cm of the soil profile (Macdonald *et al.* 1997). Further, Macdonald *et al.* (1997) caution that within this wheat belt a sizeable loss of fertilizer N is to be expected. Nearly 24% of it could be lost through leaching, volatilization, nitrification and denitrification.

Carranca *et al.* (1999) state that a greater number of studies involving ^{15}N label fertilizer have been reported from central and northern European wheat belts. For example, Pilbeam (1996) has summarized and listed large number of such studies about fertilizer N recovery using ^{15}N label. Whereas, fertilizer N recovery studies in southern European plains emanate mainly from Portugal and Spain, wherein wheat is cultivated on Haplic Luvisols (Carranca *et al.* 1999; Abreu *et al.* 1993; Carvalho and Basch, 1994). In these locations fertilizer N recovery into the wheat crop ranges from 22 to 40%, irrespective of the source, be it either urea or calcium nitrate. Higher N recovery occurred during the first year of the evaluation compared with 2nd or 3rd year. A good fraction of fertilizer N, 15 to 60% was retained in the soil after harvest of wheat crop. Top-dressed N formed the major component of N retained in soil. A significant fraction of N retained in soil, actually was held as NH_4 fixed in clay fraction, especially in surface horizon. Loss of fertilizer N occurred through leaching. Leaching losses were consistent throughout the 3 year period of evaluation. Other losses were through NO_3 leaching. Over all, total N losses ranged between 35 to 55% of fertilizer- N applied. Considering the fertilizer N dynamics as deciphered in the Portuguese wheat belt, Carranca *et al.* (1999) conclude that, fertilizer N inputs should be avoided during autumn and winter periods. It then reduces leaching losses.

Fertilizer N Inputs, Soil N Pools and Recovery by Wheat: In the field, recovery is determined to a good extent, by the complex interactions among the level of N-inputs, transformation of soil N, mainly mineralization/immobilization, the resultant changes in soil N pools and physiological status of wheat crop. Many of these aspects about N dynamics verses crop growth have been abundantly reported, and major share of our knowledge on it is derived utilizing ^{15}N label fertilizer.

Let us consider one such report by Blankenan *et al.* (2000), wherein fertilizer-N recovery by plants grown in controlled conditions range from 74 to 80%, if based on ^{15}N assay, and from 74 to 83% if based on field test using fertilized and unfertilized plot for comparison. Using a similar approach, Recous *et al.* (1988) reported 49% ^{15}N fertilizer recovery in control conditions, and 51% N recovery using difference in fertilized and unfertilized plots. Generally, fertilizer-N recovered into crop increased with N rates, indicating that fertilizer N inputs enhanced availability of soil N to plants. Clearly, the magnitude of mineralization/immobilization processes are crucial factors determining available N pools in soil. The effective soil N mineralization (Nmin) is actually indicated by soil Nmin minus fertilizer-N immobilized, and that lost. In the above trial, fertilizer-N recovered into organic-N (Norg) and microbial-N (Nmic) was only 5 to 9% and 6 to 7% respectively. Obviously, larger share of immobilized N is held in non-extractable condition within organic fraction. However, increasing N inputs did not enhance immobilization of fertilizer N. It was attributed to better root growth that enhanced fertilizer N-use efficiency. Incidentally, 60% of wheat root emergence occurs within first 4 weeks, and its spread confines to upper 60 cm layer. Unlike field trials, fertilizer N loss via leaching gets effectively avoided under controlled conditions. Hence, non-recovered fertilizer-N could be lost via denitrification leading to N-emission. At this stage we may realize that denitrification depends on organic C in soil and microbial activity. Hence, there is no doubt that fertilizer N recovery varies depending on soil characters, crop genotype and level N input.

Cropping Sequences and Nitrogen Dynamics: Cropping sequences and their influences on nutrient dynamics in European wheat belt has been studied extensively. In practical situations, wheat farming is only part of a complex cropping sequence, rotations and other soil/nutrient management decisions that relate to individual farms as a unit. In this regard, there are several computer-based simulation models that consider a wide range of such options and then evaluate the ensuing N dynamics (eg. CENTURY, DAISY, NASA-CASA, AFRCWHEAT-2). SUNDIAL, is one such dynamic simulation model that considers cropping sequences, carry forward of N' or N credits of previous crops, soil characters, crop genotype, of course weather etc. It then allows us to evaluate the pros and cons of adopting various complex rotation systems, so that best possible N-use effect could be achieved. In fact, a decision support system (DSS) helps farmers/policy makers to judge N dynamics better, and accordingly explore better cropping sequences and N management strategies (Smith *et al.* 1997; Smith and Barraclough, 1998). Table 3.4 provides an example wherein different cropping sequences that include wheat has been evaluated using SUNDIAL. Use of such simulation models is becoming frequent. In one such example, the overall N off-take from the cropping system actually improved by 21% through the use of N simulators.

Table 3.4: Simulating cropping sequence and N dynamics—overall estimates of potential N savings in Northern Europe (England)

Evaluations	Region in England		
	South West	Central	East Anglia
N lost in worst-case rotation (kg ha^{-1})	67.3	633	554
N lost in best-rotation (kg ha^{-1})	61.7	609	504
Total N savings (kg ha^{-1} 6 years^{-1})	56	24	50
N savings per crop (kg ha^{-1} y^{-1})	9	4	8
Economic value of better N dynamics (M £ ha^{-1} y^{-1})	3.4	1.5	31

Source: Smith et al. 1997; (IACR) Reports, Rothamsted, United Kingdom

E. Phosphorus Dynamics

Major categories of soil P encountered within this cropping belt, can be grouped into organic and inorganic forms. Organic forms occur as mono- and diesters (eg. inositol hexaphosphate, phospholipids, nucleic acids); as biomass P and humic P. Inorganic P is encountered in ionic (PO_4, H_2PO_4) or mineral (eg. apatite) forms (Addiscott and Thomas, 2000). A variety of inorganic physico-chemical processes affect P availability to wheat roots, as well as its dynamics in soil—mainly leaching, fixation and accumulation. The primary source of P for wheat roots is inorganic P. Dissolution of PO_4 available in apatite is a major source in addition to inputs of soluble P fertilizer such as super phosphate, mono-ammonium or di-ammonium phosphate. Phosphate precipitation is of interest, mainly because, European farms utilize higher levels of P fertilizer, and precipitation can clearly lessen the vulnerability of excess P applied, to leaching. Actually, adsorption of PO_4 which is influenced by oxides and hydroxides of Fe and Al seems more important in terms of lessening leaching loss (Addiscott and Thomas, 2000). Although, organic matter in soil does not sorb phosphate directly in all circumstances, still it can have a strong influence on sorption and desorption. Desorption processes enhance P availability to plants. Desorption of P is initiated whenever P concentration in soil solution dips, because of its withdrawal (uptake) by plants roots. Desorption may also occur immediately after rainfall event when P is leached out in run off through percolation. In many instances desorption process can enhance vulnerability of soluble P to leaching.

Phosphorus Leaching: Del Campilo *et al.* (1999) state that wheat farmers in western and northern Europe have been constantly applying all major nutrients, including P at high levels. With regard to P, it is clear that orthophosphate readily reacts with soil components, but the capacity to sorb/bind is limited depending on the soil type. The cultivated layers of soil get saturated with P because of inputs beyond plant's demands.

Beyond this stage, substantial amount of P could be leached to deeper layers of soils and ground water, or lost via drain flow and surface runoff. Analysis of Broadbalk continuous wheat experiment has shown that critical value or "change point" for leaching through drainage water begins at 60 mg Olsen's P kg^1 soil. Below this change point very little P was lost, however if P concentration in soil was increased, then proportionately increasing levels of P were leached (Leigh, 1996). Matching results were obtained in soil samples incubated in laboratory. A similar type of relationship occurred between Olsen's P and $CaCl_2$-soluble P. The $CaCl_2$-soluble P was very small up to "change point", but increased linearly above this limit. Clearly, assessing P change point for soils under wheat crop will be useful in predicting the onset of P losses through leaching, and that via drainage water. We may realize that "change point" varies with soil type. For example, the change point was only 25 μg P kg^1 soil for a sandy loam from Suffolk, England (IACR, 1996). Some of the other suggested methods that thwart P leaching via drainage water from wheat field is to prevent/or lessen drain flow itself by employing U-bend in pipes. Developing a fine tilth in soil could also restrain P loss to lower horizon by making pathway for water from cultivated layer more tortuous. For example, in experiments at Brimstone, the concentration of soluble P in drain water or surface runoff were beyond 20 μg P^{-1}, which is thought to cause eutrophication. However restricting the drain flow, so that water is held for longer duration in contact with soil decreased the soluble P losses by 29 to 52% (Tunney et al. 1997).

Mineralization: Replenishment of available P that is easily absorbed by wheat root depends on addition of inorganic phosphates (Pi), on desorption of Pi from particle surface, on solubilization of precipitated Pi and on solubilization of organically bound P (Po) (Stewart and Tiessen, 1987). Mineralization of Po may play a role in replenishing plant available Pi in many soils. A recent summarization of results from long-term trails at Askov, Denmark suggests that Po primarily comprises microbially derived tiechoic acid-P and diester-P species. This Po pool in soil is affected by seasonal changes in microbial activity and environmental influences. Fertilization generally increases labile Po, however because of rapid transformations, the effect of P fertilizer inputs on Po may not be perceived significantly. This labile Po in soil is mainly associated with clay-sized (<2 μm) organo-mineral complexes, and this fraction plays an important role in short-term turnover of soil Po (Guggenberger et al. 2000). However unlike with N in soil, mineralization is not a very significant process that contributes to phosphate availability in the soil. Addiscott and Thomas (2000) found that largest releases of P through mineralization may satisfy one half of the cereal crop's annual requirement at grain yield level of

5 Mg ha^{-1}. It is believed that mineralization may contribute as much P that is supplied through fertilizers. Among soil organic P fractions, both diesters and tiechoic acid P which are easily mineralized contribute most to available P pool. Hence, after long-term cultivation (70 to 80 years), only monoesters may be found within humic portions. Generally, mineralization or immobilization reactions occur simultane-ously, and their rates are dependent on C:P ratio of the organic amend-ment, and extent of inorganic P fertilizer inputs.

On an average, an intensively cultivated wheat crop in northern Europe, that yields 7.5 to 8.5 t grains ha^{-1} removes 32 to 36 kg P ha^{-1}, and partitions approximately 40% of it into grains. Wheat recovers nearly 60% of its P requirement before anthesis, greatest absorption rates occur between tillering and anthesis. The P utilization efficiency of wheat ranged between 250 and 350 kg grain per kg P absorbed. Overall, to produce 1 t grain, a wheat crop utilizes 4 to 5 kg P depending on soil type, fertility status, environment and genotype. The phosphate physiology of wheat genotype therefore has a say in P dynamics.

Long-term trials conducted at Humbold-University, Berlin, involving cereals (wheat, barley) have indicated that in 60 years, in plots not fertilized with P and K fertilizers, soil immobilized 16 kg P ha^{-1} and 15 kg K ha^{-1}. On plots that received additional P and K inputs through FYM, stubble and other organic sources, it led to a nutrient decrease of 19 kg P ha^{-1} and 99 kg ha^{-1} K through immobilization (Ellmer, 2000a).

Sulphur Dynamics: Sulphur inputs into wheat fields are derived from leaching of S bearing minerals, then a good fraction through atmospheric deposition and use of S containing fertilizers especially single super-phosphate. Analyses of archived soil samples using stable isotope ratio $^{34/35}$S indicated that in England, peak S deposition occurred in 1970's. At present, it has reduced by over 40% (Zhao and McGrath, 1994). Within the Broadbalk long-term wheat trials (130 years), it is estimated that 30 to 40% of S in top soil has originated from atmospheric deposition (IACR, 2001). Human activity has also affected S cycling within the European cereal belt. To a large extent, the S cycle in the soils of wheat cropping systems is regulated by soil organic matter. Sometimes, S deficiency may also influence the rates of N or C cycle. Therefore, triple labeling with stable isotopes ^{34}S, ^{13}C and ^{15}N has been suggested to arrive at quantitative relationships between S, organic-C or Organic N (IACR, 2001). Minerali-zation contributes to S uptake by crop, practically during all stages of growth. Whereas, leaching is an important aspect of S cycle that lessens its availability to wheat roots. Leaching loss of S is determined by soil type and rainfall (IACR, 1997). Therefore protecting S inputs through better timing is important.

Sulphur deficiency has increased considerably in the European wheat belt, mainly because of intensive cultivation, and use of S-free fertilizers. In Britain, nearly 30 to 50% of cereal fields are partially deficient in S. A risk assessment model for S deficiency predicts that by 2005, 22% of cereal farms in UK may develop S deficiency (IACR, 1996). Field trials indicate that S deficiency can severely threaten potential wheat yield, sometimes leading to 50% reduction in wheat grain yield. Physiologically, S containing pools in wheat plants respond to a change in external S supply. There are simple methods to ascertain plant S status. In this regard a ratio of malate to sulphate is a useful indicator of S status in plants. A ratio >1 indicates deficiency, but < than 1 indicates S sufficiency (IACR, 2001). Elemental S, if used as fertilizer needs to be oxidized to sulphate-S before it is extracted by wheat roots. Physical size of particles, either in powder or particulate form is also a crucial factor that regulates S oxidation to SO_4. Generally, S fertilizers in sulphate form were prone to leaching, consequently, there was little residual value from S inputs in the previous year. For example, annual application of >50 kg S per ha^{-1} as sulphate did not result in significant S accumulation in Broadbalk long-term wheat trials (IACR, 2001). Actually sulphate applied in excess of crop recovery was lost via leaching and drain flow.

F. Long-term Nutrient Balance in Wheat Fields

There are several locations within the European wheat belt wherein long-term nutrient balance is being monitored since mid-1800s. Among them studies at Rothamsted (since 1843), Grignon (1875), Halle (1879), Askov (1894), Bad Lauschtad (1902) Bonn (1906) and Moscow (1912) have immensely improved our knowledge about soil nutrient dynamics and wheat productivity. Such nutrient balance studies indicate that wheat crop utilizes larger quantities of N and K than P (Table 3.5). At the same time, loss of N and K from soil and the ecosystem in general, are higher because they are comparatively more mobile elements in the soil matrix. Now let us consider in detail one such long-term nutrient balance study conducted at Skeirnewiece in Poland by Mercik *et al.* (2000). They report that during the past 35 years wheat grain yield in Poland has averaged at 3 to 4 t ha^{-1} y^{-1}. For an input of 60–75 kg N $ha^{-1}y^{-1}$, the net nitrogen balance showed a loss of 11 to 14 N ha^{-1} y^{-1}, which corresponds to 15% of ammonium nitrate applied to fields without FYM. In the absence of any fertilizer-N inputs, the wheat crop extracts 30 kg N $ha^{-1}y^{-1}$, and the N content in upper soil layer stabilized at 0.033% N. Clearly, in the absence of external N input, N extracted by crop and that estimated in soil must have been derived via atmospheric inputs, symbiotic N-fixation and mineralization.

Table 3.5: A summary of long-term nutrient (N,P,K) balance and wheat grain productivity recoded at Skierniewice, Poland, between 1962 and 1996

	Nitrogen	Phosphorus	Potassium
Average crop yield (t ha^{-1} y^{-1})	3.8	4.2	3.6
Nutrient inputs (kg ha^{-1} y^{-1})			
Mineral fertilizer	59	26	91
FYM	29	NA	NA
Crop uptake			
Total (kg ha^{-1} y^{-1})	124	19	102
From fertilizer (kg ha^{-1} y^{-1})	36	8	53
Percentage	60	30	58
Nutrients retained in soil (kg ha^{-1} y^{-1})	+10	+17	+37
Balance losses (kg)	−14	−1	−1
percentage	23	4	1.5

Source: Excerpted from Mercik *et al.* 2000
Note: All the fields received mineral fertilizer, FYM and lime. Inputs were N at 90 kg ha^{-1} y^{-1}; P at 26 kg P ha^{-1} y^{-1} since 1976. Nitrogen balance calculated on data from 32 years, while that for P and K from 20 years.

The P balance for plots not fertilized with minerals nor FYM indicate a decrease of 14 kg P ha^{-1} y^{-1} in the upper soil (0.70 cm) layers. Out of this 14 kg P decrease, about 11 kg P ha^{-1} y^{-1} could be recovered in crop, meaning that 3 kg P ha^{-1} y^{-1} were lost via leaching, chemical fixation or to weeds. Long-term P balance trials at Bonn in Germany indicated that wheat crop recovered 16 kg P ha^{-1} y^{-1} and a similar value was reported from the analysis of Broadbalk wheat trials (Ellmer *et al.* 2000a, b). In plots with K replenishment, wheat crop recovered nearly 50 kg K ha^{-1} y^{-1}, derived mainly from 0.70 cm soil layer from the slow release pool. However, available K pool did not alter significantly. Over the 20 years period of testing, the slow release fractions of K decreased by 41 kg ha^{-1} y^{-1}.

2. The Wheat Agroecosystem of Australia

> *The Australian wheat agroecosystem represents historically, yet another recent human effort in expanding the limits of Agrosphere into the Southern Hemisphere. Throughout the evolution of this moderately intensified cropping ecosystem, nutrients have been key determinants of its sustenance and productivity. At present, it contributes significantly to the global wheat produce. Both, expansion and intensification of this agroecosystem are likely, but depending on soils, environment and global demand for wheat grains.*

A. Expanse, Agroclimate, Soils, Cropping System and Tillage

Historically, the seeds for wheat-based cropping system in Australia were sown by the European settlers around late 18th century (Perry, 1992). The early agricultural activity in Australia, however was predominately pastoral. Large scale cultivation of wheat began in the 1850s concurrent to the gold rush. Initially, around 1870s, the average wheat grain productivity was low at 0.7 to 0.8 t ha^{-1}. It declined to 0.5 t ha^{-1}, mainly due to phosphorus depletion. This trend in grain yield was reversed through replenishments of P and N. Then, in 1890s 'long fallows' were practiced to refurbish soil fertility. Wheat grain yield spurted during 1890 to early 1900s, mainly due to introduction of better performing varieties. Further increases in early 20th century were attained through better management of nutrients, when both major and micronutrients (Zn, Co, B and Mo) were replenished. In mid-20th century, introduction of high yielding semi-dwarfs, fallows and sequences with forage legumes gave stability to wheat yield (1.2 to 1.5 t ha^{-1}). At present, the Australian wheat-based cropping system predominantly involves two forage legumes, namely, clover (*Trifolium subtarraneum*) and *Medicago* sp., both introduced from Mediterranean countries of West Asia around 1890s.

Expanse: The wheat-based cropping system in Australia extends into all states except parts of Northern territory. However, its cultivation is predominant in Western Australia (4.5 m ha), and New South Wales (3.5 m ha), which together occupy 70% of the total 12.0 m ha wheat zone. Victoria, Queensland and Tasmania together cultivate wheat on 3.0 m ha. Published literature on cropping practices, nutrient dynamics and yield efficiency is indeed vast. However, discussions within this chapter are focused mainly on wheat in Western Australia and New South Wales, and confines to dynamics of nitrogen, carbon and phosphorus.

Agroclimate: In this region with temperate to cool Mediterranean climatic patterns, the precipitation varies with regard to total rain, seasonal length, dates of beginning and last rains, as well as inter-and-intra seasonal distribution. The southern/western Australian wheat zone receives between 275 and 625 mm annual rainfall. Rainfall peaks during winter (April to October) growing season. Nearly 70% of total annual precipitation occurs during the crop season. Rapid drainage and leaching which are pronounced immediately after a rainfall event can erode soils, and along with it the valuable nutrients could be lost. Terminal drought that may occur due to sharp decline in rainfall during spring affects grain filling in wheat (Stapper and Haris, 1988). Simulations and analysis of previously observed data by Asseng *et al.* (1998) suggests that precipitation and drainage (water loss) during the wheat crop season, spanning from April to October has varied considerably during the past 18 years. In a crop season, precipitation minus the drainage loss figures for soils in wheat belt has varied between 120 and 330 mm. Precipitation alone, on its own, is a potent factor that dictates nutrient dynamics in soil, and grain yield formation. However, more commonly, soil moisture and its interaction with N status has been influencing wheat harvests appreciably (Cantero-Martinez et al. 1999). Higher levels of soil moisture, supported by higher soil N almost always provided additive effects on grain yield. French and Schultz (1984) reported that such soil moisture mediated effects on wheat yield can be appreciable. For every mm of precipitation above cumulative 110 mm, the grain yield increased by 20 kg ha^{-1}. During the wheat crop season, the main rainfall events that spread across May, June and July months play a crucial role, in terms of soil moisture levels possible, and severity of drainage losses. The drainage loss may fluctuate depending on the intensity of individual rainfall event. However, on a cumulative basis it ranges between 100 and 200 mm per crop season. The evapotranspirational losses too accentuate gradually with time, and on a cumulative basis accounts for approximately 150 mm in a crop season. The net result is that soil moisture content is held fluctuating between 100 and 180 mm during April to May ie. at seeding, until October/November when wheat grains are harvested Anderson et al. (1998a, b).

More accurately, it is the distribution of soil moisture at different soil depths which is crucial, because it affects nutrient dynamics at different soil layers, and availability of nutrients to plant root. The dynamics of soil moisture at different depths, again is regulated by interaction between precipitation, evapotranspiration and water holding capacity. Generally, soil moisture at upper layers (0 to 80 cm) depreciates as the crop progresses, but at lower layers the curve fallows a hyperbolic trend. Peak accumulations occur at 3 weeks after sowing which coincides with better precipitation initially in the season. Soil moisture at 80 to 220 cm depth then reaches a maximum at or around anthesis and decreases a bit sharply later, to become least at harvest. Obviously, along with it the nutrients accumulated in lower horizons are also affected.

The average temperature during the growing season fluctuates between 9 and 21°C. Mid-season months June and July are coolest (10°C). The average radiation received during crop season fluctuates between 12 and 355 MJ m^{-2} d^{-1}, but during June/July which are also most rainy, it may drop to less than 12 MJ m^{-2} d^{-1}.

Soils: The major soil types utilized for cereal production in Australia are coarse textured soils (Qartzipsamments, Torripsamments), earths (Calciorthids, Palexeralfs, Haplustalfs), clays (Paleusterts, Chromousterts, Torrerts), and duplex soils (Rhodoxeralfs, Natustalfs, Paleustalfs) (Perry, 1992).

In particular, the western Australian wheat belt, to a great extent thrives on duplex and deep sandy soils. These are low in C and restricted in N supply. On such soils wheat productivity varies considerably depending on fertility status, season and location (Asseng et al. 1998a, b). One of the major constraints in exploiting soil nutrients efficiently is the restriction to root proliferation (Delroy and Bowden, 1986; Gregory et al. 1992). The solenized brown soils support 25 to 30% of wheat production, in New South Wales, Victoria and South Australia. Black earths are frequent in Queensland cereal belt. The yellow and red earths are utilized for wheat production both in Western Australia and New South Wales.

Cropping Sequences and Tillage Practices: In the Australian wheat belt, major cropping sequences practiced are wheat-wheat, wheat-lupin (or suitable legume), wheat-annual pasture and wheat-fallow. The lupin-cereal two-year sequences are most popular with Western Australian farmers (Perry, 1992). Annual cropping with a wheat-summer fallow is also common. Fallows precede each crop. A few other variations are:
  (a) six crops of wheat in 6 years (wheat—fallow repeated 6 times);
  (b) four crops in 5 years (wheat-fallow-sorghum-fallow-wheat-fallow-sorghum-fallow); and
  (c) nine crops in 6 years (wheat-fallow-soybean-wheat-soybean-fallow-sorghum-fallow-soybean-wheat-soybean-fallow-wheat).

A major constraint attached to any of these rotations is that 60% of total rainfall that occurs during May-October, is not utilized efficiently. In most situations, long fallow (10 to 18 months) which provides subsequent crops with additional soil water reserves, indeed enhances precipitation use efficiency. According to O'leary and Connor (1997) alternate forms of conservation tillage and fallows are being increasingly preferred. Since 1980, wheat farmers in southern Australia have been sequencing wheat with canola and other vegetable oil bearing *Brassicas*. The advantage called 'break crop effect' may relate to increase in soil N after canola that enhances N recovery and yield during wheat phase of the cropping sequence (Bisset and O'Leary, 1996; Kirkegaard et al. 1997, 1999).

The important tillage practices followed in Australia, like elsewhere in other wheat cropping zones has changed from the initial traditional tillage (TT) practices, to reduced tillage (RT), to the presently more common direct drilling (DD) or no-tillage (NT). It is generally accepted that continuous adoption of European conventional tillage has lead to depleted organic matter, and pulverized the soil beyond necessity (Packer and Hamilton, 1993). A change from the conventional to direct drill or no-tillage system has often reduced runoff and the attached nutrient loss. Analysis of physico-chemical properties have indicated that decrease in runoff was related more to creation of micro-porosity greater than 0.75 mm diameter, than to changes in organic carbon. The current trend within wheat farming zones is to employ RT, DD or NT, add FYM/mulches and rotate frequently with legumes green manure/break crops.

B. Soil Organic Carbon

Soil organic matter is a key factor that influences nutrient dynamics within wheat cropping ecosystem. Whitebread *et al.* (2000) point out that soil organic matter changes monitored for short or even medium term for couple of years may not be sensitive and provide useful data. Hence, long-term trials are needed. Generally, organic matter decreases following clearing of native vegetation, and subsequent cereal cropping occurs for longer durations. While it is clear that SOC decreases as cereal culture continues, changes in more labile C fractions may be crucial to nutrient dynamics in soil. Further, monitoring C management index (CMI), which is indicative of changes in both labile and total C content may be highly relevant (Parton *et al.* 1987; Whitebread *et al.* 1998). Timing of crop residue incorporation and its quality affects decomposition rates. Hence, it has relevance to nutrient release for the later stages of current crop, and the one that succeeds. However, use of crop residue that decomposes slowly may add to mulching effect and protect soil fertility by reducing erosion. Dalal and Meyer (1986) found that soils cultivated with wheat lost 1% of their organic C per year during the first 20 to 30 years. Incorporation of N-fixing grain legumes grown as break crop is known to enhance both labile and total carbon (Whitebread *et al.* 2000). Legume leys are known to be highly decomposable. For example, Hosain *et al.* (1996) compared several combinations of wheat with legumes, and found that labile C and total C increased significantly in upper layers (0 to 10 cm) of soil. Wheat stubble incorporation is known to enhance labile C and total C retention by 7% and 13.5% respectively. The rate of wheat stubble decomposition, however depends on C : nutrient ratios, microbial activity, temperature and soil moisture conditions. Overall, Whitebread *et al.* (2000) suggest that wheat residue recycling has dominating effect on SOC dynamics.

C. Nitrogen Dynamics

A recent survey indicates that 25% of wheat farmers in Australia supply >50 kg N ha^{-1} to the soil, but it ranges from 9 to 130 kg N ha^{-1}, at an average of 62 kg N ha^{-1} on long fallows, and 52 kg N ha^{-1} on short fallows (Hayman and Alston, 1996).

Nitrogen mineralization: Mineralization of N during the fallow and crop period, both are important in terms of quantity of N made available to wheat roots, and while deciding fertilizer N schedules. During the fallow period, nearly 30 to 64 kg N ha^{-1} was mineralized. Such higher values for mineralized N has been attributable to deep ploughing, longer fallow seasons and late break in wheat seasons (Anderson *et al.* 1998 a, b). In addition, whenever a legume crop precedes wheat, then the inorganic N levels reached 112 to 126 kg N ha^{-1}. However, if wheat preceded wheat, then mineral N released was much less at 72 to 76 kg N ha^{-1}. On an average, during a crop season in Western Australia, the net N mineralization rate per day ranged between 0.3 and 0.4 kg N ha^{-1}. During this period soil temperature did raise above 15°C. The cumulative N mineralization in soil measured throughout a single wheat crop season in Western Australia was 61.2 kg N ha^{-1}. During the first 100 days of the crop, nearly 45 kg N ha^{-1} was mineralized and a good fraction of it could be exploited by wheat during the grain filling stage (Anderson *et al.* 1998 a, b; Asseng *et al.* 1998). Succeeding fallows are known to contribute 14 to 21% of N supplied annually via mineralization. Comparisons between fallows, either before wheat/legume or after, indicated significant differences in mineralization rates. Given a particular wheat-based crop sequence, rainfall events during the period could influence N mineralization rates. Certain observations made in the field, and in laboratory conditions suggested that a spurt in N mineralization occurs immediately after rainfall. For example, a 40 mm precipitation in Western Australia resulted in 28 kg N ha^{-1} release into soils.

Nitrogen Recovery: Knowledge about N recovery and biomass accumulation at different physiological stages (on a Zadock's scale) of wheat is helpful in many ways. It firstly provides an idea regarding N demands, so that matching fertilizer N schedule could be decided. Then, the timing of split N doses can be managed effectively. The total N recovered into wheat crop may vary between 20 and 80 kg N ha^{-1} depending on fertilizer inputs and yield goals. Nearly 75% of total N recovery occurs by the time wheat plant reaches anthesis. Thereafter accumulated N effectively retranslocates into seeds, which are major sinks. Hence, retranslocation decides the extent of N recovered via grain harvests. On an average, wheat grown in deep sandy soils at Moore, in Western Australia which yielded between 1.6 and 2.2 t grains ha^{-1}, partitioned 25 to 45 kg N into grains. This amounts to average N-use index equivalent to 53 kg grains kg N^{-1}.

Nitrate Leaching: Primarily, rainfall events affect the extent of NO$_3$ leaching. For example, in Western Australia a rainfall spurt of 94 mm during wheat phase of a cropping sequence leached out 4.2 kg N ha^{-1} from red earth. Whereas, on the same soil, a 112 mm precipitation event leached 12 kg N ha^{-1} (Smith *et al.* 1998). According to Anderson *et al.* (1998a), NO$_3$ leaching spurts were related to initiation of winter rains, and each rainfall event. Overall, in a cropping sequence wheat phase of the rotation was most prone to NO$_3$ leaching. It is partly attributable to higher levels of inorganic N, still left in soil profile after a legume or legume based pasture. Among the options tested, clover-capeweed pastures that interject wheat sequences were least prone to NO_3 leaching. The reason being excessive N absorption by capeweed restrained N leaching losses.

In view of the above observations, Anderson *et al.* (1998a, b) aptly suggest introducing capeweed as a 'catch crop' in the wheat based sequences of Western Australia. Ordinarily, in temperate agrozones, such catch crops are grown to minimize NO_3 leaching between crop phases, particularly whenever precipitation distribution allows growing it. Soil-wise, red Kandosols (Red earth) are known to suffer comparatively less NO_3 leaching than deep sands, which is easily attributable to fine texture of deep sands.

Cropping Sequence and N Dynamics: Considering the cropping sequences, net mineralization rates were generally higher with wheat-lupin 2-year rotations compared to annual pasture-wheat or annual pastures. Leaching losses were maximum with wheat-lupin sequence, and it is attributable to deep ripping. Leaching losses were also encountered with annual pastures (Table 3.6). Perhaps, such pastures removed N efficiently, plus contributed N through symbiotic N fixation. In addition, Anderson *et al.* (1998b) believe that low extent of soil disturbance during annual pasture phase of the sequence retards leaching losses. The percentage N retained in soils was greatest at 55% of initial levels, whenever legumes were included in the rotation.

Table 3.6: Nitrogen dynamics as influenced by cropping sequences at Moore, Western Australia

Cropping sequences	Inorganic nitrogen (kg ha^{-1})	Recovery by plant %	Leached nitrogen %	Retained soil N %
2AP—1AP	224	42	12	41
W—1AP	137	21	29	22
W—L	139	28	42	27
L—W	121	28	23	55

Source: Anderson et al. 1998a
Note: Inorganic N refers to soil N available to plant at the start of season;
AP = Annual pastures, W = Wheat, L = Legume

Knowledge about N budgets during different wheat-based cropping systems is crucial while deciding on N inputs, judging agronomic N-use efficiency, and possible harvest levels. The change in total inorganic N between seeding till harvest is influenced by processes such as N mineralization rates during the crop season, N leaching, total N uptake by plant, N credits from legume whenever they are part of the crop sequence, N grazed if pastures interject the sequence, and extent of N storage in the soil. Table 3.7 provides an overview of such changes in N during a lupin-wheat sequence common in Western Australia. The net N credits derived from lupin can be appreciable, if measured at harvest. Whereas, during cereal phase of the sequence, inorganic N pools are depleted, plus there is greater N loss through leaching which is shown as 'N unaccounted for' in the Table 3.7.

Table 3.7: Nitrogen budget under a lupin-wheat sequence adopted in Western Australia (all figures in kg N ha^{-1})

YEAR 1 (1995) LUPIN PHASE			YEAR 2 (1996) WHEAT PHASE		
April	→	October	April	→	October
39	NO$_3$–N	>53	53	NO$_3$–N	27
13	NH$_4$–N	>14	14	NH$_4$–N	11
	Capeweed/Lupin uptake	−42		Wheat	−43
	Mineralization	63		Mineralization	72
	N Unaccounted	−12		N Unaccounted	−58
	N Fixation	90-112		N Credits from lupin	92-112

Source: Anderson et al. 1998 a, b
Note: Data on lupin N fixation, N credit to wheat extracted from Anderson et al. 1998b

Depending on locations, actual crop sequences and experimental observations, opinions may differ regarding the ability of legumes N fixation to supply satisfactory levels of N to the succeeding cereal (wheat). In this regard Anderson *et al.* (1998b) conclude that on deep sandy soils, legumes cultivated in rotation with cereals add sufficient organic N, through N fixation and residue recycling, that can easily support wheat productivity in excess of 3 t ha^{-1}. It is believed that efficient use of legume supplied-N is hampered mostly by asynchrony between N release verses cereal crop's demand for it. Precipitation is another factor which influences legume N fixation as well as N credits derived by the succeeding wheat crop. Above average rains generally ensured better N fixation. On an average, 23 to 38 kg N was fixed per ton biomass of clover. Higher amounts of N fixation, up to 112 kg N ha^{-1} have also been reported for clover. In practice, N fixation by lupin sequenced with wheat in Australia also varies. For example, 90 to 151 kg N ha^{-1} reported by Anderson *et al.* (1998b) is well below 181 to 327 kg N ha^{-1} observed earlier by Unkovich *et al.* (1994).

Quite often, the soil N 'spared' which is denoted as 'N credits' from legumes grown in sequence is deemed to benefit cereal phase (Evans *et al.* 1989; Chalk *et al.* 1993). However, the rates of N-fixation, mineralization of organic N and losses interfere with utilization of such legume derived N by wheat. To quote an example, *in situ* estimation of N released from senesced medic or clover based pasture and its residues indicated that only 20% N was routed to first wheat crop (Thomson and Fillery, 1997).

Certain observations on wheat in Western Australia suggest that legume (lupin)-derived N may not necessarily be a good source of this nutrient, because these deep soils are prone to heavy drainage loss. In fact, Anderson *et al.* (1998a) have remarked that rapid, early N leaching in the growing season, even before the wheat roots have proliferated enough into soil profile is the major cause of inefficiency. Furthermore, after wheat phase nearly 22 to 39% of mineralized N was left unutilized in soil at 1 to 1.5 m depth. It is a clear evidence for inefficient N recovery by wheat.

Canola is a relatively new winter oil seed crop, which is sequenced with wheat as a 'break crop' (Angus *et al.* 1991; Kirkegaard *et al.* 1999). Canola, generally resulted in superior break crop effect, but its influence on soil nitrogen (N) is questionable (Kirkegaard *et al.* 1997). Canola has 25% higher N requirement than wheat (Hocking *et al.* 1997). However, in their recent study, Kirkegaard *et al.* (1999) observed highest accumulation of mineral-N (94 kg N ha^{-1}) in all seasons after a canola crop. Possible explanations put forth are:
 a) a flush of mineral-N associated with release microbial N; or
 b) a general reduction in soil microbial activity leading to lower levels of immobilization. Whatever be the reason, mineral-N accumulated, and wheat yield increased if sown after canola crop.

Crop Physiological Aspects of N Efficiency: The crop genotypic composition of the wheat belt has a say in the nutrient dynamics. Physiological traits such as N recovery rate at different phenological stages of the crop, nitrogen use efficiency, retention in roots, N translocation index all affect the extent of N lost, that retained in soil or that recovered in stover and grains. The N uptake efficiency, and its utilization in grain formation through translocation from leaves, stem and roots varies with wheat genotype. In other words, fertilizer-N efficiency achieved in a field immensely depends on genotypic nature of wheat varieties sown. The agronomic efficiency of fertilizer N is about 40 to 50% (Anderson and Hoyle, 1999). However, among the wheat varieties cultivated in Australia, since 1990, we may encounter both N-efficient and inefficient versions (Table 3.8). Most yield efficient genotypes were also N-efficient, particularly with respect to uptake and utilization. It means that N retained in soil will be proportionately less, and that removed from the ecosystem via grains will be high. Also, more N will be recycled through stubbles.

Considering the growth stages, short season genotypes are known to recover about 13 kg N ha^{-1} in the post-anthesis period which is equivalent to 20% of N inputs. Whereas, long season genotypes recovered only 8 kg N ha^{-1} equivalent to 13% N inputs. According to Anderson and Hoyle (1999) N recovered after anthesis was not entirely related to N efficiency traits. Also, N uptake at tillering stage did not vary between crop

Table 3.8: Main nitrogen efficiency traits of wheat varieties used in Australia

Cultivar	Grain yield/kg N applied (kg/kg)	N uptake efficiency (kg/kg)	Utilization efficiency (kg grain kgN)	N uptake (kg/ha)
Yield efficient	22	0.57	39	12
Yield inefficient	14	0.47	31	10

Source: Anderson and Hoyle, 1999
Note: Yield efficient genotypes produced 2.7 t ha^{-1}, and inefficient genotypes 2.5 t ha^{-1}. Uptake efficiency relates to amount of N recovered per unit of N applied.

genotypes. Therefore, it was difficult to pinpoint the importance of N uptake rate at a particular phenological stage in terms of N recovery and yield. However, high vigor lines generally had lower recovery (48%) but greater utilization efficiency (30 kg grain kg N^{-1}).

Nitrogen budgets, and the balances achieved in soil is crucial to continued productivity. In a recent study, Hayman and Alston (1999) found that at 3.5 t ha^{-1} wheat grain yield in New South Wales, nearly 69 kg N ha^{-1} was removed through grains, but only 53 kg N ha^{-1} was supplied into soil leading to a negative balance of −16 kg N ha^{-1}. Whitebread *et al.* (1998) point out that 85% of N and P, and 61% of S in a mature wheat crop exists in grains, hence are lost from the system. In contrast, 80% of K is held in straw, which is recyclable. However, wheat straw if removed for hay, ethanol production or burnt, then K loss to the ecosystem is greater. One of the most exploitative cropping systems, according to them is wheat-lucerne-fallow with legume residues removed, no fertilizer and wheat stubbles not recycled. Least exploitative systems involve recycling of legume residues and wheat stubbles along with fertilizer inputs.

D. Phosphorus Dynamics

In Australia, P deficiency has been a prominent factor affecting the wheat yield formation (Perry, 1992). As a consequence, it can alter dynamics of N, and other nutrients. Legumes grown in sequence with wheat generally require larger quantity of P for growth, grain formation and N fixation activity (Bolland *et al.* 1999b). Phosphorus input into this wheat agroecosystem is definitely crucial. In fact, yield decline perceived during 19/20th century was effectively thwarted through P inputs. However, in recent times, P inputs into the ecosystem has varied widely between 0 and 36 kg P ha^{-1}. Phosphorus inputs have actually depended on yield goals, soil P reserves, cropping sequences and environment. Phosphorus inputs to wheat and crops sown in sequence have also depended on crop genotype and agronomic procedures adopted. Wheat sown in early April in New South Wales was provided only 0 to 2 kg P ha^{-1}. Whereas on moderate soils, nearly 15 kg P ha^{-1} was needed to achieve 90% of maximum yield. Phosphorus inputs as high as 36 to 45 kg P ha^{-1} have also been practiced in this wheat belt (Batten *et al.* 1999a).

Knowledge about recovery and accumulation of P at different phenological stages is crucial. It helps in ascertaining P demand in time and space, as well to channel P inputs efficiently. A wheat crop yielding 3.9 t grains ha^{-1} recovered 7.3 kg P ha^{-1} at floral stage, and at anthesis had accumulated 17.3 kg P ha^{-1}. At harvest, 4.8 kg P ha^{-1} was traced in straw and 13.5 kg P ha^{-1} in grains. The average P harvest index was 0.7. In other words, wheat crop draws greater fraction of P needs between sowing and anthesis (Batten *et al.* 1999). Therefore, re-translocation of P from leaves to grains seems to affect P dynamics within wheat plants. Seasonal effects on P dynamics within the wheat cropping zone can be severe. Sowing time is hence crucial. For example, in their study, Batten *et al.* (1999a) observed that late sown crop which had shorter vegetative period (only 95 d), proportionately restricted P absorption by roots. In fact in New South Wales, early sown (April) wheat required only 0-2 kg P ha^{-1} external inputs to yield reasonably. Whereas, late sown crops needed higher levels of P inputs (Batten *et al.* 1999). Phosphorus efficiency of wheat crop is dependent on two major aspects, namely P uptake and utilization efficiency. The extent of partitioning, grain productivity, amount of P recycled through stover and that removed via grains are interlinked to P efficiency traits. Phosphorus efficiency ratios (kg grain kg P^{-1}) reported for Australian wheat genotypes varies from 210 to 450 (Batten, 1992).

Fertilizer P efficiency is also influenced by several soil related aspects. Physico-chemical reactions that fix P into soil matrix, or those which render it unavailable to plant roots can decrease P efficiency. In addition to reduction in P efficiency because of chemical fixation, loss of P through leaching can be severe. Phosphorus is known to leach laterally in flowing water (runoff) during pasture phase. Vertical leaching during wheat/pasture phase is prominent on sandy soils with low capacities to retain P. Generally, P leaching is greater immediately after a high rainfall event. Phosphorus gets leached through two modes, namely as P in solution, or along with fine soil particles suspended in water. One of the options to reduce P loss in solution could be to use less soluble P fertilizers. According to Bolland *et al.* (1999a) during growing season from April to November, P leaching peaks around the wettest months June to early August. Hence, to achieve better P fertilizer efficiency, it is advisable to supply P immediately after waterlogging, and when P leaching has subsided. It then coincides with raising temperature and radiation that support higher crop growth rate, thus leading to greater P recovery. Applying P in late August can enhance residual P activity, and help raising productivity of succeeding pasture. In addition it may provide sufficient P at the start of the next season. To conclude it would be appropriate to say that P dynamics within the wheat ecosystem is affected by P leaching losses and that it could be corrected by better timing of P inputs, deeper placement and by using less soluble P sources.

3. The Wheat-based Agroecosystem of Pampas in Argentina

> *The wheat-based agroecosystem in Pampas is comparatively new. This four hundred-year-old cropping system is yet another example of historically recent human endeavor aimed at developing and extending the frontiers of Agrosphere. This cropping ecosystem has experienced signifi-cant alterations in nutrient dynamics and productivity since 1950s. Intensification of this agroecosystem in future seems imminent.*

A. Expanse, Agroclimate, Cropping Sequence and Soils

'Pampas', literally means tree-less expanse of grasslands. However, with the arrival of European settlers in the early 16th century, it has been gradually converted into an extensive agricultural zone, wherein wheat is the predominant cereal (Fig. 3.1). Pampas are geographyically demarcated into four subregions that support wheat production. They are Rolling Pampas, Flat Pampas, Western Pampas and Southern Pampas. Wheat is a secondary crop in Mesopotamic Pampas. Among these subregions, wheat cultivation is pronounced in Rolling and Southern Pampas. The physiography and geological aspects of soil formation in this agroecosystem has been detailed by Soriano (1991). Soils of this Pampas region have been cropped for only a little over one century. Obviously, this temperate

Fig. 3.1 Wheat-based farming zone of the Pampas of Argentina (Shaded area represents cropping-belt with wheat as the predominant cereal. Demarcation is not to scale).

wheat zone is recent in terms of agricultural evolution and cropping history, when compared with European or Asian wheat belts. Historically, the wheat belt in Pampas compares well with that in the Great Plains of North America. The arc of Pampas is agriculturally important. It occupies about 3.4 m ha (Giberti, 1961; Hall *et al.* 1992). The present limits to cropping areas in Pampas were reached as early as in 1930s. Nutrient inputs within the cropping zone is limited to wheat. According to Hall *et al.* (1992), a striking feature of this wheat-based agroecosystem is the increase in grain harvests that began in 1950s, which do not seem to subside. Such yield increases were attributed to introduction of semi-dwarfs and incessant application of fertilizers. Pampas in Argentina contributes 5% of yearly global wheat trade.

Agroclimate: In general, the climate of Pampas where wheat cultivation is predominant can be categorized as 'temperate, humid without a dry season'. The annual mean temperature fluctuates between 17°C in north to 14 to 15°C in south. The average annual rainfall ranges from 600 mm in the south and west, to about 100 mm in north and east. Maximum precipitations occur during summer, with decreasing amounts in spring, autumn and winter. Highest levels of precipitation are received between December and March, and the lowest in July and August. The coefficient of variation in rainfall for the whole area is around 15%. Variations in rainfall pattern do influence cropping sequences, nutrient inputs and yield (Gonzalez-Montaner *et al.* 1997). For example, in Rolling Pampas 42% of variability in wheat yield was attributable to variations in soil moisture during September and October (Hall *et al.* 1992). Also, response to N fertilization is tightly linked to precipitation levels. The eastern and southern Pampas are considered wet during autumn to winter because precipitation is greater than evapotranspiration. The drawdown of stored moisture begins around January. The diurnal variations at the extremes range between 14 and 10.2 h sunlight in northern Pampas, and 15 to 9.3 h in southern Pampas. The radiation receipts are high in December at 26.3 MJ m^{-2}, which dips to 8.9 MJ m^{-2} in June.

Cropping Sequences: Wheat is the major cereal crop in the Pampas, and it occupies 4.1 m ha. Other cereals are maize (1.8 m ha), grain sorghum (0.4 m ha) and pearl millet (0.18 m ha). Wheat is sown widely in the Pampas at large, but predominant cultivation confines to Rolling and Southern Pampas (Table 3.9). Earliest of wheat cropping booms occurred in 1857, then lately in 1950s. Of the 'pioneer crops', variation in the total area of cultivation has been least with wheat since 1930s. A typical cropping sequence in Rolling Pampas begins with ley for 3 to 4 years, followed by a single maize, then rotated frequently with wheat-soybean for several years. Wheat is planted in mid-July, and harvested early in December. In the southern Pampas, one wheat crop per year is a rule. It

Table 3.9: Crops and cropping sequences in Pampas of Argentina

Sub-region	Dominant crops	Secondary Crops	Cropped zone (%)	Mean wheat yield (t ha^{-1})
Rolling Pampas	Wheat-Soybean-Wheat	Soybean	65	1.97
Southern Pampas	Wheat-Sunflower	Maize-Oats	28	1.96
Flat Pampas	Wheat-Maize-Sunflower	Oats-Sorghum	25	1.85
Western Pampas	Wheat-Sorghum-Rye	Sunflower	20	1.59
Mesopotomic Pampas	Sorghum	Wheat-Maize	16	1.73

Source: Hall et al. 1992; Cirio, 1984.
Note : Wheat yields are overall means for the sub-region of the Pampas during the past decade.

could be either bread wheat or durum. A typical sequence begins with wheat sown in July/August and harvested late in December or mid-January, followed with a summer crop of maize/sunflower. In the west and southern Pampas, wheat becomes increasingly conspicuous, at times it is almost the only crop cultivated leading to a mono-cropping ecosystem. For a wheat-wheat mono-cropping sequence, soils are first prepared in April, disced and sown in July, and the mature crop harvested in December. Stubbles are allowed to be grazed until next April. Wheat is the only crop that receives N replenishments, but it is nominal at 20 to 30 kg N ha^{-1}. Major weeds that divert soil nutrients are *Polygonium, Brassica, Avena fatua, Lolium, Sorghum helopense* and *Cyanodon dactylon*. Alternate strategies for rainfed wheat cropping systems are being searched. In this regard, Savin *et al.* (1995) evaluated options using the CERES-Wheat model for the monsoon climate in Pampas. They suggest including combinations of two varieties of different maturity (intermediate and short) and two sowing dates (early and late). The higher yields derived by this system was related to greater water consumption and better nutrient recovery.

Soils: Mollisols are the most frequently cultivated soils in Pampas. Typic Argidolls are also common (Table 3.10). Entic Haplustolls occur in the western fringes of this wheat zone. Entic Hapludolls, Typic Argidolls of Rolling and southern Pampas possess black top soils of medium texture, with a brown subsoil enriched in clay and yellow loess. Top soil thickness varies between 18 and 80 cm. Illites predominate the clay fraction, with small portions of montmorillonite. Soils are generally acidic with organic matter approximately 3%, C:N ratio of 10, CEC 250 mg kg^{-1} and base saturation 70%. The bulk density of 1.1 g cm^{-3} in top soil may range up to 1.4 g cm^{-3} in the subsoil.

Sillanppa (1982) considers soil fertility of Pampean wheat belt as high, if compared with world soils. This high fertility is contributed by moderate acidity, high base saturation, organic matter dominated by humic acid and humin. Organic matter is a key factor in maintaining optimum N and P availability, a stable granular structure and porosity of top soil.

Table 3.10: Main soil associations of cropping zones of Pampas

Geographic region	Soil classification
Rolling Pampas with deep loess	Typic Argiudolls; Vertic Argiudolls; Aquic Argiudolls; Typic Chromousterts
Mesopotamian Pampas:	Typic Paleusterts; Typic Ochraqualfs, Calcareous silty clay
Southern Pampas:	
i) Plains and table lands	Petrocalcic Paleustolls; Typic Haplustolls.
ii) Terraces	Aridic Haplustolls; Entic Haplustolls, Udic Haplustolls, Typic Natraqualfs, Typic Argiudolls
Parana Flood Plains	Alluvial Complex

Source: Hall *et al.* 1992

The major constraint in Rolling Pampas is N deficiency encountered by winter wheat, whereas in southern Pampas widespread P deficiency affects crop yield. Potassium status of Pampean soils is sufficiently high at > 10 m eq kg^{-1}. Similarly, micronutrients Mn, Mo, Fe, B, and Zn occur at satisfactory levels. Loss of top soil, along with it the decrease in soil fertility status is a major problem in the cropping zone.

Irrutia (1984) reported 11 to 18 t ha^{-1} y^{-1} loss in top soil due to gentle gradients in Rolling Pampas. Nearly 25% of A horizon has been eroded during the past 80 years of cultivation of wheat. Hall *et al.* (1992) highlighted that during the past 20 years, nearly 21 to 56% of organic matter and total N, and 10 to 84% of available P have been lost. Concomitantly, pH decreased by one half of a unit making soils more acidic. Soil structure stability has depreciated by 40 to 73%. Overall, continuous wheat cropping has impoverished soils. The changes in crop husbandry practices towards strip cropping and reduced tillage seems to thwart erosion and loss of fertility.

Tillage and Soil Nutrients: Continuous cultivation can impair soil structure, affect nutrient dynamics and crop productivity (Glave, 1988). Most commonly employed tillage practices are Conventional tillage (CT), Restricted tillage (RT) and No tillage (NT). Area under no tillage and direct drilling has been ever-increasing, and currently it is practiced on 2.5 m ha (AAPRESID, 1995). Conservation tillage or no tillage delays soil deterioration, but leads to excessive compaction with a negative impact on crop productivity. The shallow compaction results mainly from the collapse of the larger soil pores under the action of internal and external stresses (Hill *et al.* 1985). Ferreras *et al.* (2000) compared wheat grown under both CT and NT for 3 years and noted marked changes in physicochemical aspects of soil nutrients. Firstly, in fields under NT and CT structural deterioration occured, and the stability indices were 36% for NT but 26% for CT. The saturated hydraulic conductivity was significantly lower in NT (3.5 $\times$ 10^{-7}m S^{-1}), than CT (10.9 $\times$ 10^{-7} m S^{-1}). Soil water contents in the top soil was higher in NT plots in the early growth stages, but from anthesis onwards until maturity water contents were similar for NT and CT plots. Overall, results indicated that mechanical resistance perceived under NT may impede root growth and dry matter accumulation. In another study, Tabaoda *et al.* (1998) compared sandy loams (Typic Hapludoll) by imposing conventional and zero tillage. Soil organic matter content (8.8 to 10.7 g kg^{-1}), bulk density (1.0 to 1.2 Mg m^{-3}) and relative compaction (0.7 to 0.8) were not affected significantly by zero tillage, but penetration resistance increased markedly. Quite often compaction was localized in subsoil layers. Excessive compaction can decrease biological activity and nutrient transformations. Cropping sequences followed in Pampas has also impacted several other soil physico-chemical properties. Generally, annual wheat cropping sequences (W-W-W) have resulted in lowest levels of soil organic carbon, total N and extractable P (Miglierina *et al.* 2000). Fertilizer inputs increased pore size, but wheat-natural grass (W-Gr) sequence decreased medium pore size and water holding capacity. Available water levels decreased in the following order wheat-legume (WL) > W-W > W-Gr. Overall, the W-L systems enhanced SOC and total N. Also, soil organic C and N increased as frequency of fallows increased. It was attributable to crop residue input and lower level of disturbance to soil. Miglierina (1999) has earlier reported that during the 80 years of cultivation in Pampa Humeda, C, N, and P in soil decreased by 40, 50 and 48 respectively. Such changes in soil nutrients were prominent in the coarser fraction of soil. Hence, coarse fraction of soil could play an important role in the nutrient dynamics and crop yield (Galantini and Rosell, 1997).

Soil Organic Carbon: Managing favorable C dynamics is a crucial aspect that affects ecological stability and determines productivity of the wheat cropping system in Pampas. A wheat-soybean-corn sequence in Rolling Pampas produces around 15 t ha^{-1} dry matter as straw and roots, equivalent to 5 to 8 t ha^{-1} carbon which is recyclable effectively (Alvarez *et al.* 1995). Recycling organic matter is a very important aspect because external sources of C and other nutrients are meager in many locations. There is no doubt that continuous cropping decreases SOC, especially under CT system. CT affects C dynamics in soil through its effect on crop residue decomposition and soil aeration, and through the exposure of SOC fractions protected within soil aggregates (Doran and Smith 1987). However, loss of SOM can be minimized by including pastures in rotation, and increasing the amount of wheat residues returned to soil. A study of 16 different wheat-based cropping sequences in Pampas revealed that after 11 years, SOC decreased by 4 to 8 g kg^{-1} in the absence of supplemented N, and by 2 to 7 g kg^{-1} if N was supplied. The SOC loss was accentuated if soybean occurred in the sequence. The extent of residue produced and SOC was attributable to tillage and fertilizer inputs. Fertilizer N inputs generally enhanced residue formation (Echevarria and Ferrari, 1993; Studdert and Echevarria, 2000).

Conservation tillage, especially zero tillage procedures tend to allow organic matter accumulation in the upper layers (0 to 15 cm) of the soil. As a consequence, microbial biomass too gets stratified closely matching the organic matter localization. In contrast, organic matter and microbial C is evenly distributed in soils under CT. Hence, Alvarez et al. (1995) believe that monitoring microbial biomass could be indicative of availability of organic C. Therefore, higher biological activity at soil surface induces higher CO_2-C production in NT system than under CT. Basing it on enumeration of cellulolytic microbes, which localize in the top soil layers, Toresani et al. (1998) also concluded that microbial population dynamics/respiration measurements are indicative of SOM localization under NT. The difference in carbon substrate availability, and metabolic changes could also induce differences in microbial activity and microbial C contents (Saffigna et al. 1989). For example, in a 12-year evaluation of wheat field in INTA, Buenos Aires, NT lead to 2 to 3 fold greater accumulation of coarse plant debris at soil surface (0 to 5 cm). Correspondingly, microbial biomass level at 0-5 cm soil layer was twice that under conventional plow-tilled fields. But under zero-tillage this ratio was higher in top layers and decreased with depth. The intensity of SOM mineralization is regulated by quality (C:N) of the substrate (Diaz-Ravina et al. 1988). Alvarez et al. (1995) reported that organic matter in the top 5 cm layer of soils subjected to no-till or reduced-till contains more easily decomposable materials than under deep-plowed fields. Studies on soil physical fractionation or granulometric size separation have shown a close relationship with SOC dynamics. The coarse fraction of organic matter is important because several chemical transformations are localized in it. In addition, erosional/leaching losses of N, P, S, and SOC were highest through the coarse fraction. Hence, coarse fraction of SOM plays an important role in nutrient dynamics (Galantini and Rosell, 1997)

B. Nitrogen Dynamics and Crop Productivity

Fertilizer inputs on wheat has always has been low in Argentina. Among the different crops grown in sequence in Pampas, wheat is the only crop fertilized with nitrogen. Normally, N inputs have ranged from 25 to 39 kg N ha^{-1}. The total area fertilized with N, increased from 0.1 m ha in 1970s to 1.6 m ha in 1980s (Obschatko and Del Bello, 1986) then to 2.3 m ha in 1990s. Nitrogen inputs, particularly in southern Pampas have ranged from 25 to 125 kg N ha^{-1} depending on soil fertility status and yield goals (Gonzalez-Montaner et al. 1997). One of the reasons for low fertilizer nitrogen inputs is that soil within this ecosystem still supports good crop of wheat without external inputs. Economically, 8 to 10 kg wheat grains are needed to supply 1 kg N, hence costs too forbid excessive N inputs. On an average, wheat cultivated in Pampas extracted 17.5 kg N, 3.2 kg P and 1.5 kg S to produce 1.0 Mg of wheat grains (Galantini and Rosell, 1997). According to Gonzalez-Montaner (1997) wheat cultivated in Pampas recovers N efficiently, especially in soils with low residual nutrients. At physiological maturity, a wheat crop extracts between 31 and 69 kg N ha^{-1} depending on grain yield. Normally, grain yield is positively correlated to N inputs/recovery. The mean N-use efficiency reported is 28 kg grain kg N^{-1}, and ranges between 17 and 43 kg grain kg N^{-1}. In comparison, N-use efficiency attained by soft winter wheat crop in Europe was 33 kg grain kg N^{-1}. In their trials the residual N in soil could however be enhanced by fertilizer N inputs ranging between 45 and 150 kg N ha^{-1}, Actually, an exponential increase in residual N was easily discernible. Thus indicating that all the N supplied was not recovered. Interaction with N and water was important. Nitrogen recovery was efficient under optimum water conditions, but inefficient under water deficit condition, thus leading to accumulation as residual N in soil. Immobilization of fertilizer-N may also enhance residual N.

The mean soil N mineralization during the growing period was 58 kg N ha^{-1}, but ranged between 5 and 125 kg N ha^{-1}. Hence, mineral N derived through mineralization may satisfy up to 38% of crop's demand for this element. Gonzalez-Montaner *et al.* (1997) suggest that N mineralization rates are site-specific, and vary enormously. Hence it requires careful estimation at each field. Normally, the soil temperature, SOC contents, C : N ratio, moisture and timing of fertilizer-N inputs affect N derived through mineralization (Echavarria *et al.* 1994). Variations in N mineralization rates were also attributable to organic carbon and clay formation. Soil N mineralization was negatively affected by soil mineral N content. However, in fields supplied with N levels beyond the crop's demand, it is believed that immobilization processes are important (Zourarakis, 1983). In their efforts on modeling wheat grain yield verses N inputs, Gonzalez-Montaner *et al.* (1997) found that highest grain yield (4.17 t ha^{-1}) occurred at 100 kg N ha^{-1}, and control fields supplied with 50 kg N ha^{-1} yielded lowest at 2.27 t ha^{-1}. Specific wheat-based cropping sequences adopted may also have significant role in N recovery, N-use efficiency attained and grain yield obtained (Table 3.11). Location effects, especially N deficiency affects crop responses to external inputs. In a test by Senigagliesi (1983) wheat responded to N inputs in as many as 57% of 65 locations in Pampas. Similarly, 'fallowing' as a factor influenced response to added N (Gambouda and Vivas, 1986).

Table 3.11: Nitrogen recovery and wheat productivity in different sequences in Rolling Pampas

	Wheat-Wheat (WW)	Wheat-Grass (WG)	Wheat-Legume (WL)
Total dry matter (kg ha^{-1})	10920	9340	12960
Grain yield (kg ha^{-1})	3580	3060	4050
Straw (kg ha^{-1})	7340	6290	8910
Total N recovery (kg ha^{-1})	105	91	147

Source: Galantini et al. 2000

Note: All plots received 64 kg N ha^{-1} as urea, and 16 kg P ha^{-1} as dia ammonium phosphate applied at seeding time. Nitrogen recovered was partitioned between grain:straw with ratios 86:19, 73:18 and 113:34 under W-W, W-G and W-L cropping sequences respectively.

4. Concluding Remarks

The vast European wheat belt supports both, moderate and highly intensive farming zones, and the grain productivity varies between 3 and 7 t ha^{-1}. During the recent past, nutrient inputs (N,P,K) have been excessive in many countries of this region, leading to their imbalances, excessive accumulation in soils, pollution of drainage and ground water, and disturbance to atmosphere. Hence, researchers in most European nations have focused on devising agronomic methods that thwart nutrient imbalances and ecological disturbances. Techniques, which enhance C sequestration into soil, and improve nutrient-use efficiency have been sought vehemently.

The Austaralian wheat belt is only moderately intensive. Thus, leaving us with a scope to intensify the cropping ecosystem further through nutrient and water inputs. Expansion of Australian wheat zone is a possibility, which depends on global wheat needs. The Argentinian wheat belt in Pampas is again moderately intense. Currently, the nutrient inputs are minimal in most locations, and confines to N, or at best N, P and K. The inherent soil fertility is good enough to support the moderate grain yield around 1.5 to 3.0 t ha^{-1}. However, if nutrient and water resources permit, intensification and to a certain expansion of this wheat belt is a clear possibility. A sizeable fraction of wheat produce from Pampas is exported, hence global wheat demand and economics of cultivation may hold the key with regard to intensification and expansion of this cropping ecosystem.

REFERENCES

AAPRESID (1995). Associon Argentina de productores en Siembra Directa. *Gazetilla Informativa.* Ano **6:** 18-22.

Abreu, P.P.M.E., Flores, I., Areu, F.M.G., and Madiera, M.V. (1993). Nitrogen uptake in relation to water availability in wheat. *Plant and Soil,* **154**: 89-96.

Addiscott, T.M. and Powlson, D.S. (1992). Partitioning losses of nitrogen fertilizer between leaching and denitrification. *J. of Agricultural Sciences, Cambridge,* **118**: 101-107.

Addiscott, T.M. and Thomas, D. (2000). Tillage, mineralization and leaching: phosphorus. *Soil and Tillage Research,* **53**: 255-273.

Alvarez, R., Diaz, R.A., Barbero, N., Santanatoglia, J. and Blotta, L. (1995). Soil organic carbon, microbial biomass and CO_2-C production from three tillage systems. *Soil and Tillage Research,* **33**: 17-28.

Anderson, C.C., Fillery, I.J.P.,Dunin, F.X., Dolling, P.J. and Asseng, S. (1998a). Nitrogen and water flows under pasture-wheat and lupin-wheat rotation in deep sands in Western Australia. 2. Drainage and Nitrate leaching. *Austral. J. of Agricultural Research,* **49**: 345-361.

Anderson, G.C., Fillery, I.R.P., Dunin, F.X., Dolling. P.J. and Asseng, S. (1998b). Nitrogen and water flows under pasture-wheat and lupin-wheat rotations in deep sands in Western Australia. 1. Nitrogen fixation in legumes, net N mineralization, and utilization of soil derived nitrogen. *Austral. J. of Agricultural Research,* **49**: 329-343.

Anderson, W.K. and Hoyle, F.C. (1999). Nitrogen efficiency of wheat cultivars in a Mediteranian environment. *Austral. J. of Experimental Agriculture,* **39**: 957-965.

Angus, J.F., van Herwaanden, A.F., Howe, G.N. (1991). Productivity and break-crop effect of winter grown oil seeds. *Austral. J. of Experimental Agriculture,* **31**: 669-677.

Asseng, S., Fillery, I.R.P., Anderson, G.C., Dolling, P.J., Dunin, F.X., Keating, B.A. (1998). Use of the APSIM-wheat model to predict yield, drainage, NO_3 leaching from a deep sand. *Austral. J. of Agricultural Research,* **49**: 363-377.

Barrow, E.M. and Semenov, M.A. (1995). Climate change scenarios with high spatial and temporal resolution for agricultural operations. *Forestry,* **68**: 349-360.

Batten, G.D. (1992). A review of phosphorus efficiency in wheat. *Plant and Soil,* **155/156**: 297-300.

Batten, G.D., Fattel, N.A., Mead, J.A. and Khan, M.A. (1999). Effect of sowing date on the uptake and utilization of phosphorus by wheat (cv Osprey) grown in central New South Wales. *Austral. J. Experimental Agriculture,* **39**: 161-170.

Beyer, L., Kristina, P. and Sieling, K. (2002). Soil organic matter in temperate arable land and its relationship to soil fertility and crop production. *In*: Soil fertility and crop production. K.R. Krishna (ed.), Science Publishers, Enfield, NH, USA. pp. 189-212.

Bissett, M.J. and O'Leary, G.J. (1996). Effects of conservation tillage and rotation on water inter filtration in two soils in southeastern Australia. *Austral. J. of Soil Research,* **34**: 299-308.

Blankenan, K., Kuhlmann, H., and Olfs, H.W. (2000). Effect of increasing rates of 15N-labelled fertilizer on recovery of fertilizer N in plant and soil N pools in a pot experiment with winter wheat. *J. Plant Nutr. Soil Sci.,* **162**: 475-480.

Bolland, M.D.A., Clark, M.F. and Boetal, F.C. (1999a). Effectiveness of single coastal super-phosphate applied either in autumn or spring. *Nutrient Cycling in Agroecosystems,* **54**: 133-143.

Bolland, M.D.A., Siddique, K.H.M., Loss, S.P. and Baker, M.J. (1999b) Comparing responses of grain legumes, wheat and canola to applications of superphosphate. *Nutrient Cycling in Agroecosystems,* **53**: 157-175.

Borin, M., Menini, C. and Sartori, L. (1997). Effects of tillage systems on energy and carbon balance in northeastern Italy. *Soil and Tillage Research,* **40**: 209-226.

Cantero-Martinez, C., O'Leary, G.J. and Connor, D.J. (1999). Soil water and nitrogen interaction in wheat in a dry season under fallow-wheat cropping system. *Austral. J. of Experimental Agriculture,* **39**: 29-37.

Carranca, C., de Varennes, A. and Rolston, D.E. (1999). Variation in N recovery of winter wheat under Mediterranean conditions studied with 15 N labeled fertilizers. *European J. of Agron.,* **11:** 145-155.
Carvalho, M. and Basch, G. (1994). Optimization of nitrogen fertilization. *In:* Fertilizers and environment. Rodriguez-Barmeco. (ed.), Kluwer, London, pp. 195-198.
Catt, J.A., King D.W. and Weir, A.H. (1974). The soils of Woburn experiment farm. Rothamsted report for 1974, part 2, pp. 5-28.
Chalk, P.M., Smith, C.J., Hamilton, S.D. and Hopman, P. (1993). Characterization of the N benefit of a grain legume (*Lupinus angustifolius*) to a cereal (*Hordeum vulgare*) by an *in situ* 15N isotope dilution technique. *Biology and Fertility of Soils,* **15:** 39-44.
Chan, K.Y and Heenan, D.P. (1993). Surface hydraulic properties of a red earth under continuous cropping with different management practices. *Austral. J. of Soils Research,* **31:** 13-24.
Christensen, P.T. (1989). Askov 1894 to 1989: Research on animal and mineral fertilizers. Proceedings of the Sanborne Fields centennial papers. University of Misssouri-Columbia SR-415, 28-48.
Cirio, F. (1984). La Agriculture Pampena: su potential de Crecimiento. CISEA, Buenos Aires, 72 pp.
Clough, T.J., Jarvis, S.C. and Hatch, D.J. (1998). Relationships between soil thermal units, nitrogen mineralization and dry matter production in pastures. *Soil Use and Management,* **14:** 65-69.
Conrad, R. (1996). Metabolism of nitric oxide in soil and soil microorganisms and regulation of flux into atmosphere. *In:* Microbiology of atmospheric trace gases, sources, sinks and global change processes. NATO ASI Series. Springer Verlag, Berlin, **39:** 167-203.
Dalal, R.C. and Meyer, R.J. (1986). Long term trends in fertility of soils under continuous cultivation and cereal cropping in Southern Queensland. 2. Total organic carbon and its rate of loss from the profile. *Austral. J. of Soil Research,* **24:** 281-292.
Del Campilo, M.C., Van Der Zee and Torrent, J. (1999). Modeling long term phosphorus leaching and changes in phosphorus fertility in excessively fertilized acid sandy soils. *European J. of Soil Science,* **50:** 391-399.
Delroy, N.D. and Bowden, J.W. (1986). Effect of deep ripping, the previous crop, and applied nitrogen on the growth and yield of a wheat crop. *Austral. J. of Experimental Agriculture,* **26:** 469-479.
Diaz-Ravina, M. Carballas, T. and Acea, M.J. (1988). Microbial biomass and metabolic activity in four acid soils. *Soil Biology and Biochemistry,* **20:** 817-823.
Doran, J.W. and Smith, M.S (1987). Organic matter management and utilization of soil and fetilizer nutrients. *In:* Soil fertility and organic matter as critical components of production systems Follet, R.F. (ed.), Soil Science Society of America Special Publications, Madison, WI, pp. 53-72.
Echevarria, H., Bergonzi, R. and Ferrari, J. (1994). Un modelo para estimar la mineralizacion de nitrogeno en suelos del sudesto de la provincia de Buenos Aires, Argentina. *Ciencia del suelo,* **12:** 56-62.
Echevarria, H. and Ferrari, J. (1993). Relevamiento de Algunas caractricas de los suelos agricoles del sudesto bonerensa. INTA, Ce.R.B.A.A., E.E.A. Balcarce, *Boletin Technico,* **112:** 23-31.
Ellmer, F., Peschka, H., Kohn, W., Chmielewski, F.M. and Baumecker. M. (2000a). Tillage and fertilizing effects on sandy soils. Review of selected results of long term experiments at Humbolt-University, Berlin. *J. Plant Nutr. Soil Sci.,* **163:** 267-272.
Ellmer, F., Pescheke, H., Kohn, W., Chmielewski, F.M. and Baumecker. (2000b). Long-term experiments at the Humbolt-University of Berlin—Review and selected results. *J. Plant Nutr. Soil Sci.,* **163:** 310-314.

Evans, J., O'Connor, G.E., Coventry, G.L., Fettel, D.R., Mahoney, J., Armstrong, E.L. and Walsgot, D.N. (1989). N Fixation and its value to soil N increase in lupin, field peas, and other legumes in Southeastern Australia. *Austral. J. Agriculture Research*, **40:** 791-805.

Ferreras, L.A., Costa, J.L., Garcia, F.O. and Pecorari, C. (2000). Effect of no-tillage on some soil physical properties of a structural degraded petrocalcic Paleudoll of the Southern Pampa of Argentina. *Soil and Tillage Research*, **54:** 31-39.

French, R.J. and Schultz, J.F. (1984). Water use efficiency in wheat in a Mediterranean type environment. 2. Some limitations to efficiency. *Austral. J. of Agriculture Research*, **35:** 765-775.

Friedel, J.K. (2000). The effect of farming systems on labile fractions of organic matter in Regosols. *J. Plant Nutr. Soil Sci.*, **163:** 41-45.

Friedel, J. and Gabel, D. (2001). Nitrogen pools and turnover in arable soils under different duration of organic farming. 1. Poll sizes of total soil nitrogen, microbial biomass nitrogen and potentially mineralizable nitrogen. *J. Plant Nutr. Soil Sci.*, **164:** 415-419.

Gambouda, S. and Vivas, H (1986). Resultados preliminares de fertilizcio nitrogenada en trigo en la sbregion. 1 INTA Rafeela. Informe Technico 23, Santa Fe, Argentina, pp. 28-35.

Galantini, J.A., Landricini, M.R., Iglesias, J.O., Maglierina, A.M. and Rosell, R.A. (2000). The effects of crop rotation and fertilization on wheat productivity in the Pampean semi-arid region of Argentina. 2. Nutrient balance, yield and grain quality. *Soil and Tillage Research*, **53:** 137-144.

Galantini, J.A. and Rosell, R.A. (1997). Organic fractions, N, P and S changes in an Argentinian semi-arid Haplustoll under different crop sequences. *Soil and Tillage Research*, **42:** 221-228.

Gataulina, G.G. (1992). Small-grain cereal systems in the Soviet Union. *In:* Field crop ecosystems. Pearson, C.J. (ed.), Elsevier, Amsterdam, pp. 385-400.

Gelbaly, I.E. and Johansson, C. (1989). A model relating laboratory measurements of rates of nitric oxide production and field measurements of nitric oxide emissions from soils. *J. Geophys. Res.*, **94:** 6473-6480.

Giberti, H.C.E. (1961). Historia Economica de la Ganaderia Argentina. Solar/Hachette, Buenos Aires, 217 pp.

Glave, A. (1988). Manejo de suelos y agua en la region semiarida pampena. *In:* Erosion: sistemas de production manejo y concervaciondel suelo y del agua. Cargill Foundation, Buenos Aires, pp. 1-69.

Gonzalez-Montaner, J.H., Madonni, G.A. and DiNapoli, M.R. (1997). Modeling grain yield and grain yield response to nitrogen in Spring wheat crops in the Argentinian Southern Pampas. *Field Crops Research*, **51:** 241-252.

Goss, M.J., Colbourne, P., Harris, G.L. and Howse, K.R. (1988). Leaching of nitrogen under autumn sown crops and the effects of tillage. *In:* Nitrogen efficiency in agricultural soils. D.S. Jenkinson and K.A. Smith (eds.), Elsevier Applied Science, London, pp. 269-282.

Goulding, K.W.T. (1990). Nitrogen deposition to land from atmosphere. *Soil Use and Management*, **5:** 61-63.

Gregory, P.H., Tennant, D., Hamblin, A.P. and Eastham, J. (1992). Components of the water balance on duplex soils in Western Australia. *Austral. J. of Experimental Agriculture*, **32:** 845-855.

Guggenberger, G., Christensen, B.F. and Ruback, G.H. (2000). Isolation and characterization of labile organic phosphorus pools in soils from the Askov long tem field experiments. *J. Plant Nutr. Soil Sci.*, **163:** 151-155.

Gut, A., Neftel, A., Staffelbasch, T., Riedo, M. and Lehman, B.E. (1999). Nitric oxide flux from soil during the growing season of wheat by continuous measurements of the NO soil-atmosphere concentration gradient: A process study. *Plant and Soil*, **216**: 165-180.

Hall, A.J., Vivella, F., Trapani, N. and Chimenti, C.A. (1992). The effects of water stress and genotype on the dynamics of pollen-shedding and silking in maize. *Field Crops Research*, **5**: 349-363.

Hayman, P.T. and Alston, C.N. (1999). A survey of farmer practices and attitudes to nitrogen management in the Northern New South Wales grains belt. *Austral. J. of Experimental Agriculture*, **39**: 51-63.

Hill, R.L., Horton, R. and Cruse, R.M. (1985). Tillage effects on soil water retention and pore size distribution in two Mollisols. *Soil Sci. Soc. Am. J.*, **49**: 1264-1270.

Hocking, P.J., Kirkegaard, J.A., Angus, J.F., Gibson, S.M. and Koetz, E.A. (1997). Comparison of Canola, Indian Mustard and Linola in two contrasting environments. 1, Dry matter production, seed yield and quality. *Field Crops*, **52**: 162-178.

Hossain, S.A., Dalal, R.C., Waring, S.A., Strong, W.M. and Weston, E.J. (1996). Comparison of legume based cropping ecosystems at Wara, Queensland. 1. Soil nitrogen and organic carbon accretion and potentially mineralizable nitrogen. *Austral. J. of Soil Science*, **34**: 273-287.

Hutsch, B.W., Webster, V.P. and Powlson, D.S. (1993). Long-term effects of nitrogen fertilization on methane oxidation in soil of the Broadbalk wheat experiment. *Soil Biology Biochemistry*, **25**: 1307-1315.

IACR (1996). Annual report. Institute of Arable Crops Research, Rothamsted, Harpenden, England, pp. 37-43.

IACR (1997). Annual report. Institute of Arable Crops Research, Rothamsted, Harpenden,, England, 35 pp.

IACR (2001). Annual report. Institute of Arable Crops Research, Rothamsted, Harpenden, England, pp. 37-39.

Irrutia, C.B., Musto, J. and Cubot, J.P. (1984). Caracteristicas y Delimitaion Cartografica de factores generadores de procose erosivos en el sector Argentino de la Cuenca del Plata.Publication No 174 CIRN/INTA, INTA, Buenos Aires, 32 pp.

Jambert, C., Serca, D. and Delmas, R. (1997). Quantification of N-losses as NH_3, NO and N_2O, and N_2 from fertilized maize fields in Southern France. *Nutrient Cycling in Agroecosystems*, **48**: 91-104.

Jarvis, S.C., Stockdale, E.A., Sheprd, M.A. and Powlson, D.S. (1996). Nitrogen mineralization in temperate agricultural soils. Process and measurement. *Advances in Agronomy*, **57**: 187-239.

Kaiser, E.A. and Ruser, R. (2000). Nitrous oxide emissions from arable soils in Germany—An evaluation of long term field experiments. *J. Plant Nutr. Soil Sci.*, **163**: 249-260.

Kirkegaard, J.A., Howe, G.N. and Mele, P.M. (1999). Enhanced accumulation of mineral-N following Canola. *Austral. J. of Experimental Agriculture*, **39**: 587-593.

Kirkegaard, J.A., Hocking, P.J., Angus, J.F., Howe, G.N. and Gardner, P.A. (1997). Comparison of Canola, Indian mustard and Linola in two contrasting environments. 2. Break-crop and nitrogen effects on subsequent wheat crops. *Field Crops Research*, **52**: 179-191.

Klier, J., Kubat, J. and Pova, D. (1995). Stickstoffbileanzen der Dauefeldversuche in Prague. *Miteilgn. Dtsch. Bodenkundel. Gesellsch*, **67**: 831-834.

Kormmann, M. and Koller, K. (1997). Ecological and economical effects of different tillage systems. Fragmenta Agronomica. Tom2B, Proceedings 14th ISTRO conference, Pulawy, Poland, pp. 391-394.

Knezevic, M., Durkic, M. Anotnic, O. and Zugec, I. (1999). Effects of soil tillage and nitrogen on winter wheat yield and weed biomass. *Cereal Research Communications,* **27:** 197-204.

Leigh, R.A. (1996). Crop nutrition and environmental protection. Institute of Arable Crops Research, Rothamsted, Harpenden, England, pp. 37-44.

Lipeic, J. and Stepniewski, W. (1995). Effects of soil compaction and tillage systems on uptake and losses of nutrients. *Soil and Tillage Research,* **35:** 37-52.

Lopez-Bellido, L (1992). Mediterranean cropping systems. *In:* Field crop ecosystems. C.J. Pearson (ed.), Elsevier, Amsterdam, p.p 311-370.

Lopez-Bellido, L., Lopez-Bellido, R.J., Castillo, J.E. and Lopez-Bellido, L. (2000). Effects of tillage, crop rotations and nitrogen fertilization on wheat under rain-fed Mediterranean conditions. *Agronomy Journal,* **92:** 1054-1063.

Macdonald, A.J., Poulton, P.R., Powlson, D.S. and Jenkinson, D.S. (1997). Effect of season, soil type and cropping on recoveries, residues and losses of 15N-labelled fertilizer applied to arable crops in spring. *Journal of Agricultural Science,* Cambridge, **129:** 125-154.

Maglierina, A.M. (1999). Materia organica y sistemas de production en la region semiarida bonerasa 2. Cambios de algunas propiedades quimicas del suelo. *Rev Facultad de Argonomica,* **15:** 9-14.

Maglierina, A.M., Iglesias, J.O., Landricini, M.R., Galantini, J.A. and Rosell, R.A. (2000). The effects of crop rotation and fertilization on wheat productivity in Pampean semi-arid region of Argentina. 1. Soil physical and chemical properties. *Soil and Tillage Research,* **53:** 129-135.

Mercik, S., Stepien, W. and Labetowicz, J. (2000). The fate of nitrogen, phosphorus and potassium in long term experiments in Skerniewice. *J. Plant Nutr. Soil Sci.,* **163:** 273-278.

Mikhailova, E.A., Bryant, R.B., Vassenev, I.I., Schwger, S.J. and Post, C.J. (2000). Cultivation effects on soil carbon and nitrogen contents at depth in the Russian chernozem. *Soil Sci. Soc. Am. J.,* **64:** 738-745.

Mouchova, H., Klier, J. and Lippold, H. (1996). Effect of weather conditions and time of N application on the uptake of soil and applied N by winter wheat. *In:* Progress in nutrient cycling studies. O. Van Cleemput, G., Hoffman, and A. Vermoesen (eds.). Proceeedings of 8th Nitrogen workshop. Kluwer, Netherlands, pp. 237-241.

Nalborczyk, E. and Czember, H.J. (1992). Cereal and root-crop systems in Central Europe. *In:* Field crop ecosystems. C.J. Pearson (ed.). Elsevier, Amsterdam, pp. 373-384.

Nieder, R. and Ritcher, J, (2000). C and N accumulation in arable soils of West Germany and its influence on the environment-developments 1970 to 1998. *J. Plant Nutr. Soil Sci.,* **163:** 65-72.

Obschatko. E.S. and deBello, J.C. (1986). Tendencios prodcutiva estrategia technologica para la Agricultura Pampena. CISEA, Buenos Aires, 117 pp.

O'Leary, G.J. and Connor, D.J. (1997). Stubble retention and tillage in semi-arid environment. 1. Soil water accumulation during fallow. *Field Crops Research,* **52:** 209-219.

Orlov, D.S. and Birukova, O.N. (1995) Reserves of carbon of organic compounds of the Russian federation. *Pochvovedeme,* **1:** 21-32.

Packer, I.J. and Hamilton, G.J. (1993). Soil physical and chemical changes due to tillage and their implications for erosion and productivity. *Soil and Tillage Research,* **27:** 327-339.

Parton, W.J., Schimel, D.S., Cole and C.V. Ojima, D.S. (1987). Analysis of factors controlling soil organic matter levels in Great Plains grasslands. *Soil Sci. Soc. Am. J.,* **51:** 1173-1179.

Perry, M.C. (1989). Farming systems of the Southern Australia. *In:* Proceedings of the 5th Australian Agronomy Conference. Western Australia, pp. 167-180.

Perry, M.C. (1992). Cereal and fallow/pasture systems in Australia. *In:* Field crops ecosystems. C.J. Pearson, (ed.), Elsevier, Amsterdam, pp. 451-481.

Pilbeam, C.J. (1996). Effect of climate on the recovery of crop and soil of 15 N-lebelled fertilizer applied to wheat. *Fertilizer Research*, **45**: 209-215.

Powlson, D.S., Brookes, P.C. and Christensen, B.T. (1987). Measurement of soil microbial biomass provides an early indication of changes in total soil organic matter due to straw incorporation. *Soil Biology Biochemistry*, **19**: 159-164.

Powlson, D.S., Goulding, D.W.T., Willison. T.W., Webster, C.P. and Hutsch, B.W. (1997). The effect of agriculture on methane oxidation in soil. *Nutrient Cycling in Agroecosystems*, **49**: 59-70.

Powlson, D.S., Hart, P.B., Poulton, P.D., Johnston, A.E. and Jenkinson, D.S. (1992). Influence of soil type, crop management and weather on the recovery of 15N-labelled fertilizer applied to winter wheat in spring. *J. of Agriculture Science*, Cambridge, **118**: 83-100.

Powlson, D.S., Pruden, G., Johnston, A.E. and Jenkinson, D.S. (1986). The nitrogen cycle in the Broadbalk wheat experiment: recovery and losses of 15N-labeled fertilizer applied in spring and inputs of nitrogen from the atmosphere. *J. of Agriculture Science*, Cambridge, **107**: 611-620.

Powlson, D.S., Smith, P., Coleman, K., Smith, J., Glending, M.J., Korschens, M. and Franko, U. (1998). A European network of long term sites for studies on soil organic matter. *Soil and Tillage Research*, **47**: 271-282.

Pulleman, C.J. (1996). Effect of climate on the recovery in crop and soil of 15N-labeled fertilizer applied to wheat. *Fertilizer Research*, **45**: 209-215.

Pulleman, M.M., Bouma, J., van Esseen, E.A. and Meijles, E.W. (2000). Soil organic matter content as a function of different land use history. *Soil Sci. Soc. Am. J.*, **64**: 689-693.

Rasmussen, K.J. (1999). Impact of plough less soil tillage on yield and soil quality: A Scandinavian review. *Soil and Tillage Research*, **53**: 3-14.

Recous, S., Machet, J.M. and Mary, D. (1988). The fate of labeled 15N urea and ammonium nitrate applied to a winter wheat crop. 2. Plant uptake and N-efficiency. *Plant and Soil*, **112**: 215-224.

Roner, M., Heinemeyer, O. and Kaisser, E.A. (1998). Microbial-induced nitrous oxide emissions from an arable soil during winter. *Soil Biology and Biochemistry*, **30**: 1859-1865.

Saffigna, P.G., Powlson, D.S., Brookes, P.C. and Thomas, G.A. (1989). Influence of sorghum residues and tillage on soil organic matter and soil microbial biomass in an Australian vertisol. *Soil Biology Biochemistry*, **21**: 759-765.

Savin, R., Emilio, H.S., Hall, A.J. and Slafer, G.A (1995) Assessing strategies for wheat cropping in the monsoonal climate of the Pampas using the CERES-wheat simulation model. *Field Crops Research*, **42**: 81-91.

Semenov, M.A. and Brookes, R.J. (1999). Spatial interpolation of the LARS-WG stochastic weather generator in Great Britain. *Climate Research*, **11**: 137-148.

Semenov, M.A., Porter, J.R. and Delcolle, R. (1993). Climatic change and the growth and development of wheat in the UK and France. *European Journal of Agronomy*, **2**: 293-304.

Senigagliesi, C., Garcia, R., Meira, S., Galetto, M.L., Frutos, E. and Tevel, R. (1983). La fertilizacion del cultivo de trigo en el norte de la Provincia de Buenos Aires y sur de Santa Fe. INTA, Pergamino, Informe Tecnico 191, Buenos Aires, Argentina, pp. 1-5.

Sillanppa, M. (1982). Micronutrients and the nutrient status of soils—A global study. FAO Soils Bulletin No. 48, Rome, 444 pp.

Smith, C.J. and Barraclough, P. (1998). Use of field diagnostics to improve predictions of nitrogen dynamics in arable land. IACR, Rothamsted, Harpenden, pp 1-23.

Smith, C.J., Dunin, F.X., Zagelia, S.J. and Poss, R. (1998a). Nitrate leaching from a Riverine clay soil under cereal rotations . *Austral. J. of Agriculture Res.*, **4**: 379-389.

Smith, P., Powlson, D.S., Glendening, M.J. and Smith, J.U. (1998b). Opportunities and limitations for C sequestration in European agricultural soils through change in management. *In:* Management of carbon sequestration in soils. R. Lal, J.M. Kimble and B.A. Follet (ed.), CRC Press, Boca Raton, Florida, USA, pp. 143-152.

Smith, J., Powlson, D.S., Gelendening, M.J. and Smith J.U. (1997). Potential for carbon sequestration in European soils: Preliminary estimates for a five scenarios using results from long term experiments. *Global Change Biology,* **3:** 67-90.

Skinner, J.A., Lewis, K.A., Bardon, K.S., Tucker, P., Catt, J.A. and Chambers, B.A. (1997). An overview of the impact of agriculture in the United Kingdom. *J. of Environmental Management,* **50:** 111-128.

Soriano, A., Leon, R.J.C., Sah, O.E., Lavado, R.S., Deegibus, V.A., Cauhepe, M.A., Scaglia, O.A., Velazquez, C.A. and Lemeneff, T.H. (1991). Temperate sub-humid grasslands of South America. *In:* Natural grassland ecosystems of the World. R.T. Coupland (ed.). Elsevier, Amsterdam, Vol. 8, 592 pp.

Spiertz, J.H.J., Van Heemst, H.D.J. and Van Keenlen, H. (1992). Field crops systems in Northwestern Europe. *In:* Field crops ecosystems. C.J. Pearson (ed.), Elsevier, Amsterdam, pp. 357-371.

Stapper, P.J. and Haris, H.C. (1989). Assessing the productivity of wheat genotypes in a Mediterranean climate, using crop simulation models. *Field Crops Research,* **20:** 129-152.

Stewart, J.W.B. and Tiesssen, H. (1987). Dynamics of soil organic phosphorus. *Biogeochemistry,* **4:** 41-60.

Studdert, G.A. and Echevarria, H.E. (2000). Crop rotations and Nitrogen fertilization to manage soil organic carbon dynamics. *Soil Sci. Soc. Am. J.,* **14:** 1496-1503.

Taboada, M.A., Micucci, F.G., Cosentio, D.J. and Lavado, R.S. (1998). Comparison of compaction induced by conventional and zero tillage in two soils of the Rolling Pampas of Argentina. *Soil and Tillage Research,* **49:** 57-63.

Tebrugge, F. and During, R.A. (1999). Reducing tillage intensity—a review of results from a long term study in Germany. *Soil and Tillage Research,* **53:** 15-28.

Thomson, R.B. and Fillery, I.R.P. (1997) Transformation in soil and turnover to wheat of N from components of grazed pasture in South Western Australia. *Austral. J. of Agriculture Research,* **48:** 1033-1047.

Toresani, S., Gomez, E., Bonel, B., Bisaro, V. and Montico, S. (1998). Cellulolytic population dynamics in a vertic soil under three tillage systems in the Humid Pampas of Argentina. *Soil and Tillage Research,* **49:** 79-83.

Tunney, H., Carton, O.T., Brookes, P.C. and Johnston, A.E. (1997). Phosphorus loss from soil to water. CAB International, England, pp. 370-373.

Unkovich, M., Pate, J.S., Sanford, P. and Armstrong, E.L. (1994b). Potential precision of the 15N natural abundance method in field estimates of nitrogen fixation by crop and pasture legumes in southwest Australia. *Austral. J. of Agriculture Research,* **45:** 119-132.

Van Lutzow, M. and Ottow, J.C.G. (1994). Effect of conventional and organic farming on microbial biomass and its nitrogen dynamics in Luvisols of the Friedberg plains. *Z. Pflanzenbuhr. Bodenkd,* **157:** 359-367.

Weigel, A., Russow, R. and Korschens, M. (2000). Quantification of airborne N-input in long term field experiments and its validation through measurements using 15N isotope dilution. *J. Plant Nutr. Soil Sci.,* **163:** 261-265.

Whitebread, A.M., Blair, G.J. and Lefroy, R.D.B. (2000). Managing legume leys, residues and fertilizers to enhance the sustainability of wheat cropping systems in Australia. 2. Soil physical fertility and carbon. *Soil and Tillage Research,* **54:** 77-89.

Whitebread, A.M., Lefroy, R.D.B. and Blair, G.J. (1998). A survey of the impact of cropping in soil physical and chemical properties in northwestern New South Wales. *Austral. J. of Soil Research,* **36:** 669-681.

Zhao, F. and McGrath, S.P. (1994). Extractable sulphate and organic sulphur in soils and their availability to plants. *Plant and Soil,* **164:** 246-250.

Zourarakis, D. (1983). Evolucion del contenido de nitrates en un original bajo cultivo de maiz. *Ciencia del Suelo,* **1:** 53-63.

CHAPTER 4

The Intensive Rice Culture in South and Southeast Asia
Nutrients in Rice Land Ecosystem

> *The development of intensive submerged rice agroecosystem of Southeast Asia is a laudable example of human ingenuity, and a monumental achievement in agriculture and crop based food production strategies. It serves two of the most densely populated zones of the world, in India and China. At present (2000 A.D.) this agricultural marvel supplies cereal food and energy to a third of the world populace. Projections are towards further intensification of this wetland rice belt through enhanced nutrient inputs and improved rice genetic stocks, in order to nourish an anticipated one more billion people by the year 2025, then equivalent to nearly half the global human population. It is yet another challenge to human ingenuity and his expertise in agrotechniques—hopefully we will succeed.*

CONTENTS

1. Introduction
    A. Expanse, agroenvironment, cropping patterns and soils
2. Soil Chemistry and Nutrient Dynamics
    A. Puddling, submergence and Physico-chemical transformation during rice cultivation
    B. Nitrogen dynamics
        - Forms of N in Wetland soils
        - Movement of N
        - Adsorption and desorption of N
        - Volatilization
        - Mineralization
        - Mineralization-Immobilization
        - Nitrification-Denitrification
        - Denitrification
        - Fate of fertilizer N in wet land soils
    C. Phosphorus, in intensive rice ecosystems

3. Integrated Nutrient Management in Rice Land
   A. Nitrogen management
   B. Phosphorus management
   C. Nutrients and weed management
4. Physiological Genetics of Rice versus Nutrient Dynamics
   A. Plant growth and nutrient dynamics
   B. Nitrogen use efficiency and productivity
   C. Phosphate physiology and productivity
5. Concluding Remarks

1. Introduction

A. Expanse, Agroenvironment, Cropping Pattern and Soils

The seeds for a wetland rice agroecosystem in South and Southeast Asia were sown nearly 5 to 6 millennia ago. Since then, it has consistently spread across this vast agricultural zone. Today, this intensive rice ecosystem occupies nearly 10% of global agricultural soils, and dominates the humid and sub-humid tropics of the Indian sub-continent, Indo-china area, China and Far East. Within Southeast Asia, the rice belt occupies 60 to 70% of cultivable land, and approximately 60% of the human population of this zone thrives on rice as the staple cereal and the main source of carbohydrate. Rice, actually outcompetes other dietary sources in this zone, through its most acceptable palatability and nourishing features. The rice dependent population in this agrobelt is suspected to increase by 7.2% annually, over the previous levels. Hence, demographists forecast that one billion more rice eating population will be added by 2025 (Lampe, 1993; IRRI, 1995; Tribe, 1995, FAOSTAT, 1999). Projections for area expansion of this already vast agroecosystem are bleak. Hence, consensus among rice experts has been to intensify and raise productivity per unit area. During the second half of 20th century A.D., the development and adoption of high yielding, semi-dwarf rice genotypes, and high nutrient inputs through chemical fertilizers have already intensified this irrigated rice ecosystem, causing enormous changes in the nutrient dynamics. Presently, it consumes 3 to 4 million tonnes of N, P, and K, which is equivalent to 80% of total fertilizer based nutrient inputs in Southeast Asia (Table 4.1). Limits to such intensification are not yet discernible. Trends and opinions among present rice specialists point towards further intensification, through increase in rice grain yield per unit area. For this to happen, Lampe (1995) states that significant changes in plant architecture, and genetic make-up is necessary. The super varieties and hybrids being generated may hold promise through their potential to yield over 12 to 15 t grain ha^{-1} (Yuan, 1992; Amano *et al.* 1993; Kush, 1996; Fisher, 1998).

The previous spurt in productivity via intensification was actually a synergistic effect between soil fertility specialists and rice genetics. Perhaps there is need to replicate that effect at greater intensity. We must realize that in order to achieve higher yield, carefully scheduled nutrient inputs are mandatory. Equally so, during intensification nutrient dynamics and the ecosystem equilbria should be bestowed due attention and maintained appropriately. Within this chapter, some of these aspects are discussed in greater detail.

Table 4.1: Projected changes in area and nutrient consumption by rice compared with total area changes and nutrient demand by all crops cultivated in South and Southeast Asia

Crop / Year	Area (m ha)			N P K consumption ($\times 1000$ t N + P_2O_5 + K_2O)		
	1995	2000	2005	1995	2000	2005
Rice	38.7	39.3	39.0	3290	3852	4680
Total food crops	54.7	56.6	59.5	3941	4751	5896
Percentage of rice	70	69	65	83	81	79

Source: Mutert and Fairhurst, 1997
Note: Percentage denotes area occupied and NPK consumed by rice in relation to total food crops cultivated in South and Southeast Asia, including maize, cassava, soybean, groundnuts. Values for 2005 are projected figures.

The Rice Agroenvironment: Physiologically, the rice crop is highly adaptable to its environment (Mae, 1997). It can be cultivated in a wide range of agroclimatic zones, soil types and nutrient regimes. According to Rice-Almanac (1995) this rice land agroecosystem encompasses different subzones such as:

(a) Low land irrigated rice ecosystem supports intensive cultivation through moderate to high nutrient inputs. Correspondingly high yields at 6 to 9 t ha^{-1} are possible. It occupies 80 m ha equivalent to 55% of total rice area, and contributes 75% rice harvests.
(b) Low land rainfed rice ecosystem accounts for 25% of rice area at 40 m ha. It receives moderate inputs and may yield between 2 and 4 t ha^{-1}, contributing to 18% of global rice-grain production.
(c) Flooded, deep water rice ecosystem thrives on subsistence or low nutrient input, occupying 8% of total rice land and it produces 8% of global rice. Productivity is low at 1.5 to 2.0 t ha^{-1}.
(d) Upland dry rice ecosystem thrives on subsistence inputs and contributes 5% global rice yield. However, in this chapter, the focus is confined to nutrient dynamics in lowland, irrigated (submerged) paddy.

Cropping Pattern: Major rice-based cropping systems adopted in Southeast Asia are rice-rice, rice-rice-green manure legume, rice-wheat, rice-peanut/legume, or rice-sunflower. The emphasis in this chapter is on rice-rice sequence. In most locations the rice season begins with monsoon rains around May/June. Ambient temperature during crop season ranges from 18°C to 35°C in South and Southeast Asia. If congenial conditions occur, and logistics permit, then a third rice crop is a possibility in certain pockets. Annual precipitation received in this intensive rice growing belt ranges from 700-1100 mm, to zones receiving higher ranges at 1500 mm to > 2500 mm. A typical rice-rice sequence utilizes far greater quantity of water than others. For example, 232 cm was needed by a rice-rice cropping cycle, compared with 121 cm for rice-wheat, 144 cm for rice-legume or 115 cm for rice-sunflower. The water-use efficiency is least at 44 kg ha^{-1} cm for a rice-rice sequence compared with 82 to 85 kg ha^{-1} cm sequenced with wheat, legume or sunflower. The energy equivalent of NPK inputs into intensive rice-rice sequence is also high at 12.3 MJ $\times$ 10^3 ha^{-1} (Prihar *et al.* 1999). Consequently, the grain productivity, calculated as rice grain equivalent are better for rice-rice sequence at 10.62 t ha^{-1}, compared with 9.83 t ha^{-1} for rice-sunflower, or 9.97 for rice-wheat.

Soils: Soils and their nutrient status in this intensive rice belt vary enormously depending on location. Alfisols (Boralfs, Ustalfs, Udalfs) predominate in southern and eastern India. Vertic Inceptisols (Aquapts) are common on the eastern Gangetic plains. Ultisols, mainly Udults or Ustults are frequent in Southeast Asian locations in Indo-China and China. Again, in the Far East, Udic and Ustic soils predominate (Brady, 1995).

2. Soil Chemistry and Nutrient Dynamics

A. Puddling, Submergence and Physico-chemical Transformation during Rice Cultivation

To the farmers in South and Southeast Asia, who adopt intensive rice cultivation, puddling is one of earliest land management procedures. It is both labour and capital intensive. Puddling mainly reduces percolation loss of water and dissolved nutrients. It loosens top soil allowing easy transplantation of rice seedlings, and facilitates ponding of water. Puddling destroys soil aggregates and non-capillary pores, but increases capillary pore space. This leads to increased moisture retention, decreased water loss through evaporation and percolation. It also creates a hard pan in soil. Quite often, the soil physico-chemical modifications that occur due to puddling may turnout uncongenial to the dry arable crop grown in sequence. However, optimum puddling levels are not easy to standardize. Commonly, puddling index, soil dispersion ratio, bulk density, percolation rates and root length density are used as indicators while mechanically working on paddy soils. Clearly, extent of puddling through its influence in soil physico-chemical dynamics partially determines extent of nutrient recovery, losses and productivity. Generally, intensive puddling becomes mandatory, if percolation losses are to be minimized, thus allowing good root growth, and nutrient absorption (Aggarwal *et al.* 1999; Prihar *et al.* 1999).

In addition to physical changes stated above, puddling and soil submergence brings several chemical changes. Many of these are not encountered commonly in arable soils. Physically, a flooded rice soil possesses an oxidized surface layer, below it is a chemically reduced soil layer, which is the plough layer sandwiched between yet another layer of partly oxidized sub-surface soil (Fig. 4.1). Rice roots proliferate into this anaerobic plough layer to absorb nutrients that are held in a chemically

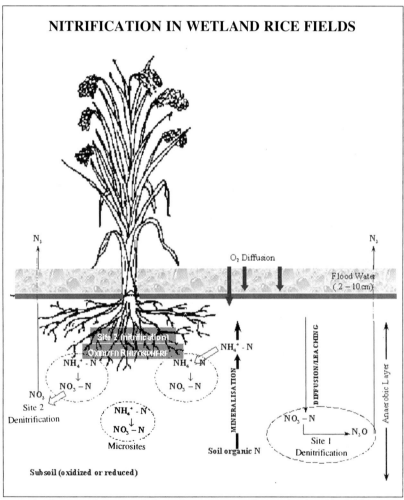

Fig. 4.1 Nitrification and denitrification sites within wetland paddy fields
Note: Soil microsites refer to oxidized local areas that may support nitrification, or anaerobic pockets that allow denitrification.

reduced state. However, several soil experts have suggested that, physico-chemically, the plough layer is not entirely uniformly anaerobic, but is a complex mosaic of anaerobic and aerobic micro-sites (pockets) (Sanchez, 1978; Sanyal and De Datta, 1991). It is estimated that within 24 hrs after flooding, oxygen supply into soil is effectively obstructed. This situation induces series of physico-chemical changes. Primarily, it retards gaseous exchange between flooded soil and air, reduces soil redox potential (Eh), increases specific conductance (Ec) and ionic strength. It affects soil microbial components through changes in oxygen tension. Aerobic microflora fades away and anaerobic microbes take over decomposition of organic matter and nutrient transformations. They utilize oxidized soil components as electron acceptors (Iyamuremya and Dick, 1996). During decomposition a specific thermodynamic sequence can be discerned depending on the redox potential. Redox potential itself changes from positive to negative, and usually decreases in a sequence as each set of reduction reactions are attained. For example, at redox potentials +800 mv oxygen is reduced to H_2O ($O_2 + 4H^+ + 4e^- — H_2O$); around + 400 to +430 mv NO_3 is denitrified to N_2, and MnO_3 is reduced to Mn_2; at 300 mV Fe $(OH)_3$ is converted to Fe $(OH)_2$; then at –180 mV organic acids and at –200 mv alcohols are reduced. Very low redox potential beyond –300 mv are detrimental to rice roots because sulphites (SO_3^{-2}) are produced.

Submergence increases hydrogen activity (pH) in acid soils, but decreases pH in calcareous soils. Several mechanisms may be operative that bring about pH changes in flooded soils. Reduced conditions that develop in submerged soils cause rapid decomposition of organic matter that liberate OH^+, thus consuming H^+ (Cang, 1985). Hence, increases in pH in flooded acid soil could be attributed to OH^- release. Reduction in pH in alkaline submerged soils may happen due to CO_2 accumulation which forms carbonic acids (Ponnemperuma, 1972). Release of organic acids by roots and soil microbes may also induce lowered pH although temporarily.

Some of the other physico-chemical changes that occur due to flooding, and relevant to rice nutrition are summarized below:
- Larger quantities of CO_2 are released during the first few weeks after submergence;
- Denitrification of nitrate is accomplished within a month after flooding, without any accumulation of nitrate;
- NH_4^+–N is more stable in reduced conditions, hence it accumulates, but is dependent on organic matter content and extent of leaching;
- Mineralization of organic C is retarded because of anaerobic conditions, whereas organic N mineralization might be hastened due to higher C : N ratios;

- Phosphorus in solution increases to more than 0.1 to 0.2 ppm during flooding, but decreases gradually. Phosphorus fixation reaction involving Al and Fe are also affected;
- Calcium, magnesium and potassium ions are displaced into solution by NH_4^+, Fe^{+3} and Mn^{+2};
- Fe^{+2} increases rapidly initially, but decreases gradually;
- Again, Mn levels increase rapidly initially due to greater solubility, but later decrease due to precipitation as $MnCO_3$;
- B, Cu, and Mo is increased, but Zn availability decreases.
- Under flooded conditions, the end products of organic matter decomposition are CO_2, NH_4^+, CH_4, NH_2, merceptons, H_2S and partially humified residues and resistant humified materials.

In South and Southeast Asia, rice is also grown on sodic and saline soils. Such problem soils are widely distributed within this intensive rice belt, occupying nearly 18 m ha (Ponnemperuma and Bandhyopadyay, 1981; Bandhyopadyay and Sarkar, 1999). Under such sodic/saline conditions rice roots experience excessive salts which results in electrical conductivity (EC) more than 4 m ha cm^{-1} at 25°C. The dominant salt is usually sodium chloride, or sodium sulphate, and the pH ranges between 6.5 and 8.5 or even 9.0. Hence, nutrient acquisition gets severely hampered. Generally, salt accumulation in flooded soils effectively displaces K^+, NH_4^+, Fe^{+2}, Mn^{+2}, Ca^{+2} and Mg^{+2} from the exchange sites. However, inherently rice is moderately tolerant to salinity. Any injury to roots or retardation in growth occurs due to high osmotic strength. In fact, rice is one of the most preferred crops, whenever salinity or sodicity problems are suspected in soils. There are some rice cultivars which are salt tolerant. They are cultivated in saline soil tracts of Bangladesh, Burma, India, Indonesia and Philippines (Ponnemperuma, and Bandhyopadyay; 1981; Chauhan et al. 1999).

B. Nitrogen Dynamics

Forms of N in Wetland Soils: The nitrogen content of wetland soils may vary widely, and whatever be the proportions of different fractions, nearly 60 to 80% of N acquired by rice crop is derived from native N pool in soil (Broadbent, 1979). A major fraction of soil N in lowland paddy fields is encountered as organic N, whose chemical nature may vary, therefore not clearly known. The subfractions of organic N pool contain amino sugars, amino acids, humin-N etc. The percentage concentration of these organic-N fractions vary depending on organic components added into soil, weathering processes, cropping, and soil microbial component. For example, green manuring affects soil N fractions, their contents and chemical transformations. Green manure incorporation into rice-rice sequence may preferentially add N into fulvic than humic fractions. If Azolla was used as green manure, upto 48% residual ^{15}N originated from

Azolla. Whereas, if Milk-Vetch was incorporated in wetlands, then 35% of ^{15}N was traced in residual humin and fulvic fractions. Green manuring also affects microbial N. Singh *et al.* (1996) estimated that nearly 10% of microbial biomass N in rice fields was derived from green manure. The other major N fraction in rice soil is the inorganic-N component. It is composed of NH_4^+–N, NO_2–N, and N_2, found in an adsorbed state to clay, organic colloids, lattice structure, in soil solution and soil air. This fraction is important in terms of N recovery by rice roots because they preferentially absorb NH_4^+–N.

Movement of N: In submerged conditions, transport of N ionic species occurs through phase diffusion, soil solid phase movement and mass flow. All of these three physico-chemical processes affect the net nitrogen transformation in submerged soils. More accurately, knowledge about diffusion coefficient for different N species, the actual extent of N movement, especially its direction either away or towards root zone is crucial. In undisturbed flooded soil which is not yet planted, the downward movement of NH_4^+ ranges from 12 to 14 cm in 4 weeks (Savant and De Datta, 1980). In coarse textured soils, comparatively low CEC tends to preserve NH_4^+ in soil solution. This accentuates downward convective transport of NH_4^+–N in flooded soils. The ^{15}N analyses indicate that horizontal shifts of deep placed NH_4^+–N could be as much as 4 adjacent rice hills planted at 20×20 cm spacings (Ventura and Watanabe, 1978). Puddling during rice cultivation retards water movement beyond the puddled layer, hence it decreases leaching losses of NH_4^+. Therefore, puddling is a significant factor in improving fertilizer N efficiency in rice zones. Soil compaction beyond bulk densities 1.72 to 1.75 g cm^{-3} reduces leaching losses of N beyond root zone. It decreases N percolating rates. However, higher inputs of organic matter may increase NH_4^+–N in the leachates. Quantity, source and timing of N inputs to paddy fields could be dependent on such data. Placement of N is also important. For example, a 10 cm deep placement of NH_4^+–N, in a transplanted paddy field resulted in greater downward movement than lateral or horizontal movement. Savant and De Datta (1982) have cautioned that a downward displacement of N ionic species beyond 5 mm day^{-1} can be detrimental to fertilizer N efficiency. It is predicted that on an average 4 to 6 kg N ha^{-1} per week could be lost due to improper rates of N movement/leaching in flooded soils.

Adsorption and Desorption of N: Within wetland rice ecosystem, knowledge on the dynamics of NH_4^+–N in soil is of great significance, because, rice crop preferentially absorbs and responds to NH_4^+–N inputs. Fertilizer N inputs, and mineralization of soil organic matter liberates NH_4^+–N ionic species. The resultant NH_4^+ pool interacts physico-chemically with wetland soils maintained under submergence. Most importantly, a sizeable amount of NH_4^+-N may be assimilated and immobilized in microbial and phytic fractions. Apart from these biological interactions, NH_4^+-N may be adsorbed in a chemically exchangeable ionic state on clay fraction, and it can be chemisorbed by humic fraction or fixed within clay lattice. Ammonium-N not extractable by KCl is termed 'fixed-N' in soil. The extent of sorption, chemisorption or fixation depends on variety of factors related to soil texture, nature of clay minerals, soil environment, submergence, chemical nutrient and organic manure input, and intensity of cropping etc. for example, soils with montomorrillonite exerts adsorption power greater than kaolinitic, then kalonities > allophanes. Chemically a reduced soil environment, normally resulted in lower levels of NH_4^+-N adsorption, compared with oxidized layers in the rice field.

Volatilization: Nitrogen loss from the rice ecosystem that occurs through volatilization as NH_3 and N_2O or N_2 emissions can be a severe factor that diminishes fertilizer N-use efficiency. If not checked at appropriate time, volatilization can alter N dynamics in wetland soils significantly. The extent of N loss through this mechanism varies widely, depending on a large set of factors related to soil, irrigation (flood water), atmosphere, crop and its management. For example, when soil was flooded N_2O emissions were through rice plants, while in the absence of flood water, N_2O was emitted through the soil surface (Yan et al. 2000).

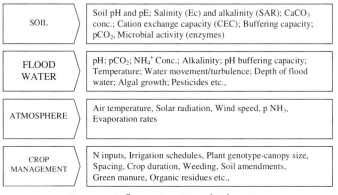

Factors influencing N volatilization:

Details on kinetics of NH_3 volatilization, chemical equilibrium and factors that influence such N loss have been discussed in great detail elsewhere by several authors (Ponnemperuma, 1972; Hagin and Tucker, 1988; Bandhyopadyay and Sarkar, 1999; Rao, 2002). To quote examples, under typical air temperature, aerodynamics and other meteorological parameters prevalent in this rice belt, volatilization may increase by 0.25% for every degree Celsius rise (Vleck and Crasswell, 1979). Actually, NH_3

volatilization in submerged rice field follows a definite diurnal pattern. It is generally observed that rice genotypes which rapidly develop a thick canopy are useful in avoiding undue air movement. It also reduces ambient and soil temperature presumably through standing water, therefore reducing NH_3 volatilization. For example, a rice crop with sparse/small canopy lost 6.1 mg NH_3–N cm^{-2} hr^{-1}, but larger canopies lost only 2.8 mg NH_3–N cm^{-2} hr^{-1}. Recently, Gill et al. (1998) reported that wetland soils amended with organic matter exhibited marked increase in urease activity. It means that higher quantities of urea are catalyzed, therefore aggravating N loss through ammonia volatilization. In fact, N loss via volatilization was directly related to urease activity, and the amount of organic matter added in certain rice fields of eastern India.

Mineralization: Mineralization of nitrogen, in simplest terms, is the conversion of organic forms of nitrogen into inorganic mineral forms. Whereas immobilization is a chemical process that runs antiparallel, resulting in conversion of inorganic NO_3–N or NH_4^+ ions into organic components in paddy soils (Brady, 1995). Both of these soil biochemical processes affect nitrogen nutrition of wetland rice crop, its productivity, and net N dynamics in the ecosystem.

Under anaerobic conditions created by submergence, organic nitrogen conversion proceeds only up to ammonia formation. Hence, it is commonly termed as ammonification whenever it concerns flooded rice ecosystems. The microbiological and biochemical reaction sequences and intricacies of regulation of mineralization versus the opposite process—immobilization have been discussed in great detail at periodic intervals by soil researchers (Ponnemperuma, 1972; Broadbent, 1979; Hagin and Tucker, 1988; Sanyal and De Datta, 1991; Iyamuremya and Dick, 1996). Ammonification is a catabolic, deamination of amino acids existing in wetland soils. The oxidative degradation of such amino acids derived from microbial and phytic fractions results in formation of amino acids, which are then split into ketoacids and gaseous NH_3. The oxidative deamination occurs in oxidized upper layer of low-land rice fields, perhaps nearer to rhizosphere, as well as in microsites with oxidized environments. On the other hand, reductive deamination may occur within the reduced soil layer that occurs below flooded zone. This process liberates gaseous NH_3 and forms saturated acids.

Ammonification reaction follows a first order kinetics in unplanted submerged soils, which however gets altered whenever rice crop is established. Immediately after N fertilizer inputs, ammonification increases and peaks in 2 weeks after submergence. This phase is followed with decrease in ammonification rates upto 8 weeks. A similar pattern could be noticed for ammonification within acid soils, or soils amended with organic matter. However, when ^{15}N tagged fertilizers were introduced into rice fields, soil mineralization appeared to peak twice during the cropping season, which was attributed to other soil chemical processes that result in N loss, such as denitrification. In the intensive rice zone, application of fertilizer N is known to provide a positive priming effect for ammonification. Similar priming effect, but of low magnitude can be noticed whenever organic nitrogen sources are incorporated into lowland rice soils. In general, during a 12-week rice crop duration, the trends of ammonification in submerged soils could follow different patterns such as linear, rectilinear or sigmoidal.

Factors affecting ammonification rates in submerged paddy soils are:

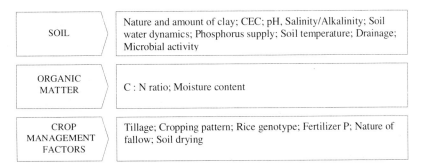

Mineralization-immobilization: In the wetland rice fields, mineralization-immobilization reactions, which are either coupled or uncoupled chemical reactions can be crucial in terms of N availability and turnover. As the process of immobilization proceeds mediated by soil microbiota, reammonification of this immobilized N follows a rate and pattern different from that obtained on native soil nitrogen. Initially, such an ammonification process can be rapid with native soil N, and it prolongs for a period. Such enhancements in N availability through reammonification may or may not get reflected as increased plant N uptake. Progressively as immobilized N gets incorporated in greater proportions into high molecular weight organic compound, which complex with clay fractions, the ammonification of immobilized N reduces. At this stage, it needs to be clarified that mineralization of organic N and immobilization of inorganic N in wetlands are not mutually exclusive. They may proceed simultaneously at different rates in pockets (microsites) within the soil.

Immobilization: The assimilation of inorganic nitrogen into organic forms mediated by soil microbes is an important aspect of soil N dynamics, mainly because, it affects N availability to rice crop. Again, several factors related to soil, inorganic and organic N inputs, and agronomic procedures adopted during rice culture affects the immobilization process. Firstly, C:N ratios of the soil, and plant residues that may be added markedly influence the pattern of immobilization. Larger the C:N ratio greater is the extent of N immobilization. The ^{15}N labeled studies indicate that the extent of added N lost through immobilization can vary between 0 and 80% (Hood, 2001). Biochemically, easily degradable substances allow rapid immobilization process, whereas those with greater proportion of crude fibers may support slower rates of immobilization. Immobilization is clearly a temperature and moisture dependent process. Higher rates of immobilization are expected at 37°C > 30°C > 20.6°C in that order. Under submerged conditions, added inorganic ^{15}N was rapidly immobilized within first 2 weeks, but it slows down later with lapse of time.

Nitrification-Denitrification: The native soil nitrogen in wetland rice fields that occurs as NH_4^+, which is derived from mineralization of soil organic N, gets oxidized biologically to NO_3. This strictly aerobic, biochemical process termed nitrification is mediated by microbes—*Nitrosomonas* and *Nitrobacter*. It confines to aerobic microsites in soil, and soil layers wherein oxygen is not limiting. It could be detrimental to N availability to rice roots, since rice roots prefer NH_4^+ over NO_3.

In flooded soils, an initial lag period that occurs before nitrification begins is variable. Simulations of submerged soil ecosystem indicate that such a lag period could extend upto 6 days. Sometimes, in well fertilized flooded soils, NH_4^+–N conversion to NO_2^{-2} and NO_3^- could peak in 7 to 10 days. The nitrification reactions occurring in oxidized layers of submerged paddy fields (Fig. 4.1) are deemed to be comparatively slow paced, that follow a zero order kinetics. Yet another site, where an oxidized microsite supports nitrification is in the rhizosphere surrounding rice roots/rootlets. Steady diffusive flux of oxygen of approximately 10^{-8} g O_2 cm^{-2} root surface min^{-1} at 23°C, from rice roots into the wetland soil alters, both chemical and microbiological environment in the surrounding zones. The rhizosphere alterations in pH, Fe ionic forms, Eh (350 – 450 mV), pe (+2 to 15), microbial flora (from anaerobic to aerobic), all of which are dependent on oxygen diffusion rates have been studied and reported good detail. Rhizosphere nurtures chemically most favorable soil micro-environment for nitrification in submerged soils. Since, nitrification in the rhizosphere is influenced strongly by organic exudates and diffusion of O_2, it is indirectly dependent on physiological stage, and genetic traits of rice plant.

Denitrification: This process occurs within the anaerobic soil environment provided copiously by the wetland soils. Our understanding of this important process which affects N-use efficiency has been periodically reviewed during the past four decades (Hauck, 1979; Broadbent, 1979; Savant and De datta, 1982, Hagin and Tucker, 1988; Sanyal and De Datta, 1991; Iyamuremya and Dick, 1996; Krishna and Rosen, 2001).

Firstly, a gradient of NH_4^+–N gets established between aerobic and anaerobic layers. Next, the NH_4^+–N in aerobic layers of flooded rice is converted to NO_3 through nitrification. It then diffuses into lower layers of soil wherein anaerobic conditions prevail. Here, it gets denitrified leading to gaseous N_2O or N_2 evolution. This is clearly a net N loss to the wetland soil ecosystem. During this coupled nitrification-denitrification process, the upward translocation (diffusion) of NH_4^+–N and comparatively slower rates of nitrification—i.e. NO_3 formation are indeed the rate limiting steps. These biochemical steps actually affect the ratio of NH_4^+–N/NO_3–N availability to plant roots and extent of reduction in fertilizer N-use efficiency, or N loss as gaseous N_2. Obviously, wide range of factors related to soil, fertilizer management practices and crop genotype influence nitrification-denitrification mediated N loss from wetland soils. According to rice researchers, well stabilized wetlands with pH between 6.5 and 7.2 and pE between 1 and 3 are generally conducive to denitrification. Sometimes, redox systems related to conversion of Fe^{+3}–Fe^{+2} and chemical energy derived from it could be involved in denitrification process. At the bottom line, critical O_2 concentration less than 4×10^{-4} M which corresponds to 1% of partial pressure O_2 supports denitrification. Under such anaerobic soils, denitrification rates between 10^{-9} and 10^{-8} mol N gas g soil^{-1} m^{-2} are possible. It means considerable amount of soil N could be lost as gaseous N in a day. Also, the native organic matter, and near neutral pH (7.0) is crucial to the activity of denitrifying bacteria. The above details on nitrification and denitrification actually relates to the process that occurs at site-I, that is out in the wetland soils and away from rhizosphere. It is believed that denitrification at site-I may not be of significance in terms of N loss, at least in some parts of eastern India, Bangladesh, Burma and Vietnam, because soils are acidic laterites with very low organic matter. Generally, soils with <1% organic C suffer least loss of N in gaseous form because denitrifiers are scanty.

The site-II for potential nitrification-denitrification is identified around the oxidized rice-rhizosphere and reduced soil microsites (Fig. 4.1). Chemically, NO_3-N that may be produced at the rhizosphere through nitrification, should be immediately absorbed by plant roots. If not, its diffusion to reduced layers of soils will support denitrification and loss as gaseous N. The nitrogen loss through denitrification at site-II i.e. oxidized rhizosphere-reduced soil micro-site system seems more conspicuous even in soils with low organic matter. However, plant factors related to root growth, its activity and redox environment are also crucial factors for nitrification-denitrification at site-II. Keeney and Sahrawat (1983) estimated that such N loss due to nitrification at site-II in rice fields can account for 18% of applied fertilizer N, and added losses are to be expected through gaseous N_2 loss.

Fate of Fertilizer N in Wetland Soils: Fertilizer nitrogen supplied to submerged paddy soil immediately becomes part of soil N dynamics, and takes part in soil physico-chemical and biological interactions. Such transformations may either be useful or detrimental in terms of N availability to rice roots, and the N-use efficiency achieved. In particular, chemical transformations of N fertilizer and their consequence on rice crop production has been reviewed in great detail (Savant and De Datta, 1982; Hagin and Tucker, 1988; Bandhyopadyay and Sarkar, 1999; Krishna and Carl Rosen, 2001; Hood, 2001). Overall, the expected fate, either physical or chemical transformation of N fertilizers in wet submerged rice soils are:

(a) Volatilization of NH_3 derived through NH_4 based fertilizers;
(b) Nitrification and denitrification of ammonium/nitrate fertilizers; loss through N_2O emissions and gaseous N_2 evolution;
(c) Immobilization of available N (NH_4^+ or NO_3) in biomass;
(d) Freely available and soluble NO_3 can be leached or lost to lower layers of soil-where it is denitrified.
(e) Uptake by roots, translocation to seeds, loss through transhipment outside field/ecosystem.

Quantifications of different N transformations mentioned above have been utilized to fix actual N fertilizer dosages to rice crop and to assess their long-term effects on N recovery index, soil organic N pool, N-use efficiency etc. During intensive cultivation of wetland rice, higher levels of N fertilizer requirements by the crop are satisfied either as a larger basal application and/or split application (top dressings). Such N inputs initially enhance NH_4^+ level in puddled soils, both in floodwater and soil surface (plough layer). For example, a typical N dosage between 40 and 60 kg N ha^{-1} results in 58 ppm NH_4^+–N in floodwater and 8 ppm NH_4^+–N in soil surface (Savant and De Datta, 1982). As NH_4^+–N diffuses to lower layers, both nitrification and denitrification processes result in N loss to system, and reduces NH_4^+ availability to plant roots. Whenever, such N loss is severe, N recovery into rice plant tissue could drop to 17 to 27% of basal N. Recovery of fertilizer N supplied as top dressing is usually better compared with that achieved through basal application. A well spread, established root system and lowered levels of N loss via ammonia volatilization are the causes for better N recovery from top dressings. Loss of fertilizer N due to percolation to lower horizon and leaching can be significant in submerged paddy fields. The fate of fertilizer N is also influenced much by the method of its application. For example, ammonia volatilization can be minimized to a great extent by incorporating N fertilizer at a depth of >6.0 cm in soil. Nitrification–denitrification losses too could be reduced considerably, if fertilizer N is placed deeper than 10 cm. Deeper placement of N fertilizer can enhance N recovery by plants to a tune of 10% over surface broadcasted ones. For example, deeper urea placement gave 30.2% N recovery, compared with 19.5% due to surface application. Also, it reduced N loss by 13% over surface broadcasted treatments (Sharma, 1995; Bandhyopadyay and Sarkar, 1999). Deeper placement is equivalent to gaining 0.7 t rice grain ha^{-1}, or between 20 and 25% gain in N-use efficiency and N recovery (Mishra, 1999). Considering the several ^{15}N based analyses, we may summarize that between 19 and 50% fertilizer N is recovered by a rice-rice sequence in Southeast Asia. Variations in N recovery from inorganic fertilizers are mainly caused by formulation, method of application, crop genotype and the season (Sharma, 1995; Mohanty and Sharma, 2000; Hood, 2001). To quote an example, during intensive rice culture in southern India, out of 100 kg ^{15}N applied, nearly 24.3% of N was recovered into rice grains, volatilization and leaching losses accounted for 17.2%, and 27.3% ^{15}N was retained in soil. Upto 31% ^{15}N could be lost through denitrification. Although, considerable ^{15}N was retained, only 5% of it seemed to flow into the second rice crop sown in sequence (Sharma, 1995).

C. Phosphorus in Intensive Rice Ecosystem

Dynamics of phosphorus in the South and Southeast Asian rice belt is often affected by the widespread deficiency of this element in the soil (De Datta *et al*. 1990). Phosphorus deficiency is conspicuous on major soil types such as Ultisols, Oxisols, Vertisols, Inceptisols (particularly Andepts) and acid sulphate soils. The problem gets accentuated due to a high degree of phosphate fixation associated with these soils. The result being, sub-optimal or inappropriate levels of P recovery into rice crops, and reduced productivity. To overcome it, larger initial inputs of P are practiced by rice farmers.

Forms of P in Flooded Rice Soils: The total P in soils of south Asian rice belt, particularly within the Indian rice zone could vary between 200 and 400 kg P ha^{-1} in the plough layer. However, these figures have little significance, because they do not correlate well with available P levels in soil, which is more crucial to rice productivity. Flooded, lowland rice soils contain P in both organic and inorganic forms. Inorganic forms encountered are iron phosphate (Fe-P); aluminum phosphate (Al-P), calcium phosphate (Ca-P) and reductant soluble or occluded Fe-P and Al-P. However, Fe-P seems the dominant form, accounting for 75 to 80% of inorganic P.

Phosphorus Availability in Submerged Soils: In general, submergence enhances available P levels, which is attributable to Fe–P (Sanyal and De Datta, 1991). Major reasons for such an increase in the availability of P as summarized by Iyamuremya and Dick, 1996; Sah *et al*. 1989, and Sanyal and De Datta, 1991 are:
 a) Reduction of hydrous ferric compounds, particularly Fe-oxides. Liberation of sorbed and/or co-precipitated P increases solution or extractable P in flooded soils (Lu Ru-Kun *et al*. 1982; Willet, 1989;

De Datta et al. 1990). Reduction and dissolution of Mn do not seem to affect P release during flooded rice culture.

b) In acid soils, higher solubility of $FePO_4 \cdot 2H_2O$ and $AlPO_3 \cdot 2H_2O$ could result due to increased soil pH.

c) Under submergence, organic acids released during anaerobic decomposition of organic compounds complexes Ca^{+2} to increase their solubilities. Sometimes level of organic matter could influence P release due to its effect on ferric reduction. However, mineralization of organic P in flooded soil as a mechanism for enhanced P availability is considered less significant, because it is too slow (Tate, 1984; Uwasawa et al. 1988; Iyamuremya and Dick, 1996).

d) PO_4 ion release can occur through exchange between organic anions and phosphate ions in Fe-P and Al-P compounds.

e) In calcareous soils, fall in pH due to flooding caused by CO_2 accumula-tion results in increased Ca–P solubility (Ponnemperuma, 1985).

f) Flooding enhances P diffusion and buffering of soil P. Such increased buffering capacity is attributable to P adsorption from soil solution by ferrous hydroxides or carbamates.

g) Microbial activity around the oxidized zones of rhizospheres can mobilize phosphorus (Sanyal and De Datta, 1991).

h) In flooded soils with highly reduced conditions, yet another chemical process involving Fe-P and sulphides can result in enhanced P availability via formation of H_3PO_4.

Even within the severely intensive rice belts of South Asia, which practice a continuous rice-rice sequence, a conspicuous drying phase occurs in between the two crops. Such drying causes re-oxidation, thus changes Eh, pH and Fe^{+2} ions. Flooding and draining affect P sorption. Phosphorus sorption capacity is also dependent on the extent of drying phase prior to submergence (Sah et al. 1989). Therefore, drying a flooded soil reduces P availability. Also, P replenished at the beginning of flooding often gets immobilized to a greater extent than if applied during oxidized phase. It is believed that alternate flooding and drying phases increase activity of ferric hydrous oxides in sorbing P. This results in P immobili-0zation whenever flooded soils are drained, dried and aerated.

Green Manuring and P Availability: Green manure crops are a common interject in a rice-rice sequence. Its incorporation and mineralization release P in available form to rice roots. Such P release is obviously influenced by soil and environmental factors. Foremost, P contents of green manure itself, and its C:P ratio regulates P mineralization. Critical C:P ratio of green manure for mineralization during a wetland rice culture ranges between 55 and 300. Both, P content and C:P are however dependent on green manure species, its age at harvest and available P status of soil (Singh *et al.* 1996). Soil pH is an important factor that influences P release from green manures.

Potassium in Submerged Soils: Potassium is a key element in intensive rice ecosystem. Like other nutrients, physico-chemical transformations and dynamics of K are influenced by soil submergence. Potassium availability increases immediately after flooding, because it favors better nutrient mobility. The chemically reduced environment that ensues also increases K availability. Hence, rice plants may absorb higher levels of K leading to its exhaustion in wetland soils. Irrigation water contains traces of K, but it does not match K requirements of the crop. At best, K^+ input through irrigation water may offset leaching losses or run off. Fertilizer based potassium input into wetlands is transformed into water soluble, exchangeable and non-exchangeable fraction of K. The release of water soluble K in submerged soils may range between 2 and 3 ppm if FYM was applied, 2.5 to 5.5 ppm when blue green algae was incorporated, and 3 to 73 ppm when inorganic K fertilizer was employed (Prasad and Rokima, 1991). Potassium derived from organic matter inputs can be useful to rice crop. For example, Mohanty and Singh (1997) observed that organic matter applied to wetland soils in eastern India, released 0.1% exchangeable K, 0.8 to 1.0 non-exchangeable K and 5.2 to 7.0% HNO_3 soluble K (slowly soluble K). These fractions sum upto 6 to 8% total K^+ in the organic source. Split applications of potassium plus FYM have also been useful (Thaker *et al.* 1999).

Sulphur in Wetland Paddy Soils: Intensive cultivation of rice, and consistent use of high analysis fertilizers in south and southeastern Asia, without commensurate replenishments of secondary elements has resulted in appearance of sulphur deficiency (Hussain, 1990; Biswas and Tewatia, 1990). Sulphur status of wetland rice soils varies widely and total S contents range between 0 and 0.6%. For example, soils in Philippines used for wetland rice production contained between 125 and 446 ppm total S, but SO_4^{-2} ranged between 3 and 58 ppm. Sulphur inputs to the rice eco-system can occur through atmosphere as SO_2 depositions, which is a natural source. Atmospheric S is usually encountered as H_2S, SO_2 or in particulate S forms. Intensive rice cultivation practiced in southeast Asian coast line may receive S deposits through sea water evaporation, and SO_4 possessing aerosols. Knowledge about sulphur content of irrigation water used for ponding rice fields is also important. It is suggested that irrigation water containing dissolved S between 4 and 6 mg l^{-1} suffices to support a moderately good rice crop. However, such S sources may turn out insufficient, if intensive rice-rice sequence is practiced for more than couple of years. Generalizations indicate that 3 to 5 ppm SO_4–S in soil could be sufficient to achieve optimum growth. Sulphur occurs in both organic and inorganic forms in flooded soils. Organic-S can be traced in ester forms, or bonded in S containing amino acids or within residual S. Organic-S could account for 65% of total S in submerged soils. Isotopic analysis indicate that majority of S absorbed by rice plant arises from this fraction. Residual S in wetlands is yet another fraction which is resistant to acid and alkali hydrolysis. Major quantity (65 to 98%) of soil S occurs as organic S (Tabatabai and Bremner, 1972) which is not is extractable unless mineralized. Nearly 10% of organic S could be mineralized within a single rice crop season. Under flooded situation, SO_4^+-S released through mineralization of organic-S, quickly gets transformed to sulphide forms because of chemically reduced environment. Under submerged conditions, even high levels of SO_4^{-2}–S (upto 1500 ppm) could be reduced to unavailable form within a span of 5 to 6 weeks (Ponnemperuma, 1972). Therefore, immobilization of S inputs as SO_4^{-2} fertilizer is quite common.

Sorption and desorption of SO_4 in rice soils of southern Asia can be explained in different ways. In volcanic soils of the Far East, sorption of large quantities of SO_4 is ascribed to hydrous oxides of Fe and Al. The SO_4 is absorbed on the positive charges by displacing hydroxy or argo ligand. Sulphate sorption, that varies with horizons may actually be related to organic matter status and pH of soil solution. Inhibition of SO_4 sorption is often related to oxygen containing functional groups in soil, for example PO_4.

Since rice roots prefer to absorb SO_4^{-2} forms, application of SO_4^{-2} bearing fertilizers might be efficient. Water logging, which creates anaerobic conditions, usually reduces S fertilizer efficiency by converting SO_4^{-2} to sulphides. Immobilization as organic S, or precipitation as Fe or Al–S reduces S fertilizer efficiency. Possibility for oxidation of S in submerged soils is confined to oxygenated area on rhizoplane and rhizosphere. Hence, timing of S bearing fertilizers, depending on the nature of source, and duration of crop seem crucial. Obviously, dynamics of S transformation, crops' demand and absorption patterns need appropriate synchronization. Basal inputs generally achieve better efficiency with early rice genotypes, whereas for medium and late maturing genotypes, a belated application 30 days after transplantation provides high S fertilizer efficiency.

Zinc: Submergence during paddy cultivation, chemically transforms both native and applied Zn. Potential availability of Zn seems to decrease due to reducing conditions, and proportionately based on duration of submergence (Singh *et al.* 1999d; Savithri *et al.* 1999). Submergence induces formation of different forms of Zn, namely a) Water soluble exchangeable Zn (WSEX); organically complexed Zn (OC); manganese oxide bound Zn (Mn Ox); amorphous sesquioxides (AMOx) and Crystalline sesquioxides

(CRYOx). Concentrations of all the different Zn containing fractions decrease gradually with submergence. As a consequence, low land rice crop recovers proportionately lowered levels of Zn. Zinc recovery by rice roots is also influenced by soil factors such as pH, organic C, clay content, CEC and temperature.

3. Integrated Nutrient Management in Rice Land

Nearly 80% of fertilizer based nutrient inputs in Southeast Asia are allocated to this intensive rice zone (Mutert and Fairhirst, 1997). This trend of lavish nutrient inputs to rice may continue for many years. However, efficient management of soil nutrients, their dynamics and resultant rice crop productivity is important. Several procedures are adopted, but among them integrated nutrient management systems are advocated more frequently. Let us consider a few salient features of this nutrient management system.

A. Nitrogen Management

Nitrogen-use Efficiency: The main focus of integrated nutrient management programs is to improve fertilizer N efficiency, grain productivity and maintain optimum N dynamics in rice fields. Improved N-use efficiency is crucial to sustenance of intensive rice-rice sequence. Potential benefits due to higher N-use efficiency is large. However reliable information on N demand, supply, and turnover within the different subzones of this rice land is meager (Singh *et al.* 1995; Kundu and Ladha, 1999). Rice researchers believe that complexity of nutrient transformations, uptake, and influences of interacting factors are immense. Hence, it may demand new approaches that maximize N-use efficiency. In this regard, computer-based simulation on nitrogen dynamics could be useful in obtaining better insights.

Nitrogen is generally the most limiting nutrient in this rice ecosystem. Thus, efficiency/inefficiency related to N may also influence utilization of other nutrients. Perhaps, careful simulations of such effects will lead us to better decisions on N inputs, and its management. Recovery of Urea-N applied to wetland rice is extremely low, owing mainly to N losses. The use of complex fertilizers, slow-release formulations and nitrification inhibitors, under the integrated system is known to increase N recovery. Percentage N recovered in different locations averaged 26 for sole N inputs, whereas complex NPK formulation gave 38% N recovery. Generally, combining NPK formulations with FYM gave best N recovery at 45 to 48% (Mohanty and Sharma, 2000). Clearly, obtaining suitably high N recovery is crucial to proper management of N dynamics in rice fields. Slow-release N fertilizers tend to improve N recovery by rice crop. Urea-G (Phosphogypsum urea—36.8% N) and neem-coated urea (NCU—35.4% N) often provide better productivity. Soil analysis proved that, while urea (untreated) released higher amounts of NH_4^+–N which was vulnerable for volatilization and leaching losses, the slow-release N formulation released comparatively lower levels of NH_4^+–N (Deshmuk and Tiwari, 1996; Purkhayatstha and Katyal, 1998). Physical placement of N fertilizer could be crucial. Among the variations, mechanical placement at 5 cm depth provided best fertilizer N use efficiency (Table 4.2) (Mishra *et al.* 1999). Olk *et al.* (1999) suggest that in addition to agronomic efficiency, changes in native soil fertility should also be considered while interpreting fertilizer-N efficiency.

Table 4.2: Nitrogen recovery and N-use-efficiency obtained using different methods of placement of urea

Method of application	NUE	RE (%)	ANR (%)
Broadcasting			
Prilled Urea (PU) broadcast	18	100	21
Urea super granules (USG) broadcast	22	122	30
Placement			
USG manual placement at 5.0 cm depth	26	146	41
USG mechanical placement at 5.0 cm depth	25	140	63

Source: Mishra *et al.* 1999
Note: NUE = kg grain/kg N fertilizer applied; RE = Relative efficiency; ANR = Apparent Nitrogen Recovery

Nitrogen Balance: Nitrogen input and output equation, that is nitrogen balance is influenced by crop productivity, N-use efficiency, the nature of soil, and characteristics of the N fertilizer. Based on 14-year evaluation of rice-rice sequence in different locations in India, Nambiar *et al.* (1992) and Nambiar (1994) reported that, annually 50% of N inputs was extracted by rice crop, and out of the other half a portion is retained in soil, rest is lost. This could invariably lead to build up of N in soil, but is dependent on extent of N loss through volatilization, seepage etc. (Patnaik *et al.* 1989). Incidentally, estimation of other nutrients indicates that intensive rice-rice sequence has resulted in their build-up. In particular, organic C and available P levels increase with time (Mohanty and Sharma, 2000; Table 4.3).

Similarly, long term (10-year) evaluations of rice-rice sequence of eastern Indian locations have shown that incessant N, or NPK complex fertilizers can result in build-up of N, plus disproportionately increase Al^{+3} and Fe^{+3} activity. In acid soils, it even resulted in drastic rice grain yield reductions. Hence, Mohanty and Sharma (2000) concluded that productivity of rice-rice cropping sequence is influenced by residual and cumulative effects of N inputs, and it is always better to apply N, in both

Table 4.3: Residual soil fertility after long term (14-year) rice-rice sequence in southern India (Hyderabad)

Location	Initial value	Residual nutrients after 14 years of rice-rice sequence		
		N	NPK	NPK + FYM
Soil pH	8.40	8.20	8.30	8.00
Organic carbon (%)	0.51	0.69	0.82	1.25
Available N (kg ha^{-1})	27.20	214.00	250.00	254.00
Available P (kg ha^{-1})	9.00	10.00	23.00	53.00
Available K (kg ha^{-1})	315.00	234.00	269.00	253.00

Source: Mohanty and Sharma, 2000
N, NPK and NPK + FYM are manurial treatments

organic and inorganic forms. Infact, considering residual soil fertility during intensive culture helps in avoiding nutrient imbalances if any.

The productivity of rice-rice ecosystem could be enhanced through integrated N management system, but grain output and nutrient removals were also affected by other factors, for example, location and its agro-environment (Hedge, 1996). An 8-year evaluation in the Indian rice belt indicated that grain yield hovered around 4 to 5 t ha^{-1} in Assam and Bengal plains in eastern India, whereas it exceeded 8 to 9 t ha^{-1} in peninsular and coastal India. Whatever be the location-related effect on potential grain productivity, managing N inputs through both inorganic fertilizers and organic manures was most effective. Use of sizeable quantities of organic manures was most congenial considering the variations in thermal and moisture regimes (Yadav et al. 1999).

B. Phosphorus Management

Phosphorus management is an important aspect during intensive rice culture. In this regard De Datta et al. (1990) have excellently summarized, over 13,000 field trials, that relate to P recovery, P-use efficiency and their relationship to rice productivity. On an average 17 to 26 kg P were applied into a single rice crop, and it resulted in grain formation between 480 and 940 kg ha^{-1} in excess over control. The efficiency of P inputs varied between 18 and 54 kg grain kg P^{-1}. Location trials in Southeast Asia indicate that crop genotype, P source, timing, method and amount, all of these factors affect P recovery and productivity of rice crop (IRRI, 1996, 1997).

Applying P to soils in one stretch at the beginning seems best. Mainly because, rice crop extracts P rapidly during early seedling stage. Adequate P supply during early stages supports better root development and tillering. At lower temperature common in Far East, more P is needed during early growth stage. On an average, a rice crop that accumulates 12 to 15 t ha^{-1} stover and 6 to 7 t grain ha^{-1} depletes 31 kg P ha^{-1} from the soil. This could be derived from P fertilizer, P in recycled stubbles, through green manure incorporation, and inherently available soil P (De Datta and Morris, 1984). Integrated management procedures adopted at IRRI, Philippines resulted in agronomic efficiency between 53 and 91 kg rice grain per kg P, whenever grain yield levels fluctuated between 3.8 and 6.0 t ha^{-1} and stover between 7.5 and 12 t ha^{-1} (Medhi and De Datta, 1997). Physiologically, the efficiency of stover production was 232 kg stover per kg.

Lu et al. (1982) state that whatever be the improvements in fertilizer P management or in rice genetic stock with reference to P efficiency, only 8 to 20% of P applied is utilized, and the rest 80 to 90% may remain in soil to provide residual effects on rice-rice sequence. Long-term evaluation in different countries provided evidence that rice grain yield increased by 22% due to P inputs. The agronomic P-use efficiency ranged from 0 to 114 kg grain per kg P applied. Of course, it varied depending on location and crop genotype. Doberman et al. (1996) have summarized that between 16 and 33 kg P needs to be recovered per ha to produce 8 t grain ha^{-1}, and at the minimum, 1.8 to 4.2 kg P are consumed to produce 1.0 t grain, at a physiological P efficiency ranging between 220 and 600 kg grain per kg P absorbed. In many of the 11 locations utilized by Doberman et al. (1996), they found that P supplementation was mandatory in order to obtain a positive P balance in soils (Table 4.4).

Table 4.4: Phosphorus inputs, recovery, net balance and rice productivity in eleven Southeast Asian locations

Treatment	Olsen's P mg kg^{-1}	Grain yield kg ha^{-1}	Fertilizer P input kg ha^{-1}	Recycled P in stubble kg ha^{-1}	Total uptake kg ha^{-1}	Net P balance kg ha^{-1}
All sites control[1]	3.9	3341	0.0	0.6	7.6	−7.0
Phosphorus	9.4	6189	20.0	2.0	18.3	+3.7

Source: Doberman et al. 1996
Note: Total P uptake = P in grain + P in straw; Net P balance = Fertilizer P input−(P uptake− P recycled).Values are means of 11 sites in Southeast Asia namely IRRI (Philippines), Phil Rice (Philippines), Maros (Indonesia), Lanagang (Indonesia), Cauliang (Vietnam), Shipai (China), Jinxian (China), Quin Pu (China), Pantnagar (India), Coimbatore (India).

Based on these long term, multilocational trials, Dobermann et al. (1996) suggest that present recommendation of 20 to 25 kg P ha^{-1} input is inadequate to maintain rice productivity levels between 5 and 6 t ha^{-1}. Annual P balance can turn out to be negative, if P inputs do not match the crops demand. Such a situation occurred frequently in many locations in India, Vietnam, China, wherever P inputs were less than 30 to 40 kg P ha^{-1} (Medhi and De Datta, 1997; Mohanty and Sharma, 2000). However, with the advent of rice hybrids and super hybrids with ability to yield between 7 and 13 t g ha^{-1}, the P dynamics ought to differ. Greater P inputs are needed, but only after careful considerations about physiological and agronomic P efficiency, P cycling and net P balance. At higher yield levels (13 to 15 t ha^{-1}), which are already possible in some locations in China, P recovery per rice crop will be 40 to 50 kg ha^{-1} (Kush, 1993). Most importantly as this rice belt gets further intensified through super hybrids, management of P inputs recovery and recycling needs to be more efficient.

Rock phosphates are not a preferred source of P for submerged paddy culture. However, in certain areas of intensive rice culture in eastern India, China and other Southeast Asian locations, phosphate rocks have been utilized on acid soils (Chien *et al.* 1990; Rajkhova and Bharoova, 1998;). Relative fertilizer efficiency and net P recovery levels are generally lower with phosphate rocks used in rice fields.

Sulphur during Rice Production: Dynamics of sulphur is gaining importance in the rice belt of Southeast Asia. It is being bestowed greater priority by soil fertility experts, because its deficiency and imbalance are becoming pronounced due to intensification of rice culture. Hence, S is being called the 'Fourth Major Element' after N, P, K. On an average, rice cultivars producing between 4 and 9 t grain, and proportionate straw, at harvest indices between 0.4 and 0.5 draw up between 3 and 6 kg S ha^{-1}. Intensive sole rice crop (IR64) grown for three years at Batangas, in Philippines recovered 3 to 6 kg S ha^{-1} during the first year. Slightly lower levels of S at 2 to 3.5 kg S ha^{-1} were recovered during second year. Rice productivity responses to residual S during second and third year ranged between 4.5 and 6 t grain ha^{-1}. It was 1.5 to 3 t grain ha^{-1} in excess over rice fields not receiving S inputs. Clearly, while assessing S requirement, and managing S dynamics in rice land, residual effects need due consideration. Quite often, use of S bearing compounds such as ammonium sulphate, gypsum, urea-sulphur rectifies S dearth if any (Singh, 1999).

Green Manure and Biofertilizers: It is generally argued that higher costs and greater loss of inorganic nutrients are attached with chemical fertilizers, particularly N and P. Hence, biofertilizers such as Azolla, Blue green algae, and organic N sources have been gaining importance in southern Indian rice belt. They tend to provide better nutrient dynamics in the rice ecosystem. The situation is similar in Vietnam, China, Philippines and other countries. Indeed, a wide range of combinations involving chemical and biofertilizers have been utilized by the rice farmers in Southeast Asia. Firstly, admixtures of chemical and biofertilizers such as Azolla-blue green algae, enhances N recovery considerably. In addition, Mishra *et al.* (1998) found that even after 2 years of rice-rice sequence, pH, organic C, K, were held at optimum levels in soils, plus a positive balance was noticed for N and P. A significantly large doze of organic manure and Azolla is sometimes opted, because it can reduce requirements of inorganic N and P fertilizers (IRRI, 1997, Gopalswamy *et al.* 1997). In eastern India, nutrient gains through Azolla-BGA incorporation ranged between 8 and 41 kg N ha^{-1}, 2 and 13 kg P ha^{-1} and 10 and 18 kg k ha^{-1} over control (Roy and Pederson, 1992; Singh and Mandal, 1997). A variation that involves intercropping P enriched Azolla provided 10 to 12% increase in N recovery over control (Singh, 1998). Strains of Azolla vary with respect to growth rate and their ability to support nitrogen fixation. For example, with Azolla hybrids nitrogen gains were consistent and ranged between 8 and 51% over control (Gopalswamy and Kannaiyan, 1998).

A green manure crop after rice-rice sequence is a common prescription under integrated nutrient management procedures. Green manure species such as *Sesbania* spp, *Crotalaria juncia, Vigna, Gliricidia* etc. produce between 5 and 16 t biomass ha^{-1}, accumulating N at 2 to 4% on fresh weight basis (Kush and Bennet, 1992; Soliappan *et al.* 1996; Hiremath and Patel, 1996; 1998; Mehta *et al.* 1996; Choudhary and Thakaria, 1998; Chapale and Badole, 1999; Rai *et al.* 1999). Across different locations in India and other South Asian countries, green manuring enhanced N recovery by 16 to 33% over farmers' conventional practices (Jana and Gosh, 1996). More recent summarizations indicated that green manure crops absorb up to 100 kg N ha^{-1}, which is contributed by residual N in soil and symbiotic N fixation. Out of which, 40 to 60 kg N ha^{-1} could be traced in the succeeding rice crop (Mohanty and Sharma, 2000; Lata *et al.* 2001). Comparisons made by Kush and Bennet (1992) across several locations suggest that, if asymbiotic microbes in the rhizosphere contribute 30 kg N ha^{-1} per crop, rice-blue green alga/Azolla contributed nearly 80 kg N ha^{-1} per crop, whereas, legume-rhizobium contributed high levels of N averaging 230 to 360 kg N ha^{-1}. Fishmeal, common in coastal rice belt enhanced soil N by 26 kg ha^{-1}, available P$_2$O$_5$ by 2.0 kg ha^{-1} and K by 10 kg ha^{-1}. The corresponding productivity increase may range between 5 and 10 q grain ha^{-1} (Talashilkar *et al.* 1999).

C. Nutrients and Weed Management

Weeds that proliferate during intensive rice-rice sequence can immensely affect nutrient dynamics. Basically, it lessens nutrient recovery by rice crop and reduces fertilizer use efficiency. In a crop season, approximately 18 kg N, 3 kg P$_2$O$_5$ and 16 kg K$_2$O could be lost to weeds, if unchecked (Rao *et al.* 1995; Choubey *et al.* 1999). Hence, careful weed management means reduced nutrient diversion from the immediate rice crop. Transplanted paddy, in the general course, suffers less due to weed competition. Hand weeding drastically reduces nutrient loss to weeds, but is labour intensive. A wide variety of chemical weedicides are used in Southeast Asian countries. Still, on an average, if N uptake by rice crop in a season ranges between 35 and 55 kg ha^{-1}, then weeds may have recovered between 2 and 12 kg ha^{-1} despite control measures. Singh et al. (1999a) caution that nutrient diversion to weeds should also be considered while judging and estimating nutrient needs of a rice-rice sequence. Further, they noticed that if a season-long weedy rice field recovered 35 kg N, 15 kg P$_2$O$_5$ and 45 kg K$_2$O per ha, a weed free field in the rice belt in India absorbed 62 kg N, 26 kg P$_2$O$_5$ and 80 kg K$_2$O per ha. Overall, a 45-day weed-free situation beginning at transplantation seems necessary to achieve better nutrient recovery and obtain satisfactory productivity of rice (Singh et al. 1999b; Table 4.5).

Table 4.5: Average recovery of major nutrients by rice, and weeds associated with rice culture in the Indian subcontinent (1993–1995)

Rice culture	Rice productivity		Nutrient recovery by rice (kg ha^{-1})			Nutrient depletion by weeds (kg ha^{-1} at 100-day after transplant)			
	Grain (q/ha)	Stover (q/ha)	N	P	K	Weed dry matter (g m^{-2})	Nutrients depleted by weeds (at 100 day after transplant) (kg ha^{-1})		
							N	P	K
Transplanted	32.5	91.2	96.0	14.0	160	6.0	5.7	0.62	7.9
Weed control									
Weedy check	18.2	48.1	48.2	6.8	81	16.8	42.7	4.52	63.1
Hand weeding	35.5	92.0	96.1	14.1	205	5.9	5.6	0.63	7.2
Weedicides	24.0	58.6	64.1	10.8	105	10.4	16.7	1.85	24.5

Source: Chander and Pandey, 1997

4. Physiological Genetics of Rice versus Nutrient Dynamics

Genetic manipulation of physiological traits, such as crop duration, plant stature, harvest index, biomass and nutrient allocation have been constantly occurring within this South and Southeast Asian rice land. Such modifications immensely influence dynamics of major nutrients N, P and K in the ecosystem. Let us consider a few pertinent relationships between the physiology of rice plant and nutrient dynamics, with emphasis on nitrogen and phosphorus.

A. Plant Growth and Nutrient Dynamics

During early stages of rice crop growth nitrogen is contributed by both, fertilizer N inputs and that available inherently in soil, or that left as residual N by previous crop. Mae (1997) states that chief source of N to rice roots is NH_4^+-N. Nearly, 70% of it is derived from soil, and relatively low levels of N originated from fertilizer N. It is generally observed that basal N inputs at transplantation sustains rice plant through the seedling phase. Mineralization of soil organic N too provides for N required by rice plant at seedling stage, and N-use efficiency may fluctuate between 40 and 60% of fertilizer N (Mae, 1997). Basal N inputs get almost exhausted by the time rice crop traverses through seedling stage and reaches maximum tillering (Table 4.6). That is, it satisfies N requirement until the end of vegetative phase. Obviously, splitting N inputs and applying it at right stages, whenever its consumption is greater, enhances nutrient-use efficiency. Fertilizer N provided as top dress is important in achieving maximum tillering potential, and proper panicle primordial initiation. Hence, it affects grain formation levels, net N recovery and N-use efficiency. Sometimes, basal plus two split doses of fertilizer N is advocated (Singh *et al.* 1999c). The underlying principle is to match N supply with crops demand. Nearly, 40 to 60% of basal N is exhausted by the time crop reaches flowering, and during panicle stage about 75 to 80% of top dressed N is consumed. During ripening stage, i.e., until harvest 20% of basal N and 50 to 80% of second top dress (if practiced) is utilized by the rice crop. Actually, immediately after panicle initiation, grains which are largest sinks dictate N absorption and retranslocation. These two processes in turn determine grain yield and N-use efficiency. Since, vegetative phase would have subsided, N top dressed will be efficiently translocated for seed growth, and will not cause lodging or increase canopy size.

Table 4.6: Average nitrogen recovered and accumulated by low land rice at different growth stages

Stage	Vegetative phase		Reproductive phase		
	Seedling	Tillering	Flowering	Ripening	Harvest
	(20 days)	(Variable)	(30 days)	(25 days)	Grand Total
N inputs (kg ha^{-1})	Basal N 40–120		Top dress 15 to 45 N		55–165
N recovery from soil (kg ha^{-1})	40–110 N		7–20N	5–12N	52–142 N
N recovered from fertilizer	40–60%		75–80% of TDN	20% of BSN 50-80% TDN	

Sources: Vergara, 1993; Mae, 1997
TDN = Top dressed N; BSN = Basal N; d = days

Clearly, N needs are greatest at grain filling stage, but rates of nutrient uptake, particularly N and P by the already senescing roots do not match such requirements. In fact, retranslocation of N already absorbed and stored in leaf tissue accounts for greater portion of N accumulated in grains. Remobilized-N is crucial, since 60-70% of seed N is contributed through this source. Whereas, instantly absorbed N may constitute the rest ranging between 10 and 30% (Mae and Shoji, 1984; Amano *et al.* 1993). Therefore, ability to remobilize N into seeds is an important genetic trait, because it helps to overcome vagaries, if any, in soil N availability during different stages of rice crop. Also, it determines the extent of N and/or other nutrients removed from the rice ecosystem through grains. Rice grains may contain between 1.0 and 1.5% N by dry weight basis at 14% moisture content. We may realize that changing rice genotypes, to hybrids with potential to form 6 to 13 t grains ha^{-1} in Japan, or even higher at 12.5 to 16 t ha^{-1} as projected in China, or similar increases elsewhere in Southeast Asia, will need sufficient build up of remobilizable-N in the leaf tissues of these genotypes. Appropriate increases in translocation index, N-use efficiency and harvest index are a necessity. Efficient and long lasting roots active in absorbing nutrients even during ripening stage may be helpful. Mae *et al.* (1997) suggest that parameters such as specific leaf N, CO_2 fixation rates, and kinetics of CO_2 fixing enzymes are crucial to proper biomass accumulation and N (nutrient) accumulation. Rice, as a plant species, ranks high among C_3 plants, in terms of photosynthetic and N-use efficiency. Overall, it should be clear that any modification in physiological traits such as crop duration, growth rate, harvest index, or nutrient retranslocation index will enormously alter the net nutrient demand, recovery, grain and stover production. Consequently, extent of nutrient recycling achieved through stover incorporation, and nutrient removed through grain harvest will also differ. A good example to study will be the overall changes in nutrient dynamics, and rice ecosystem productivity that will happen in future due to change from presently used; conventional moderately yielding rice cultivars to very high yielding rice super hybrids, with potential to yield 12 to 15 t grain and 22 t stover per ha (at harvest index over 5.6 to 6) as professed by Kush (1996). Net N and P demand during different stages, extent recycled through stover and that removed away through grain will be enormously different due to adoption of super hybrids of rice. Consequently nutrient turnover rates will be enhanced within the rice ecosystem. General productivity of rice in this belt has so far increased steadily and is expected to increase further through super hybrids. However, the extent of nutrients recycled via stover has not shown proportionate increases, because semi-dwarf genotypes and hybrids possess high harvest index, translocate larger fraction of nutrients into seeds, and retain less in leaves at maturity.

Harvest Index and Nitrogen Allocation: Let us consider the ramification of modifying HI and nitrogen allocation on rice crop productivity and net nutrient dynamics in the ecosystem. Actually, the total biomass potential of rice cultivars in a given environment, and harvest indices are crucial parameters that decide net nutrient demand, recovery and allocation. The transition from traditional tall rice varieties with harvest indices around 0.3 to 0.35, to the present day high yielding, semi-dwarf, high tillering genotypes with harvest indices between 0.42 and 0.50 has already affected nutrient dynamics of this intensive rice culture zone enormously. In particular, greater levels of N, P, K inputs, proportionately larger recovery into plants, and removal through grains have occurred due to this transition. Presently, nearly 10 to 20 t of biomass generated per ha, allow grain formation between 3 and 10 t ha^{-1}, depending on the genotype. For this to happen, harvest indices have ranged between 0.35 and 0.50. Rice breeders have been ever active, and we may find this situation changed in the near future (see Kropff *et al.* 1994; Lampe, 1995; Kush, 1996). Rice varieties and hybrids being developed for future consumption may further affect nutrient dynamics, because HI could further improve from 0.60 to 0.65 (Amano *et al.* 1996). With grain yield potential at 15 to 16.6 t ha^{-1} (Kropff *et al.* 1994), farming community in southeast Asia may adopt it, rather rapidly. Therefore, increases in nutrient inputs, particularly N, P, K, their recovery rates per unit area and time, nutrient allocation to grain and net nutrient removal from ecosystem will change drastically. In general, nutrient turn over rates in soils and above ground, and in the ecosystem at large will be enhanced proportionately. Managing such changes in nutrient dynamics in this rice ecosystem will need attention from policy makers, rice researchers and soil fertility experts, so that we may achieve a gradual change, congenial to the environment. Computer-based simulations about soil nutrients versus productivity, and accurate forecast may help us in arriving at better judgments.

B. Nitrogen Use Efficiency and Productivity

Nitrogen use efficiency is currently an important genetic trait receiving attention from rice scientists, because it has relevance to nitrogen economy and grain productivity of this cropping ecosystem. In addition, it has great relevance to N dynamics, and the ecology of the rice land (see De Datta and Braodbent, 1993; Mae, 1997; Singh *et al.* 1998; Fisher, 1998).

In simple terms, genetic improvement in N-use efficiency, allows production of higher amounts of biomass and grain at the same levels of N input. Alternatively, similar yield levels could be achieved at lower N inputs. It then delays NO_3 accumulation and toxic effects in areas where excessive N inputs have been practiced previously. In any case, a genetic change in N-use efficiency of a genotype, which is adopted for mass cultivation potentially induces vast changes in N dynamics of this intensive rice ecosystem. Essentially, physiological traits such as total N uptake, N translocation index, and grain N index can regulate N-use efficiency

(Jennings, 1991; Singh et al. 1998). Some rice specialists believe that genetic improvements in N-use efficiency during the past decades has been a major contributor to higher grain productivity of this rice-based ecosystem (Fisher, 1998; Singh et al. 1998). In certain locations N-uptake efficiency may be crucial. Higher N absorption is actually decided by an interplay between N-availability in soil with root morphogenetics and physiological activity. Normally, better results are possible by synchronizing N availability at root-soil interphase with the pattern of root physiological activity and crop demand for N, which is easier said than achieved.

C. Phosphate Physiology and Productivity

On an average, rice seedlings absorb between 0.5 and 1.0 kg P ha^{-1} before the tillering, and this constitutes around 2.5% of the total P demand by a crop that yields between 5 and 8 t ha^{-1}. Although a small fraction, it is crucial since it stimulates rapid root and shoot growth, which forms the basis for tillering, further growth and productivity. During panicle initiation and flowering phase, rice crop absorbs 5 to 6 kg P ha^{-1} (Table 4.7). Incidentally, flowering is dependent on phosphorus status of rice plants. Once again like N, crop demand for P is greatest during grain filling and ripening. It accounts for 75% of P accumulated in rice crop (Medhi and De Datta, 1997). Now, unlike N inputs, P is often applied to soil at the beginning. Usually, submerged paddy receives between 20 and 60 kg P ha^{-1} depending on yield goals. Splitting P inputs is uncommon. In tune with low mobility of P in soil, a basal dosage is practiced for rice. Whenever, breeders aim at deriving genotypes with vastly branched root systems that enhance P scavenging ability, it actually suits P dynamics in soil (AICRPR 1993; Ni et al. 1996). Remobilization of P is an important physiological phenomenon that ensures optimum grain formation, and to a certain extent determines amount of P removed away from the rice ecosystem via grains. A high-yielding, semi-dwarf rice cultivar that produces 8 to 10 t grain and 10 t straw per ha, extracts 30 to 33 kg P ha^{-1}, and partitions 3 to 11 kg P into grains. Once again, this equation will change, the moment super hybrids yielding 12-15 t grain and 18 to 22 t stover are adopted across this rice belt. Rapid and higher removal of P from soil needs to be appropriately replenished. Phosphorus removal from the cropping ecosystem is expected to increase considerably. Again, simulations and accurate predictions may be helpful in avoiding untoward effects. Yet another crucial aspect is the P-use efficiency of rice genotypes. Rice specialists suggest that physiological P-use efficiency might have contributed significantly towards yield gains in wetland rice (Dobermann et al. 1996; see Figure 4.2).

Table 4.7: Average P recovered and accumulated by low land crop at different physiological stages (in kg ha^{-1})

Nutrient recovery	Vegetative		Reproductive			
	20 d Seedling	Variable Tiller to panicle	30d Flowering	35d Grain formation Straw	Grain	Harvest Grand total
P recovery (kg ha^{-1})	0.6	0.9	4.0	4.4	14.1	24
Percentage P recovered	2.5	3.7	16.6	77.2		100

Sources: Vergara, 1993; Medhi and De Datta, 1997
Note: d = days

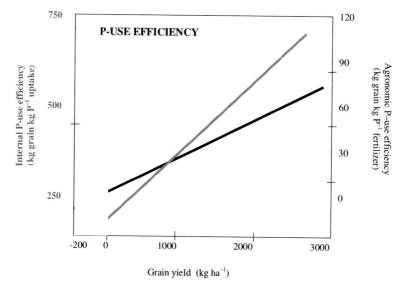

Fig. 4.2. Relationships between physiological (internal) P-use efficiency, and agronomic P-use efficiency with rice grain yield.
Source: Drawn based on data reported by Dobermann *et al.* 1996;
Note: Similarly, linear relationships between N-use efficiency of rice genotypes and grain yield have been reported by Fischer, 1998)

5. Concluding Remarks

The rice land of Southeast Asia, which is one of the most intensive agricultural belts, is being incessantly impinged with high rates of fertilizer-based nutrients. Consequently, the grain yield and biomass productivity of this ecosytem is high, ranging between 4 and 9 t grain ha^{-1}. Still, the yield levels attained seem insufficient. Already countries in Far East such as Japan, South Korea and Taiwan use >300 kg fertilizer-based N, P, K per ha. Forecasts by rice experts indicate a clear trend towards further intensification using higher levels of nutrient inputs and higher yielding genetic stocks. For example, super hybrids that may yield between 10 and 12 or 15 t grain ha^{-1} will demand higher nutrient inputs. As a precaution, during this process drastic changes or upheavals in nutrient dynamics and other ecosystematic functions need to be avoided. Firstly, intensification through higher nutrient inputs will directly affect the soil ecosystem. Then this soil microflora and transformations mediated by them may vary. As we know, higher use of nutrients will result in greater loss of nutrients. Emissions such as NH_3 may reduce fertilizer use efficiency, and CH_4, CO_2, N_2O can be deleterious to atmosphere. Use of super rice hybrids and excessive nutrient rates may also influence other crops and natural flora encountered within this ecosystem.

Overall, demographic forecasts indicate that a billion more rice eaters will be added in the next 25 years. Hence, the trend to intensify the rice ecosystem is clear and inevitable. Such intensification has to be achieved gradually, correctly and harmoniously without affecting other ecospheres. The upper limits to intensification, if there is one, particularly with regard to fertilizer inputs, and improvements in genetic stocks need to be ascertained. In this regard, simulating and modeling using computers to learn before hand, the consequences of greater intensification should be helpful.

REFERENCES

AICRP Rice (1993). Annual report—All India Co-ordinated Research Project on Rice. Indian Council of Agricultural Research, New Delhi, pp. 129-130.

Aggarwal, P., Garg, R.N., Das, D.K. and Sharma, S.M. (1999). Puddling, soil physical environment and rice growth on a typic Ustochrept. *J. Indian Soc. Soil Sci.*, **47**: 355-357.

Amano, T., Shi, C., Quin, D., Tsudu, M. and Matsumato, Y. (1996). High yielding performance of paddy achieved in Yunan province, China. 1. High yielding ability of Japanese F_1 hybrid rice—Yu Za 29. *J. J. of Crop Science*, **65**: 16-21.

Amano, T., Zhu, Q., Wang, Y., Inone, N. and Tanaka, H. (1993). Case study of high yields of paddy rice in Jiangxce Province, China. 1. Characteristics of grain production. *Jap. J. of Crop Science*, **62**: 267-274.

Bandhyopadyay, K. and Sarkar, M.C. (1999) ^{15}N Balance in rice-wheat cropping system. *Fertilizer News*, **44**: 39-47.

Biswas, B.G. and Tewatia, R.K. (1990). Role of sulphur in balanced plant nutrition: Indian experience. *Fertilizer News*, **36**: 13-17.

Brady, N.C. (1995). The nature and properties of soil. Prentice Hall of India, New Delhi, 467 pp.

Broadbent, F.E. (1979). *In:* Nitrogen and Rice. International Rice Research Institute, Manila, Philippines, pp. 105-118.

Cang, D.G., Jing, H.W. and Zang, X.N. (1985). Acidity. *In:* Physical chemistry of paddy soils. T.R. Yu (ed.), Springer Verlag (Science Press), Beijing, China, pp. 131-154.

Chander, S. and Jitender Pandey, J. (1997). Nutrient removal by scented basmati rice (*Oryza sativa*) and associated weeds as affected by nitrogen and herbicides under different rice cultures. *Indian J. of Agron.*, **42**: 256-260.

Chapale, S.D. and Badole, W.P. (1999). Effect of green manuring and NPK combinations on soil health and yield of rice (*Oryza sativa*). *Indian J. of Agron.*, **44**: 448-451.

Chauhan, R.P.S., Singh, B.B., Singh, R.K. and Singh, V.P. (1999). Strategic nutrient management for sustained rice production in sodic soil. *Fertilizer News*, **44**: 13-26.

Chien, S.H., Sale, P.W.G. and Hammond, L.L. (1990). Comparison of the effectiveness of phosphorus fertilizer products. *In:* Phosphorus requirements for sustainable agriculture in Asia and Oceania. International Rice Research Institute, Manila, Philippines, pp. 143-156.

Choubey, N.K., Tripathi, R.S. and Ghosh, B.C. (1999). Effect of fertilizer and weed management of direct seeded rice (*Oryza sativa*) on nutrient utilization. *Indian J. of Agron.* **44**: 313-315.

Choudhary, J.K. and Thakuria, R.K. (1998). Effect of green manuring on transplanted rice (*Oryza sativa*). *Indian J. of Agron.* **41**: 151-153.

De Datta, S.K., Biswas, T.K. and Charoenchamratcheep, C. (1990). Phosphorus requirements and management for low land rice. *In:* Phosphorus requirements for sustainable agriculture in Asia and Oceania. International Rice Research Institute, Manila, Philippines, pp 307-323.

De Datta, S.K. and Braodbent, F.E. (1993). Development changes related to nitrogen use efficiency in rice. *Field Crops Research,* **34**: 47-56.

De Datta, S.K. and Harris (1984). Integrated nutrient management in relation to soil fertility in low land rice based cropping system. *In:* Rice farming systems: New directions. International Rice Research Institute, Manila, Philippines, pp. 38-78.

De Datta, S.K. and Morris, R.A. (1984). Systems approach for the management of fertilizers in rice and rice based cropping sequences. Proceedings of the seminar and system approach to fertilizer industry. Fertilizer Association of India, New Delhi.

Deshmuk, S.C. and Tiwari, S.C. (1996). Efficiency of slow-release fertilizers in rice (*Oryza sativa*) on particularly reclaimed sodic Vertisols. *Indian J. of Agron.*, **41**: 586-590.

Dobermann, A., Cassman, K.G., Physiological Sta-Cruz, P.C., Adviento, M.A.A and Panipolino, M.F. (1996). Fertilizer inputs, nutrient balance and soil nutrient supplying power in intensive, irrigated rice systems 3. Phosphorus. *Nutrient Cycling in Agro-ecosytems,* **46**: 111-125.

FAO STAT (1999). Food and Agricultural Organization of the United Nations. http://.FAO.org.

Fisher, K.S. (1998). Toward increasing nitrogen-use efficiency in rice cropping system; the next generation technology. *Field Crops Research,* **56**: 1-6.

Gill, J.S., Khind, C.S., Singh, B. and Yadavinder Singh (1998). Ammonia volatilization under flooded conditions as affected by Urease activity of soils amended with crop residues on long term basis. *J. Indian Soc. Soil Sci.,* **46**: 448-450.

Gopalswamy, G., Anthoni Raj, S. and Ranganathan, T.B. (1997). Biofertilizer application strategy to rice (*Oryza sativa*). *Indian J. of Agron.*, **42**: 68-73.

Gopalswamy, G. and Kannaiyan, S. (1998). Utilization of Azolla hybrids for bio-reclamation of soils and rice (*Oryza sativa*). *Indian J. of Agron.*, **43**: 511-517.

Hagin, J. and Tucker, B. (1988). Fertilizers for dryland and irrigated soils. Springer Verlag, Heidelberg/New York, 435 pp.

Hauck, R.D. (1979). *In:* Nitrogen and Rice. International Rice Research Institute, Manila, Philippines, pp. 73-93.

Hedge, D.M. (1996). Integrated nutrient supply on crop productivity and soil fertility in rice (*Oryza sativa*) rice system. *Indian J. of Agron.* **41**: 1-8.

Hiremath, S.M. and Patel, Z.G. (1996). Influence of seeding data and level of green-manure during winter season on performance of summer rice (*Oryza sativa*). *Indian J. of Agron.*, **41**: 149-151.

Hiremath, S.M. and Patel, Z.G. (1998). Effect of winter green manure and nitrogen application on summer rice (*Orzya sativa*). *Indian J. of Agron.*, **43**: 71-76.

Hood R. (2001). The use of stable isotope in soil fertility research. *In:* Soil fertility and crop production K.R. Krishna (ed.) (in press).

Hussain, A. (1990). Sulphur in Bangladesh Agriculture. *Sulphur Agric.*, **14**: 25-28.

IRRI (1995). Rice facts. International Rice Research Institute, Los Baños, Philippines.

IRRI (1996). Annual Report. International Rice Research Institute, Manila, Philippines.

IRRI (1997). Annual Report. International Rice Research Institute, Manila, Philippines.

Iyamuremya, F. and Dick, R.P. (1996). Organic amendments and phosphorus sorptions by soils. *Advances in Agron.*, **56**: 139-185.

Jana, M.K. and Gosh, B.C. (1996). Integrated nutrient management in rice (*Oryza sativa*) rice crop sequence. *Indian J. of Agron.*, **41**: 183-187.

Jennings, P.R. (1991). Rice crop improvement. International Rice Research Institute, Manila, Philippines, pp. 175-176.

Keeney, D.R. and Sahrawat, K.L. (1983). Nitrogen transformations in rice puddle. *In:* Nitrogen economy of flooded rice. S.K. De Datta and K.W.H. Patrick (eds.), Martinus Nijhoff Publishers, Amsterdam, pp. 15-38.

Krishna, K.R. and Rosen, C. (2002). Nitrogen in soil: Transformations and their influence on productivity. *In:* Soil fertility and crop production. Krishna, K.R. (ed.) Science Publishers, New Hampshire, USA. pp. 91-108.

Kropff, M.J., Cassmeng K.G., Peng, S., Mathews, R.B. and Setter, T.C. (1994). Quantitative understanding of yield potential. *In:* Breaking the yield barrier. K.G. Cassman (ed.), International Rice Research Institute, Los Baños, Philippines, pp. 21-38.

Kundu, D.K. and Ladha, J.K. (1999). Sustaining productivity of low land rice soils: issues and options related to N availability. *Nutrient Cycling in Agroecosystems* **53**: 19-93.

Kush, G.S. (1993). Filling the Rice bowl. International Rice Research Institute, Manila, Philippines, 53 pp.

Kush, G.S. (1996). Breaking the yield frontier of rice. *Geo Journal* **35:** 329-332.

Kush, G.S. and Bennet, J. (1992). Nodulation and nitrogen fixation in rice: potential and prospects. International Rice Research Institute, Manila, Philippines, 136 pp.

Lampe, K. (1995). As coated in Tribe 1995. Feeding and Greening the World. CAB International, Wallingford, Oxford, England, p. 143.

Lampe, K. (1993). Foreword. *In:* Manual for hybrid rice production. Virmani, S.S. and Sharma, H.L. (eds.). International Rice Research Institute, Manila, Philippines, pp. 1-57.

Lata, Salexena, A.K. and Tilak, K.V.B.R. (2001). Biofertilizers to augment soil fertility and crop production. *In:* Soil fertilizers and crop production. Krishna, K.R. (ed.) (in press).

Lu Ru-kun, B. Jiany, and Ching-Kwai, L. (1982). Phosphorus management for submerged soils. Proceedings of 12th International Soil Science Congress—II. Indian Society of Soil Science, New Delhi, pp. 182-191.

Mae, T. (1997). Physiological efficiency in rice: nitrogen utilization, photosynthesis and yield potential. *Plant and Soil,* **196**: 201-210.

Mae, T. and Shoji, S. (1984). Studies on the fate of fertilizer nitrogen in rice plants and paddy soils using ^{15}N as a tracer in north-eastern Japan. *In:* Soil science and plant nutrition in north-eastern Japan. *Japanese Soc. of Soil Science and Plant Nutrition,* Sondai, Japan, pp. 77-94.

Medhi, B.N. and De Datta, S.K. (1997). Phosphorus availability to irrigated low land rice as affected by sources, application level and green manure. *Nutrient Cycling in Agroecosystems*, **40:** 195-203.

Mehta, H.D., Ahlawat, R.P.S., Chaudhary, P.P. and Jadhav, Y.R. (1996). Effect of green manuring on nitrogen management in low land rice (*Oryza sativa*). *Indian J. of Agron.*, **41:** 38-40.

Mishra, B.K., Das, A.K., Dash, A.K., Jena, D. and Swain, S.K. (1999). Evaluation of placement methods for urea super-granules in wet land rice (*Oryza sativa*) soil. *Indian J. of Agron.*, **44**: 710-716.

Mishra, S., Pradhan, L., Rath, B.S. and Tripathy, R.K. (1998). Response of summer rice (*Oryza sativa*) to different sources of nitrogen. *Indian J. of Agron.*, **43**: 60-63.

Mohanty, S.K. and Sharma, A.R. (2000). Nutrient management in rice-rice cropping system. *Fertilizer News*, **45**: 45-58.

Mohanty, S.K. and Singh, T.A. (1997). Nutrient changes and dynamics under a rice-wheat system. *In:* Sustainable soil productivity under rice-wheat system. T.D. Biswas and Narayanaswamy, G. (eds.). *Indian Soc. of Soil Sci. Bull.*, **18**: 20-31.

Mutert, E.W. and Fairhirst, T.H. (1997). Changing scenario in fertilizer consumption and supply in S.E. Asia: Perspective. *Fertilizer News*, **42**: 41-47.

Nambiar, K.K.M. (1994). Soil fertility and corp productivity under long-term fertilizer use in India. Publication and information division. Indian Council of Agricultural Research, New Delhi, 144 pp.

Nambiar, K.K.M., Soni, P.N., Vats, M.R., Sehgal, D.K. and Mehta, D.K. (1992). Annual report 1987-88/89. All India coordinated research project on long term fertilizer experiment. Indian Council of Agricultural Research, New Delhi, pp. 144.

Ni, J.J., Wu, P., Lou, A.C., Zhang, Y.S. and Teo, Q.N. (1996). Low phosphorus effects on the metabolism of rice seedlings. *Commun. Soil Sci. Plant Anae.* **27**: 3073-3084.

Patnaik, S., Panda, D. and Dash, R.N. (1989). Long-term fertilizer experiments with wetland rice. *Fertilizer News*, **34**: 47-52.

Ponnemperuma, F.N. (1972). The chemistry of submerged soils. *Advances in Agron.* **24**: 29-96.

Ponnemperuma, F.N. (1985). Chemical kinetics of wetland rice soils relative to soil fertility. *In:* Wetland soils characterization, classification and utilization. International Rice Research Institute, Manila, Philippines, pp. 71-89.

Ponnemperuma, F.N. and Bandhopadyay, A.K. (1981). Soil salinity as a constraint on food production in humid tropics. *In:* Soil related constraints to food production in the tropics. Cornell University/International Rice Research Institute, Manila, Philippines, pp. 203-216.

Prasad, B. and Rokima, J. (1991). Integrated nutrient management III. Transformation of applied K into various K fractions in relation to its availability and uptake in calcareous soil. *J. Indian Soc. Soil Sci.* **39**: 722-726.

Prihar, S.S., Ghildyal, B.P., Painuli, D.K. and Sur, H.S. (1985). Soil physics and rice. Proceeding workshop in physical aspects of soil management in rice based cropping system. International Rice Research Institute, Manila, Philippines, pp. 57-61.

Prihar, S. S., Pandey, D.D., Sukla, R.K., Verma, V.K., Chaure, N.K., Chaudhary, K.K. and Pandya, K.S. (1999). Energetics, yield, water use efficiency and economics of rice based cropping system. *Indian J. of Agron.* **44**: 205-209.

Purakayastha, T.J. and Katyal, J.C. (1998). Evaluation of compacted urea fertilizers prepared with chemical additions applied to wetland rice. *J. Indian Soc. Soil Sci.*, **46**: 147-150.

Rai, A.K., Pareek, R.P. and Chandra, R. (1999). Rate of decomposition and N release and microbial biomass C content in moist and flooded soils amended with Sesbania species green manures *J. Indian Soc. Soil Sci.*, **47**: 365-368.

RajKhowa, D.J. and Bharoova, S.R. (1998). Performance of Udaipur rock phosphate and superphosphate on rice in acid soils of Assam. *Indian J. of Agron.* **43**: 637-643.

Rao, K.V., Rao, B.P. and Rama Rao, K. (1995). Weed control techniques in transplanted rice (*Oryza sativa*). *Indian J. of Agron.*, **38**: 474-475.

Rao, K.V. (2002). Seasonal conditions influencing ammonia volatilization loss in irrigated wetland rice culture. *Journal of Indian Society of Soil Science*, **50**: 151-158.

Rice-Almanac (1995). The IRRI Rice almanac—1995. International Rice Research Institute, Manila, Philippines, 145 pp.

Roy, R.N. and Paderson, O.S. (1992). *International Rice Communication News Letter.*, **41**: 118-125.

Sah, R.N., Mickelom, D.S. and Hafiz, A.A. (1989). Phosphorus behavior in flooded drained soil. II Iron transformation and phosphorus sorption. *Soil Science Society of America J.*, **53**: 1723-1729.

Sanchez, P.A. (1978). Properties and management of soils in the tropics Wiley, New York.

Sanyal, S.K. and De Datta, S.K. (1991). Chemistry of phosphorus transformation in soil. *Advances in Soil Science*, **16**: 2-120.

Savant, N.K. and De Datta, S.K. (1980). Movement and distribution of Ammonium-N following deep placement of urea in Wet Land Rice. *Soil. Soc. Am. J.* **44**: 559-565.

Savant, N.K. and De Datta, S.K. (1982). Nitrogen transformations in wetland soils. *Advances in Agronomy*, **35**: 241-300.

Savithri, P., Perumal, R. and Nagarajan, R. (1999). Soil and crop management technologies for enhancing rice production under micro-environment constraints. *Nutrient Cycling in Agroecosystems*, **53**: 83-92.

Sharma, A.R. (1995). Fertilizer use in rice and rice based cropping system. *Fertilizer News*, **40**: 29-41.

Singh, D.P. (1998). Performance of rice (*Oryza sativa*) as affected by intercropping with phosphorus-enriched *Azolla carobiniana* under ranging levels of urea-nitrogen. *Indian J. of Agron.*, **43**: 13-17.

Singh, M.V. (1999). Current status of micro and secondary nutrient deficiency and crop response in different agro-ecological region. *Fertilizer News*, **44**: 65-82.

Singh, U., Cassman, K.G., Ladha, J.K. and Bronson, K.F. (1995). Innovative nitrogen management strategies for low land rice systems. *In:* Fragile ecosystems. International Rice Research Institute, Manila, Philippines, pp. 236.

Singh, Y., Ladha, J.K., Singh, B. and Khind, C.S. (1996). Management of nutrient yields in green manure systems. International Rice Research Institute, Manila, Philippines, pp. 159-166.

Singh, Y.V. and Mandal, B.K. (1997). Nutrition of rice (*Oryza sativa*) through Azolla, organic materials and urea. *Indian J. of Agron.*, **42**: 626-633.

Singh, V., Ladha, J.K., Castillo, E.G., Punzalan, G., Tirola-padere, A. and Duquiza, M. (1998). Genotypic variation in nitrogen use efficiency in medium and long duration rice. *Field Crops Research*, **58**: 35-53.

Singh, G., Singh, R.K., Singh, V.P., Singh, B.B. and Nayak, R. (1999a). Effect of crop-weed competition on yield and nutrient uptake by direct seeded rice (*Oryza sativa*) in rainfed, low land situation. *Indian J. of Agron.*, **44**: 722-727.

Singh, G., Nayak, R., Singh, R.K., Singh, V.P. and Singer, S.S. (1999b). Weed management in rainfed lowland rice (*Oryza sativa*) under transplanted conditions. *Indian J. of Agron.*, **44**: 316-319.

Singh, S.P., Sreedevi, B. and Pillai, K.G. (1999c). Influence of seedling age, source and schedule of nitrogen application on rice (*Oryza sativa*). *Indian J. of Agron.*, **44**: 530-533.

Singh, A.K., Khan, S.K. and Nonghynrich, P. (1999d). Transformation of zinc in wetland rice soils in relation to nutrition of rice crop. *J. Indian Soc. Soil Sci.* **47**: 248-253.

Solaiappan, U., Muthu Krishnan, S. and Veerabadran, V. (1996). Effect of rainfed green crops on succeeding rice (*Oryza sativa*). *Indian J. of Agron.*, **41**: 147-149.

Tabatabai, M.A. and Bremmer, J.M. 1972 Distribution of total and available sulphur in selected soils and soil profiles. *Indian J. of Agron.*, **65**: 40-44.

Talashilkar, S.C., Patil, S.H., Mehta, V.B. and Powar, A.G. (1999). Yield of rice and nutrient availability in soil as influenced by integrated use of fertilizers and fish meal. *J. Indian Soc Soil Sci.*, **47**: 725-727.

Tate, K.R. (1984). The biological transformation of phosphorus in soil. *Plant and Soil*, **76**: 245-256.

Thaker, D.S., Patel, S.R. and Yogeshwar Lal (1999). Effect of split application of potassium with FYM in rice (*Oryza sativa*). *Indian J. of Agron.* **44**: 301-303.

Tribe, D. (1995). Feeding and Greening the World. CAB international, Wallingford, Oxford, England, pp. 137-173.

Ventura, W. and Watanabe, I. (1978). Dry season soil conditions and soil nitrogen availability to wet season wet land rice. *Soil Science and Plant Nutrition*, **24**: 535-545.

Vergara, B. (1993). Rice. International Rice Research Institute, Manila, Philippines, pp. 1-218.

Vlek, P.J.G and Crasswell, E.T. (1979). Effect of nitrogen source and management on ammonium volatilization losses from flooded soil systems. *Soil Sci. Soc. Am. J.*, **43**: 352-358.

Uwasawa, M., Sangtong, P. and Cholitkal, W. (1989), Behaviour of phosphorus in paddy soils of Thailand. Contents of inorganic, organic and available soil P in relation to rice plant P-nutrition. *Soil Science and Plant Nutrition*, **34**: 41-53.

Willet, I.R. (1989). Causes and prediction of changes in extractable phosphorus during flooding. *Australian Soil Research*, **27**: 45-54.

Yadav, R.L., Dwivedi, B.S. and Prasad, K. (1999). Current status of crop response to fertilized in different agro-climatic zone. *Fertilizer News*, **44**: 53-60.

Yan, X., Shi. S., Du, L. and Xing, G. (2000). Pathways of N_2O emission from rice pady soil. *Soil Biology Biochemistry*, **32**: 437-440.

Yuan, L.P., Yung, Z.Y. and Yung, J.B. (1992). Hybrid rice in China. Second international symposium on hybrid rice. International Rice Research Institute, Los Baños, Philippines.

CHAPTER 5

The Rice-Wheat Agroecosystem of the Indo-Gangetic Plains
Nutrients and Productivity

Historically, the seeds for agricultural activity (agrosphere) were sown in the Indo-Gangetic region several centuries ago, beginning with the domestication of important cereals (e.g. rice) and legumes by the local population. Natural savannas, scrubland, shrubs, tropical and subtropical jungles, and barren expanses along the major rivers Indus and Ganges, and their tributaries, gave way for a regular agricultural cropping ecosystem. Since then, the cropping systems in particular, and general agro-ecology has co-evolved with human population in these plains. Human ingenuity has played a dominant role in developing and maintaining the agroecosystem's equilibrium, particularly the nutrient dynamics and productivity equations. Driven by dense population pressure in recent times (beginning mid 20th century till date), this agroecosystem experienced drastic modifications towards intensification through high nutrient inputs, improved crop genetic stocks, and increased productivity. Presently, this 'emerald green belt' predominately grows two most important cereals—rice and wheat.

CONTENTS

1. Introduction
    A. Soils in Rice—Wheat cropping zones
    B. Rice—Wheat-based cropping sequence
2. Nutrient Dynamics
    A. Nitrogen in Rice-Wheat Ecosystem
    B. Phosphorus in Rice-Wheat Ecosystem
    C. Potassium in Rice-Wheat Sequence
    D. Micronutrients in Rice-Wheat Ecosystem
3. Soil Microbes, Nutrient Dynamics and Rice-Wheat Productivity

4. Sodicity versus Nutrient Dynamics
5. Integrated Nutrient Management and Productivity
6. Concluding Remarks

1. Introduction

The rice-wheat agroecosystem of the Indo-Gangetic plains, which is among the most intensive agro-belts of the world is comparatively a recent development. Its evolution began approximately 3 decades ago. In India, it occupied a mere 0.2 m ha in 1960's, then increased to 0.4 m ha in 1970's. Beginning with an exponential increase since late 70's, it has reached 11 m ha at present (Velayutham, 1997). This rice—wheat ecosystem extends into 1.6 m ha into the Indus plains of Pakistan, about 0.5 m ha in Bangladesh, and 0.4 m ha in Nepal (Tarai belt) (Fig. 5.1).

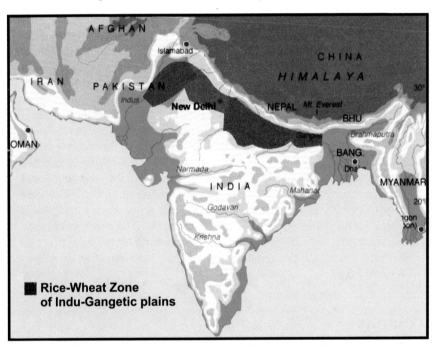

Fig. 5.1 The Rice-wheat Agroecosystem of the Indo-Gangetic Plains (Shaded area represents the rice-wheat cropping belt. Demarcation is not to scale)

So far there has been no standardized sub-classification of this cropping ecosystem. However, for convenience the Indo-Gangetic plains could be demarcated into (a) Indus plains covering southern plains in Sind and northern plains in Punjab provinces of Pakistan; (b) Trans-Gangetic plains (TGP), comprising Punjab in Pakistan and India, Haryana, Delhi and parts of Rajasthan in India; (c) Upper Gangetic plains (UGP) that occupy area between the two rivers Ganga and Yamuna; (d) Middle Gangetic plains (MGP) and (e) Lower Gangetic plains (LGP) (Pasricha and Brar, 1999). The most commonly encountered cropping sequence in this double-cereal based agroecosystem is as follows:

The rice-wheat agroecosystem extending mainly into Pakistan and India, and small portions in Bangladesh is hot, sub-humid or dry eco-region with soils derived mostly from alluvium, of either Indus and tributaries in the west or Ganges and its tributaries in the central and eastern parts. Soil could vary from red, yellow, vertic to podzolic (Swarup and Srinivasa Rao, 1999; Yadav *et al.* 1999a).

Annual rainfall in Indus plains varies between 500 and 875 mm. Throughout the plains precipitation is distributed unevenly, because rainfall events are more during July to September. A third of the annual precipitation is received during winter, when the plains support wheat crop production. Within the rain-fed wheat belt, ecological regions could be classified based on extent of precipitation levels, such as those receiving greater than 300 mm, 200 to 300 mm zones, and less than 200 mm zones during the crop season (Malik and Hassan, 1992). Water efficiency during wheat phase ranges between 10 kg ha^{-1} mm^{-1} and 14.0 kg ha^{-1} mm^{-1} depending on evapotranspiration that ranges between > 50 mm and <200 mm respectively (Chattopadhyay *et al.* 2001).

A. Soils in Rice-Wheat Cropping Zones

The rice-wheat cropping sequence is preferred by the farming community in Indo-Gangetic plains due to both natural and man-made factors. Mainly the edaphic and climatic factors, and to a lesser extent availability of infrastructure, food habits of the population and better economic consequences. The rainfall distribution during monsoon, annual ambient and soil temperature changes, humidity and inundation of soil during rainy season (Kharif) make rice the best option. Let us consider the soil-related factors. The general soil chemistry gets severely influenced by the anaerobic and reducing environment which results because of flooding. In fact, redox changes could be rampant, leading to changes in nutrient status, their availability, leaching and percolation etc. Similarly, the edaphic and climatic parameters that set in after rice is harvested, such as the soil aerobicity, low soil temperatures from November onwards, shifting of soils to dry oxidized condition, and nutrient transformation patterns make wheat the most preferred crop in this season (Rabi).

Major soil types are Inceptisols (Aquepts) and Alfisols (Ustalfs), and to a lesser extent Aridisols (Argids) and Vertisols (Usterts) are also encountered in this belt. These soils extend from Indus plains of Pakistan in the West, throughout the mid-Gangetic plains up to Bangladesh in the East. Information on classification of soils, utilized specifically for rice-wheat cropping sequence in the gangetic belt is meager or sporadic (Sawhney and Sehgal, 1989). More recently, Pannu *et al.* (1999) provided some information on morphology and physico-chemical traits of such soils. These soils in Indo-Gangetic plains have actually been used for rice-wheat cropping pattern only for over 3 decades. They are mostly clay loams, sandy loams, or silty loams. Soils in the northern Punjab (Pakistan) plains have been developed *in situ* from tertiary rock, consisting mainly of shales and sand stones. Loess and alluvium are major components. Soils in this rice-wheat belt are alkaline (pH 7.8 to 8.1), calcareous (5 to 8% free $CaCO_3$) with organic matter <1% (Khan *et al.* 1989; Khan, 1997). Crops respond to N, P, and K input since these nutrients are generally deficient.

The eastern Gangetic alluvial soils are mainly silty throughout the profile. Silty clay or clay could be dominant in several areas. Bulk densities are comparatively higher and the hydraulic conductivity is low to very low (Khan *et al.* 1998). Morphologically, soils in eastern Gangetic plains are alluvial, with uniform gray matrix. Prolonged submergence, low C : N ratio due to advanced decomposition, and hyperthermic conditions during rice phase seem to affect soil traits immensely.

Soil Physical Environment: Rice requires soil conditions that allow stagnation of water, and a soft rooting medium. Discing, leveling, and importantly puddling helps in achieving it. These farming methods allow ponding of water and reduces soil permeability, so that rice root growth, nutrient availability and recovery are favored (Ghildyal, 1978). Incidentally, puddling the soil is a major operation that distinguishes rice-wheat ecosystem from those which involve cereals other than rice. Puddling affects soil structure, reduces macrospores (aeration), decreases hydraulic conductivity and permeability. Whenever soils are stirred up while they are submerged, the soil aggregates are lost (Sharma and De Datta, 1985). Puddling increases bulk density of soil, both at surface and sub-surface layers (Sur *et al.* 1981; Chenkul and Acharya, 1990; Bushan and Sharma, 1999). It actually hastens soil compaction just below puddled layer affecting soil water retention severely (Kar *et al.* 1997). Water retention increases due to increased fraction of micropores and concomitant decrease in amount of macropores. For example, silt loams under rice-wheat ecosystem, often retain greater quantities of water at all levels of tension compared with a dry soil based crop rotation (Chenkul and Acharya, 1990). The reduction in soil aggregates decreases the transmission of micropores and hydraulic conductivity. Hardpan due to incessant puddling cycles also reduces hydraulic conductivity (Sur *et al.*

1981). However, impedance to water infiltration in puddled soils, still, depends on soil texture, structure, mineralogy, organic matter (Prihar *et al.* 1985) and the extent of puddling compaction achieved.

Puddling during rice cultivation in Indo-Gangetic plains keeps soil temperature a trifle low. Puddling favors water ponding, therefore it will fetch cooler soils during rice cultivation. The impounded irrigation water, quite often shifts from one field plot to next. During this process it causes equitable soil and water temperature regime. Soil aeration is an important factor affected by puddling during rice cultivation. The submergence results in low oxygen diffusion rates, often less than 5 μg cm^{-2} min^{-1}. Rice roots are well adopted to this micro-aerophilic or anaerobic condition. Whereas, for arable crop species, oxygen diffusion rates below 30 μg cm^{-2} min^{-1} will be detrimental. Rice roots, due to their high oxidizing power (equivalent to 1 to 3 mg day^{-1}) keep themselves, as well as rhizosphere in an oxidized state. This enables microbial activity adapted to aerobic conditions, thus allowing nutrient transformations possible through them (Ghildyal, 1982; Sarkar *et al.* 1989). However, the microenvironment in soil remains reduced due to flooding. Soil compaction due to puddling also affects redox potential (Kar *et al.* 1997). Puddling is indeed a crucial soil management procedure that distinguishes rice-wheat cropping sequence from other arable crop sequences. It affects several aspects, perhaps most conspicuously the redox changes and the ensuing alteration in soil nutrient transformations and net availability. Firstly, it reduces percolation of water and nutrients, thus making it congenial for rice root proliferation, nutrient acquisition and seedling establishment. Puddling is to be regulated, if not, roots confine excessively to top layers of soil. For example, 70 to 80% rice roots can be encountered within top 0 to 0.15 m. Whereas, wheat that succeeds is comparatively deep rooted, but will get affected by hardpan created through puddling during the rice phase. Hence, Aggarwal *et al.* (1999) suggest a careful compromise and regulation of puddling. Wheat, that follows the wetland rice in sequence also requires specific soil tilth and an oxidized state. Soil loosening is usually achieved by ploughing and pulverizing the top soil. Unlike rice, intensity of microporosity which supports better water retention should decrease for wheat root proliferation. Hardpans, if any, that impede root need to be dismantled, so that sufficiently deeper root penetration is promoted. As stated above, the rice-wheat agroecosystem is also unique in terms of seasonal changes in soil chemistry, particularly the electrochemical changes. Rapid alternations in electrochemical properties due to submergence and drying cycles affect nutrient transformations and availability. Conspicuously, the soil electro-chemical properties such as redox potential (Eh) drops to negative values during rice phase, but continues to be positive during wheat phase (Mohanty and Singh, 1997). Such drop in Eh can also be discerned transitorily, even after an irrigation event on dry soils during wheat phase. Flooding during rice phase affects pH. It is known to increase pH of an acid soil by affecting Fe^{+3}–Fe^{+2} equilibrium. Alternatively, it decreases pH of alkaline soils to neutrality through affecting CO_3–HCO_3 equilibrium (Ponnamperuma, 1972).

B. Rice-Wheat-based Cropping Sequence

The rice-wheat cropping system is currently the most common sequence adopted in the Indo-Gangetic plains. Rather, it is the mainstay of the farming community in this zone (Gangwar and Sharma, 1996). The rice-wheat cropping pattern is an important aspect of food grain supply mechanism for people dwelling in the Indo-Gangetic plains. This cropping system is known to contribute nearly 75% cereal food-grain required by India (Bandhyopadyay and Sarkar, 1999).

Overall, the major rice based cropping sequences adopted in the Indo-Gangetic plains are: (a) rice-rice-fallow; (b) rice-wheat (or any other cereal); (c) rice-pulse (mainly mung bean, black gram, soybean, cowpea etc.), (d) rice-oilseed (mustard or groundnut). Going by a survey conducted by Indian Council of Agricultural Research (ICAR), New Delhi, Kar et al. (1997) suggest that rice and wheat cropping sequence is the major rice based pattern in Gangetic plains, accounting for nearly 85% of the rice or wheat area, and amounts to 11.0 m ha in India. In Pakistan it occupies 1.6 m ha, while it is sparse in Bangladesh and Nepal (Table 5.1). The rice-wheat sequence is preferred over other combinations because of several crucial advantages to farming community. It offers better production efficiency, benefit : cost ratios, land and water use efficiency. The grain output equivalence and energy output are higher than other sequences involving legume or oilseeds. It is shorter in duration compared to rice-rice sequence (see Table 5.2). Summarization of research efforts over the past two decades, encompassing over 200 field trials in different locations in Indo-Gangetic plains indicate that cropping sequences envisaged under the improved integrated plant nutrient system (IPNS) are consistently efficient. Nutrient recovery, nutrient use efficiency, recycling, and grain biomass productivity of both rice and wheat were better. Therefore, most common and successful cropping sequence, both in India and parts of Pakistan include rice, wheat and a legume or a green manure crop during summer.

Table 5.1: Area (m ha) under rice-wheat ecosystem in the Indo-Gangetic plains

India	
1. Irrigated rice-wheat in Northwest India	**10.5**
• Scented rice (SR)-wheat	1.5
• Non-scented rice (NSR)-wheat	3.0
2. Partially irrigated NSR-wheat	2.0
3. Rice in salt affected soils-wheat	2.0
4. Lowland Rice-Wheat	1.5
5. Deep water Rice-Wheat	0.5
Pakistan	**1.6**
Bangladesh	**0.5**
Nepal	**0.4**
Total	13.0

Source: IRRI Almanac 1995; Woodhead *et al.* 1994; Modgal *et al.* 1995; Velayutham, 1997

2. Nutrient Dynamics

A. Nitrogen in Rice-Wheat Ecosystem

Mineralization-Immobilization: Nitrogen is a key element that sustains the high grain and biomass productivity levels of the rice-wheat ecosystem in the Indo-Gangetic belt. During the past 3 decades, N inputs through organic and inorganic manures have been comparatively enormous. Such manures undergo several transformations commensurate with soil conditions. Nitrogen mineralization/immobilization are some of the most intensely researched soil fertility aspects of this cropping ecosystem. Submergence allows for mineralization of organic N, but only until ammonia level. Hence, NH_4^+–N is dominant during rice phase of the cropping sequence. In contrast, the dry aerobic soils during wheat period supports mineralization reactions to continue up to NO_3–N. The inherent N mineralization potential of the soil, and that achieved through additions of either fertilizer N or cereal straw is an important parameter. It clearly affects N availability, therefore N recovery by rice and wheat. Simple rice straw amendments often have less mineralization potential. Consequently, net N release is much lower compared to urea. Admixtures of urea based N fertilizer and FYM have provided greater N release into soil. Mineralization rates were also better when straw was burnt. On the other hand, simple incorporation of straw is known to cause a fair degree of immobilization (Sarkar *et al.* 1991). Actually, immobilization could set in very early, and 95% of it could be completed within 7 days after urea application. Sometimes, the process of N immobilization may continue upto 40 days after rice seedlings are transplanted, after which mineralization becomes dominant. In general, throughout the duration of

Table 5.2: A comparison of advantages that occur through rice-wheat sequences adopted in Indo-Gangetic plains

Crop sequence	Duration (days)	Land use efficiencies (%)	Water use efficiency (kg ha^{-1} cm^{-1})	Production efficiency (kg day^{-1} ha^{-1})	Wheat grain equivalence	Energy output (MJ × 10^3 ha^{-1})	Energy efficiency (output/input)	Benefit : cost ratio
Rice-Wheat	226	62	82	43	127	260	10.6	0.67
Rice-Rice	236	62	46	42	124	279	10.8	0.62
Rice-oilseed (mustard)	250	60	73	36	88	230	11.9	0.51

Note: Production efficiency has been calculated, first by converting to rice grain equivalence.
Sources: Excerpted form Tripathi *et al.* 1999; Prihar *et al.* 1999 and several others.

rice crop, the flooded soils contain NH_4–N as the predominant available N, compared to NO_3–N. Hence, immobilization of NH_4^+–N is also possible through chemisorption to humic content, sorption as exchangeable N or could even be fixed into lattice structures of vermiculite, montmorillonite, or illite fractions of soil. To a large extent, cation exchange capacity (CEC) determines the amount of NH_4^+–N fixation, and this chemical process is dependent on soil pH, moisture, organic C content and presence of K, Al etc. Alternate submergence and drying during rice–wheat sequence affects such chemisorption and NH_4^+–N fixation levels. Remineralization of previously immobilized fertilizer-N is a clear possibility and becomes more rapid in light textured soils. On a long term basis, enhanced N inputs lead to increase in almost all forms of soil N, except hexose amine and non-hydrolyzable-N (Prasad and Rokima, 1991; Rokima and Prasad, 1991). They further noticed that, levels of exchangeable, hydrolyzable and available N were directly related to N recovered into rice plants.

Nitrogen Emissions: Fertilizer N applied during rice-wheat sequence is prone to losses through volatilization. Urea, which is the most common N fertilizer used is hydrolyzed enzymatically to $(NH_4)_2 \cdot CO_3$ in soil. Such hydrolysis could be rapid in soils treated with organic matter or in soils with high pH. Evaluations using ^{15}N indicate that the hydrolysis process is usually complete within 24 h after urea application (Bandhyopadyay et al. 1994), releasing N as NH_4^+–N or NO_3^+–N depending on redox potential of soil. Whenever, high humidity and wind velocity occur, these chemical transformations leading to NH_4^+ formation results in volatilization losses. Obviously, the extent of ammonia volatilization varies depending on N inputs, soil and environmental conditions. Such N volatilization from a basal urea application to rice could range between 13 and 17%. Splitting urea-N application and disbursing a portion of it at tillering stage reduces volatilization losses. Approximations indicate that maximum N loss through volatilization occurs during 3 days immediately after urea inputs (Prasad and Singh, 1992). In the normal course of rice-wheat sequence in central and eastern Gangetic plains, nearly 10 to 15 kg N ha^{-1} are lost through volatilization during rice phase of the sequence. About, 7 to 14 kg N ha^{-1} are then lost during wheat crop season (Mandal and Kar, 1995). On a percentage basis, about 8.4% N loss occurs during seedling stage itself. Nearly 15 to 20% could be lost as volatile NH_3 during wheat cultivation period. Thus, on an average, 23 to 35% N loss occurs during rice-wheat sequence (Prasad et al. 1999).

Urease activity, which determines extent of hydrolysis accentuates, whenever urea is added along with crop residues. Such a situation is also possible when green manure crops are included in the sequence. According to Gill et al. (1998), the green manure species, and its quality

(C:N ratio) influences urease activity during rice phase of the sequence. It was observed that wheat straw plus urea, which induced higher urease activity, resulted in correspondingly higher N loss through volatilization. However, NH_3 volatilization was appreciably less if ammonium chloride or ammonium sulphate were N sources. Among the factors that influence the extent of NH_3 volatilization, method of N input into rice-wheat ecosystem could be important. NH_3 volatilized subsided best if urea inputs were placed around roots. incorporation into soil also reduced NH_3 volatilization. However, broadcasted urea was lost to greatest extent as volatile NH_3 (Prasad et al. 1999).

Nitrification-Denitrification: Nitrification during rice-wheat sequence is immensely influenced by the aerobicity/anaerobicity of soils caused due to submergence during rice season, then draining to achieve arable condition throughout the wheat season. Denitrification is a major factor inducing loss of N applied as urea. The onset of N loss via denitrification could be quick, say as early as the next day after fertilizer application. Nitrogen loss due to denitrification is severe during rice phase of the sequence. Sometimes, it may reach nearly 23% of fertilizer N. Evaluations using ^{15}N mass isotope technique have provided variable figures for N loss through denitrification. It could even be as low as 10% (Mohanty and Mosier, 1990). Much of N loss due to denitrification in submerged rice soils, nearly 94% of it occurs as N_2, and the rest about 6% could be as N_2O (Mohanty and Singh, 1997). It is believed that, managing N inputs as urea into rice-wheat ecosystem is comparatively easier during wheat phase than during rice cultivation period. For rice, the suggested practice is to split the total N inputs, stagger application of N, use slow-N release sources and neem coated urea or add nitrification inhibitors (Prasad and Power, 1995).

Nitrogen Availability: Preferences for available nitrogen levels, their chemical forms and the duration of rapid recovery by roots differ enormously between rice and wheat. Above it, the cyclical changes in soil conditions, that occur due to submergence and drying affects chemistry of soil nitrogen and its availability. Rapid changes in chemical forms of N occur as a rule. Rice, which is cultivated under submerged soils prefers NH_4^+-N over NO_3-N for absorption through roots and rootlets. Nearly 80% of N that is absorbed and transported into rice plant is drawn as NH_4^+-N (Oury et al. 1997; Wiren et al. 1999). Incidentally, the submerged soils during rice phase allows transformation to proceed only upto NH_4^+-N, which is then preferentially recovered by rice roots. Whereas, wheat roots prefer NO_3-N, and the aerobic soil conditions during wheat growth allows mineralization to proceed beyond NH_4^+, until NO_3-N formation, which then is easily absorbed by wheat plants. NH_4^+-N like NO_3-N is soluble, and a highly mobile moiety in soils. Obviously, it is prone to rapid movement away from root zone through percolation or leaching. The long term effects on sizes of NH_4^+–N and NO_3–N pools achieved after different nitrogen inputs is shown in Table 5.3.

Table 5.3: Fractions of soil nitrogen (ppm) after 5 years of rice-wheat cropping in eastern Gangetic plains (at Pusa, Bihar)

Treatment	NO_3–N	NH_4^+–N	Hydrolyzable N	Non-hydrolyzable N	Total soil N
Control	20	49	245	135	450
Inorganic N	22	63	264	140	489
Inorganic N plus FYM	22	87	306	145	560
FYM	16	61	300	142	519

Source: Excerpted from Mohanty and Singh, 1997
Note: Control refers to plots that did not receive N fertilizer or FYM. Inorganic N and FYM treatments were adopted at levels recommended by local soil testing labs. Treatment involving FYM additions improved NH_4^+–N, hydroliziable N and Total soil N.

Analysis of rice crop grown on Hapludolls of northern and central Gangetic plains indicate that NH_4^+–N pool increased upto 30 to 50 days after seedling transplantation, that is until tillering stage, and decreased thereafter. In fact, NH_4^+–N levels measured at tillering stage seem to be a good indicator of total N recovery possible and grain yield levels. Mohanty and Singh (1997) opine that around tillering stage, during a rice crops season, a dynamic equilibirium between different forms of N, particularly NH_4^+–N, NO_3–N and nonhydrolysable-N would have been achieved. Further, NH_4^+–N versus NO_3–N ratios measured at panicle stage are possibly the dominant factors that determine extent of N recovery by rice and grain yield. Overall, the rice-wheat cropping sequence adopted in these plains seems to reduce both the forms of available NH_4^+–N and NO_3–N. There are very little chances for legume-based N inputs. Upon it, if N replenishments are inadequate in scale, then rapid decreases in available form of N should be expected (Bhandhari *et al.* 1992; Kumar and Yadav, 1993). Inorganic N inputs do enhance NO_3–N, but green manure treated rice/wheat fields show slightly higher levels of NO_3–N, which percolates upto 60 cm depth (Tiwana and Narang, 1997). Hence, the bottom line suggestion is to adopt integrated N supply scheme, that envisages application of organic N, introduction of legume or leguminous green manure, and judicious timing of inorganic N replenishments (Bharadwaj and Omanwar, 1984; Swarup, 1991).

Nitrogen Recovery: Nitrogen consumption in Indo-Gangetic agrobelt has been very high, owing to higher productivity levels aimed and intensive cultivation practice. Both the cereals, rice and wheat demand higher levels of N inputs. Between the two, nitrogen recovery by rice remains low at 20 to 40% of applied N (Goswami *et al.* 1988). Recent investigations using ^{15}N isotope too indicate that conjunctive use of inorganic N fertilizer and FYM enhances N recovery by rice, which may reach upto 31%. Wheat, which follows rice in sequence recovered between 2.2 and 4.3% of N applied to rice at the beginning of the sequence (Table 5.4). The unaccounted N in the system could be minimized to 35.5% if FYM is added (Bandhyopadyay and Sarkar, 1999). Inclusion of mineralizer, calcium monohydroxy monohydride enhances N recovery by wheat marginally, which is not reflected in grain yield. Similarly, use of 2, 4-dinitrophenyle-hydrozone or napthyl ethylene diamine resulted in better N recovery by wheat (Patil and Jain, 1994). Higher levels of initial N inputs to rice, in general, enhanced retention of N in soil. It also increased N losses (leaching) and immobilization via microbial activity. Again, ^{15}N based studies by Sharma and Gangwar (1995) indicated that though a sizeable quantity of N is retained in soil after rice harvest, its utilization by wheat was meager, sometimes slow. Hence, they suggested that this fraction be termed residual N, which may be recovered through cropping in due course.

We may realize that fertilizer N recovery during rice-wheat sequence is less than 50% owing mainly to sizeable losses through volatilization, run-off, leaching and denitrification. Reduction in such N losses could be managed by using nitrification inhibitors (N-serve, AM), neem cake, karanji cake or lac. Obviously, slow release of N will alter, both soil N dynamics as well as extent of N recovery by crops. To quote an example, Singh *et al.* (1999a) found that coating urea with neem cake, or adding N-serve or using prilled urea resulted in higher levels of available N in soil after rice harvest (268 kg N ha^{-1}) compared with uncoated Urea (216 kg N ha^{-1}). It is clear that sizeable N losses occur during rice-wheat sequence, but cropping in this agrobelt is intensive, and expected yields at 7.5 to 8.5 t ha^{-1} rice and 5 to 6 t ha^{-1} wheat necessitate excessive additions of fertilizer N. Tiwana and Narang (1997) report that at such yield levels, common in these plains, the two cereals remove on average 200 kg N ha^{-1} in one cropping cycle. Quite often, the recommended N inputs just match the recovery, and to overcome any dearth, farmers resort to organic N inputs.

B. Phosphorus in Rice-Wheat Ecosystem

Phosphorus is a crucial element in the Indo-Gangetic agrobelt, during both wetland rice culture, and equally so for the succeeding wheat crop. Once again, alternation between wet and dry soil conditions bring about

Table 5.4: Nitrogen recovery by rice-wheat sequence adopted in Northern Gangetic plains, at IARI, New Delhi, India

	Rice			Wheat			Fallow		
	June	Oct	Nov		Mar	Apr		June	
N inputs	Total N uptake (kg N ha^{-1})	Fertilizer ^{15}N kg ha^{-1}	Recovery (%)	Total N uptake (kg N ha^{-1})	Fertilizer ^{15}N kg ha^{-1}	Recovery (%)	Fertilizer ^{15}N in soil		Fertilizer ^{15}N unaccounted	
							kg ha^{-1}	(%)	kg ha^{-1}	(%)
Urea (180 kg ha^{-1})	107	25	**14**	67	1.4	**0.8**	41	**24**	113	**61**
Urea (180 kg ha^{-1} + CCC)	115	38	21	70	2.1	1.2	32	16	108	60
Urea (90 kg ha^{-1} + FYM)	95	21	24	68	0.6	0.3	28	32	40	43

Source: Excerpted from Bandhyopadyay and Sarkar, 1999
Note: All treatments under succeeding wheat crop received 50 kg N ha in two split dosages. Percentage N recovery values that should add upto 100 are in 4 different columns and highlighted (e.g. 14 + 0.8 + 24 + 61 = 99.8). Unaccounted N includes that lost via volatilization, leaching, percolation etc.

enormous fluctuation in the chemistry of soil P, the major P forms, P availability levels, the net recovery by crops etc. Overall, this alternation of wet/dry conditions bestow certain unique features to phosphorus dynamics within the rice-wheat ecosystem.

Soil P Fractions: In general, within the rice-wheat ecosystem, continued P fertilizer application, for over 10 years, caused increases in Al–P and Fe–P fractions. Such accumulations depended on soil type. In certain locations in eastern Gangetic plains, the proportion of Al–P fraction increased with rates of P replenishments. Inorganic P and organic forms added through FYM or blue green algae (BGA) enhanced Al–P, Fe–P, and saloid P. These fractions predominantly contributed to the available P pool utilized efficiently by rice and wheat (Mohanty and Singh, 1997). On a long-term basis, a variety of ecosystematic processes affect the soil P fractions. For example, conversion of Ca–P and Fe–P to Al–P and saloid-P were aided by the exudations of HCO_3 from wheat roots. Certain combinations of P fertilizer and organic manure caused accumulation of organic–P and Ca–P, which accounted for 43% of the total P. A long-term evaluation at Pantnagar (in western plains), again indicated that saloid-P, Al–P, Fe–P and available P increased due to rice-wheat sequence (Agarwal *et al.* 1987). Depending on the soils, and their redox potential Fe–P could be the most dominant fraction of P.

Soil P Transformations: Some researchers believe that transformation of native P, and the size of available P that results is of greater relevance to the productivity of rice, than to wheat sown later in sequence (Vig *et al.* 1999). Therefore let us consider some of the major physico-chemical transformations of P that occur within soils in this cropping ecosystem. Based on series of investigations, it has been summarized that acetic acid extractable P levels are enhanced whenever soils are submerged. Similarly, submergence is accompanied by increase in Fe-P. Whereas, water soluble P and Al– P might decrease (Singhania and Goswani, 1978a, 1978b). In contrast, within the oxidized soils during wheat phase of the sequence, firstly, response to applied P and native soil P differ widely. Dash *et al.* (1981) have reported that P inputs at the beginning of wheat season were transformed to Al–P and Fe–P, thus leading to a degree of chemical fixation. However, on subsequent flooding during the ensuing rice reason available P pools get enlarged. Contrastingly, if P inputs are made at the beginning of rice season, initial flooding, anaerobicity and lowered redox potential decreases Fe–P, Al–P and available P pools in soils. Hence, the general suggestion is to replenish P during wheat phase of the sequence, and exploit better residual effects of P.

Transformations related to organic-P are equally important. Again, changes in moisture regimes during rice and wheat affects chemical reactions involving organic-P. Firstly, moisture is a pre-requisite for mineralization of organic-P. Therefore, flooding supports better mineralization rates on added organic-P, resulting in higher levels of soluble and labile P pools. The alternate wetting and drying accentuates such transformations (Vig and Milapchand, 1993; Bahl *et al.* 1998). Soil aggregates that crumble during flooding actually allows better access to humic substances, so that microbial and chemical process can take effect resulting in higher P availability.

Available P Pools: The native soil P levels in this agrobelt are considered marginal to medium. Soil test values may even indicate deficiency of this element. Still, response to added P could be small. This is attributable to enhanced availability of P in submerged soils during rice cultivation. The chemically reduced conditions in soils allow better solubilizations of Fe^{+3}–P to Fe^{+2}–P. Infact, conversion of Fe^{+3}–P to Fe^{+2}–P under submerged soils is the chief source that replenishes available P pools in rice fields (Dash *et al.* 1982; Mohanty and Singh, 1997).

Incessant P replenishments done each season sustains enhanced levels of Olsen's available P. Also, periodic incorporations of green manure, FYM or burning trash for 7 to 8 cycles of rice-wheat sequence, generally resulted in enhanced $NaHCO_3$ extractable P (Bhat *et al.* 1991). Withholding P inputs obviously decreased available P pools (Swarup, 1991; Bhandari *et al.* 1992; Kumar and Yadav, 1993).

In some eastern locations, upon P fertilization, the available P pools increased upto 10 days after transplantation. Whereas, in western and central plains, increases in available P continued until 40 days, that is upto flowering stage. These variations were attributable to soil pH, soil CO_2 status, and reductant aided solubilization of Fe–P. Generally, water logging induces greater P availability. However, Hundal *et al.* (1989) suggest that in certain alluvial soils of Gangetic plains, solubility and availability of P to rice crop could decrease. Such a situation could however be alleviated by green manure incorporation. Green manuring, in this case, actually shifted soil P towards solubility by formation of octacalcium phosphate. Also, P inputs along with lime amendments on submerged soils increased P availability. Calcium application remarkably alters Al^{+3} and Mn^{+2} to lower levels and enhances P availability. Such treatments have been effective on acid soils of eastern plains, where both, deficiency of P and fixation are rampant (Mongia *et al.* 1998). Phosphorus concentration in rhizosphere soil solution may also be increased by the hydrolytic cleavage of soil organic matter, mainly the phosphomonoester. It is believed that secretion of extracellular phosphatases are regulated based on plants demand for phosphate (Chhonkar and Tarafdhar, 1981).

Clearly, available P pools in soil are crucial to productivity of rice-wheat ecosystem. The critical levels of available P should be satisfied, either through fertilizer P inputs or by manipulating soil conditions/reactions. Also, crop genotypes that tolerate low P availability could be utilized. In general, the salient aspects of soil P dynamics in rice-wheat zones to be remembered are (a) P requirement of wheat is higher than rice; (c) ability of wheat roots to utilize Fe–P is comparatively lower, and (d) rice roots are endowed with ability to absorb residual–P in soil that occurs as Al^{+3}–P, Fe^{+2}–P and Ca–P. Therefore, P inputs recommended for wheat, that follows rice is generally higher.

Timing of P inputs during rice and wheat sequence is equally crucial, because it determines the extent of P recovery by rice and/or wheat (Singh et al. 2000). The carryover effects of P application to first crop, on the subsequent ones can be appreciable (Table 5.5). Clearly, both in terms of P–use efficiency and related economics, applying P at the beginning of wheat season seems beneficial (Sagger et al. 1985). Splitting P fertilizer application provided greater P recovery from fertilizers (Table 5.5). Productivity of rice–wheat ecosystem also depends enormously on the type of P inputs. Series of evaluations indicated that P inputs as single superphosphate (SSP), and ammonium polyphosphate have consistently enhanced P recovery (AICARP, 1986/87; Hegde and Dwivedi, 1992; Meelu and Yadavinder Singh, 2000; Vig et al. 1997). Sparingly soluble sources such as rock phosphates supported lower levels of P recovery on the first crop, but residual effects were conspicuous (Goswami et al. 1996).

Table 5.5: Influence of phosphorus source on fertilizer P-use efficiency, and residual P fraction retained in soil

P Source	Rice phase			Wheat phase		
	Fertilizer P utilized (%)		Unutilized P fraction (%)	Fertilizer P utilized (%)		Unutilized P fertilizer (%)
	Basal	Split	after rice	Basal	Split	after wheat
Single super PO_4	31	38	65	21	24	77
Di-ammonium PO_4	41	45	57	22	27	59

Note: Fertilizer P was added at 40 kg P_2O_5 ha^{-1} either as basal or split (at sowing and tillering).

C. Potassium in Rice-Wheat Sequence

The alluvial soils of Indo-Gangetic plains are generally rich in potassium because of mica and illite clay minerals. Despite it, continuous intensive cultivation of rice and wheat depletes potassium levels considerably. This necessitates seasonal or yearly replenishments of this element. The situation may get aggravated, if recycling through stover incorporation or burning is not practiced. Instead, if the produce, both biomass and grain are removed out of the field/local ecosystem then potassium is lost.

During rice phase of this cropping system, waterlogging of soils increases the availability of K^+ to plant roots (Mohanty and Patnaik, 1977). Depletions in K^+ can be rapid, if N and P replenishments are solely done in inorganic form. Use of FYM, generally delays onset of K^+ deficiency, because it replenishes K^+ to a certain extent through mineralization (Bharadwaj et al. 1994). Long term (10-15 yrs) studies involving fertilizer treatments that include K^+ inputs, along with other major nutrients, indicates that rice-wheat sequence depletes K^+ severely (Bhat et al. 1991; Kumar and Yadav, 1993; Nambiar, 1994). Severest of decreases in available K^+ occurred whenever N and P were added incessantly for several cropping cycles without proportionate K additions (Agarwal et al. 1993).

Transformation of Potassium: Application of K^+ to rice and wheat fields, either in organic or inorganic form brings about chemical changes. For example, a four-year evaluation indicated that major form of K^+ encountered after due transformations in soils are water soluble K^+, exchangeable K and non-exchangeable K. The concentrations of these fractions varied depending on the source of K and its quantity. Approximately, an organic source upon transformation released 0.1% exchangeable K, 0.8 to 1.0%, non-exchangeable K^+, and 5.2 to 7.0% HNO_3 soluble K. It sums to about 6 to 8% of total K present in the source. If we considered the K^+ recovery by rice crop and its influence on productivity, then, 1N ammonium acetate extractable K, which is water soluble and exchangeable is the most important fraction. Whereas, both exchangeable and non-exchangeable K^+ fractions are equally important while evaluating wheat crop for K recovery and productivity (Mohanty and Singh, 1997).

D. Micronutrients in Rice-Wheat Ecosystem

Zinc inputs into the rice-wheat ecosystem have varied depending on extent of its deficiency in soil and the crop genotype cultivated. About 68% of the nearly 4500 field trials conducted using zinc inputs in this agrozone indicate that a 25 kg $ZnSO_4$ ha^{-1} addition can result in 2 to 5 q ha^{-1} more rice and wheat grains (Sakal et al. 1997; Singh et al. 1998; Singh, 1999). In general, zinc replenishments to these alluvial soils have been more efficient with rice crop than wheat. In certain locations within Indo-Gangetic belt, such as western Punjab, parts of Uttar Pradesh, or in Bihar the zinc inputs are mandatory in order to obtain optimum productivity (Singh, 2001). Here, the efficiency of zinc inputs are generally high, with rice grain yield levels reaching 3.5 to 5.5 t ha^{-1} and wheat 1.3 t to 2.5 t ha^{-1}. Obviously, soils deficient for Zn provide greater yield response upon replenishment (Sakal et al. 1991; Mehla, 1999).

The preferred method of Zn replenishment, which is also easy to adopt would be to add Zn salts on an yearly, bi-yearly or tri-yearly basis to soil.

For a sandy loam which is alkaline in reaction, long-term Zn addition once in 3 yr served better. In mid-eastern Gangetic plains, alternate crops were provided with Zn inputs. Whereas in the western Indo-Gangetic plains, seasonal low rates of Zn inputs were preferred (Nayyar et al. 1990; Sakal et al. 1997) because it provided better micronutrient recovery and grain productivity.

Zinc applied to submerged paddy soils undergoes chemical transformations into different forms which are preferentially utilized by roots, and the rest retained as residual Zn in soil. For example, of the total Zn applied to soil, 60 to 70% was retained as residual fraction (Singh et al. 1999b). Analysis of residual Zn after paddy harvest indicated its presence in water soluble fraction (exchangeable Zn at 3%), in amorphous sesquioxides (47%), complexed with manganese oxides or in crystalline bound form (12 to 20%). Singh et al. (1999c), further observed that rice prefers Zn that occurs complexed with Mn oxides. In general, continuous submergence reduces the concentration of different Zn fractions in soil. Zinc inputs into rice-wheat cropping system is often agriculturally inefficient, because only 5% of it seems to be recovered by the crop. Mixing organic matter, farm waste, or biogas slurry could alleviate the situation. It is believed that chelating effects produced by biogas slurry or FYM form soluble complexes with Zn, thus enhancing Zn use efficiency. Available Zn levels after rice harvests were generally lower than after wheat. Once again, submergence could have reduced zinc availability.

Dearth of Fe in the soils of rice-wheat ecosystem could be felt frequently. Mainly, the calcareous, saline and sodic soils of western Indo-Gangetic plains are low in Fe contents. Even in sandy loams, impounding of water for rice cultivation causes low redox, and this makes reduction of Fe^{+3} less feasible (Sakal et al. 1993). Puddling can offset this situation by enhancing solubility of Fe. Inputs of Fe have been 50 kg Fe ha^{-1} in western plains (Punjab in Pakistan and India). Farmyard manure added along with inorganic Fe allows better chelation of this micronutrient in soil. Hence, greater quantities of Fe are recovered into crops. Marked increases in DTPA extractable Fe are possible through green manure incorporation (Singh et al. 1992). Sometimes, Fe ($FeSO_4$ 1.0%) inputs through foliar sprays are preferred because it results in better grain and biomass formation by both rice and wheat.

Clearly, intensive cropping of rice-wheat sequence since the past 3 to 4 decades has depleted micronutrient resources of these alluvial soils. Dearth of not only Zn and Fe, but other micronutrients such as Mn, Co and B have started appearing. The imbalance in crop nutrition that they cause is severe in terms of loss in productivity. Quite often even the major nutrient recovery is impeded. Manganese inputs during wheat cultivation has been practiced on calcareous soils of Bihar and alkaline soils of Punjab in India. It is known to cause economically useful grain and biomass increases, upto 4 to 5 q ha^{-1} wheat grains (Sakal *et al.* 1993). Manganese inputs as water soluble $MnSO_4$ is more likely to be efficiently recovered, compared with Mn frits or Mn_2O_3 (Sadana *et al.* 1989). $MnSO_4$ (0.5 to 1.0%) foliar spray is 1 to 1.5 times more efficient in delivering Mn requirements to rice and wheat. Some attempts to improve crop genotypes, in order that they withstand low Mn conditions have been made. Generally, aestivum wheat is known to tolerate low-Mn better than durum wheat. Genotype dependant variations in Mn-efficiency has also been noticed (Bansal *et al.* 1991, 1994).

Copper and Boron deficiencies are sporadic in Pakistan and India, and Cu inputs are negligible. Both, foliar spray and soil applications have been tested. Soil application before wheat planting provided between 5 and 10 kg grain increase in western plains (Sakal *et al.* 1997). Boron deficiency during rice-wheat sequence has been reported in eastern plains (Bihar). Soil application of B as Borax is popular, because it allows appreciable recovery of this element by rice and wheat. Molybdenum is yet another micronutrient where deficiency is being felt during rice-wheat sequence. Molybdenum inputs have been practiced in certain locations, usually, as complex mixtures of micronutrients. Residual effects of Mo may persist for 2 to 3 years.

3. Soil Microbes, Nutrient Dynamics and Rice-Wheat Productivity

Within this intensively cultivated agrobelt, soil microbes mediated transformations of inorganic and organic forms of essential nutrients are important. Soil microbes, actually participate in a wide range of soil processes that directly or indirectly affect cereal productivity.

Residue Recycling: Innumerable reports are available on this aspect and its relevance to nutrient dynamics and productivity of rice-wheat sequence. Bharadwaj *et al.* (1997) state that intensive tillage of soils, high inputs of inorganic nutrients, and rapid decomposition of native organic matter are the major causes for low levels of organic components in soils. They further point out that, although rice-wheat cropping sequence generates large quantities of organic residues, its management may be inefficient. Main reason quoted is that, farming community in Indo-Gangetic plains prefers to burn the stubble. This exercise adds to inorganic nutrient pool in soil, but decreases organic component and subsides microbial activity. Therefore organic recycling is hampered. Basically, rice and wheat residues are cellulose and energy rich. Hence, it supports transformations through heterotrophic soil microbes in the alluvial soils (Bangar *et al.* 1999). The net advantage of microbial activity on organic residues is immensely influenced by nutrient ratios that occur in the residues. For example, if rice residue added is deficient for N, then it results in microbial immobilization of mineral. Because of high C:N ratios of wheat straw (120) and rice straw (70), a sizeable fraction of fertilizer N inputs, first gets immobilized (Patil and Sarkar, 1994). Hence, it is suggested that organic residue incorporation should occur well ahead of seeding (wheat) or transplantation (rice) to avoid nutrient immobilizations. Incorporating organic residues or green manure during fallow after wheat season seems most important in Indo-Gangetic plains (Srivatsava et al. 1988; Bharadwaj et al. 1994). Actually the soil microbial load, and its interaction with organic residues is the key to proper decomposition and release of nutrients. Sometimes, inoculation of rice or wheat residues with specific microbial species, or atleast enriching with species such as *Aspergillus, Paecilomyces* along with adjuvants (cowdung) have been advantageous (Table 5.6).

Table 5.6: Microbial population levels (number g^{-1} dry soil) achieved through rice straw incorporation, by burning, or due to totally removing straw from the field (Punjab, India)

Rice straw management	Bacteria $\times 10^6$	Fungi $\times 10^2$	Nitrogen fixers		Nitrifiers	
			Azotobacter $\times 10^4$	Rhizobium $\times 10^4$	Nitrosmonas $\times 10^4$	Nitrobacter $\times 10^4$
Removed from the soil ecosystem	43	51	36	55	131	70
Burnt in situ	11	46	35	57	78	60
Incorporated in situ	68	76	50	74	256	125

Source: Sidhu and Beri, 1985
Note: While, populations of bacteria and fungi are indicative of decomposition rates possible, those of nitrogen fixers relate to possible N credits to the ecosystem. Similarly *Nitrosomonas, Nitrobacter* populations relate to N transformation potential.

Nitrogen Fixation: Diazotrophs that occur in the cereal rhizosphere and bulk soil are free-living. They proliferate under anaerobic, micro-aerophilic and aerobic soil conditions, and fix atmospheric nitrogen asymbiotically, resulting in net N gains to the rice-wheat agroecosystem. However, the extent of N inputs derived through such asymbiotic N fixation varies widely. Nitrogen gained may range between statistically insignificant and sometimes significant levels. However, there are innumerable reports and reviews depicting the extent of positive N balance or N credits to the rice-wheat ecosystem derived through asymbiotic N fixation, both in rhizosphere and bulk soils (Kundu and Gaur, 1980; Bharadwaj et al. 1997; Saxena and Tilak, 1998; Lata et al. 2002). Recently, Lata et al. (2002) have provided detailed information about net nitrogen gain, residual N available, nitrogen use efficiency and concomitant yield advantages. To quote a few examples from the Indo-Gangetic plains, asymbiotic N fixation during the wheat season enhanced productivity by 4 to 7% over uninoculated controls in Hoshairpur, Punjab (Gill et al. 1993). Seed inoculation with *Azotobactor croococum* resulted in cereal grain yield increases between 10 and 15% (Gill et al. 1993). They suggest that between 15 and 24% N in wheat grains could have originated through N fixation by free-living N fixers.

In order to sustain appropriate soil nitrogen levels during Rice-Wheat sequence, grain legumes or leguminous green manures (e.g. Sesbania, Glyricidia) are introduced. Both, direct soil analysis and ^{15}N isotope based analysis indicate positive N balance and N credits from such legume species. Soil nitrate accumulation due to legumes in the sequence may range from 10 to 60 kg N ha^{-1}, depending on legume species. For example, N credits from pea could be +28 kg N ha^{-1}, from green gram +26 kg N ha^{-1}, black gram +38 kg N ha^{-1}, pigeon pea +15 to 60 kg N ha^{-1} (Ladha et al. 1996).

Azolla-Anabaena are free-floating aquatic ferns which fix atmospheric N_2 through symbiotic BGA *Anabaena azollae*. Use of Azolla during rice cultivation is in vogue in tropical and subtropical zones of the Indian subcontinent since several decades. Normally, repeated inoculations are made whenever dry conditions interject. Positive N balance to rice and succeeding crops have been confirmed through multilocational studies (Singh and Singh, 1987a, c). The biomass of Azolla, which doubles in 48 to 78 hrs contains between 4 and 6% N on a dry weight basis. During a rice season, N inputs through Azolla could range between 30 and 100 kg N ha^{-1}. Using ^{15}N isotope, it was deciphered that nearly 29 and 95% of N incorporated into soil as Azolla was actually derived from atmosphere— i.e., through nitrogen fixation process. Azolla decomposes rapidly, and N recovery rates range between 30 and 70% depending on method of incorporation, time and quantity (Lata et al. 2001). Sometimes, N inputs through Azolla could serve more efficiently because of reduced losses compared with that applied as inorganic N fertilizer. Further, Singh and Singh (1989) state that in a sandy loam, Azolla based succulent components are mineralized at faster rates. Advantages equivalent to 60 kg N ha^{-1} were possible using Azolla. In some Indo-Gangetic soils, Azolla incorporation enhanced NH_4^+-N and P availability (Singh et al. 1981). It also improved soil porosity (3.7% to 4.2%), enhanced organic C from 8.6 to 23% (Singh and Singh, 1987b). In summary, a crop of Azolla grown and incorporated provided 20 to 40 kg N ha^{-1} with corresponding rice grain equivalent of 500 kg ha^{-1}.

Blue Green Algae (BGA): Tropical and subtropical climates within Indo-Gangetic plains ensures appreciably high incidence of cyanobacteria in rice fields. Through a series of networked, multilocational trials within rice-wheat zones, Venkatraman (1979) inferred that algal inoculation could benefit the ecosystem with 30 kg N ha^{-1} input. These algal inoculants are essentially of BGA constituted by N fixing *Anabaena variabilis, Tolypothrix timus, Nostoc muscorum* and *Aulosira fertilissima*. The N accruals by BGA into the rice-wheat system in Indo-Gangetic plains could be as high as 1.5 to 1.7 mt of urea per season (Venkatraman and Tilak, 1990).

Productivity gains through such BGA inoculations have hovered around 10 to 15% over uninoculated control in rice fields (Singh and Bisoyi, 1989). Cyanobacterial inoculation also increases water holding capacity, provides better aeration of wheat rhizosphere and increases organic matter contents (Pabbi and Singh, 1998). Physiologically, peak N fixation by BGA occurs when rice is at tillering stage, but a certain lapse of time is required before N is released. However, there are suggestions that certain BGA strains continuously fix N and release it into soil (Lata *et al*. 2001).

Arbuscular Mycorrhizas (AM): Alternation between submergence and dry arable conditions might provide an interesting situation with reference to ecology of arbuscular mycorrhizas and related P dynamics. Mycorrhizal propagule density, root colonization levels and P dynamics will be severely affected due to submergence during rice-wheat sequences. Arbuscular mycorrhizas are aerobic fungi and colonize wheat roots efficiently. Whereas, rice cultivated under submergence may not support functionally, any degree of AM symbiosis, only, upland rice is known to be mycorrhizal. Fluctuations in AM propagules due to alternate submergence and dry conditions, their activity and influence on P recovery from soils have not been studied in Indo-Gangetic plains. However, under natural conditions, a good fraction of P recovered by wheat root traverses through AM hyphal network, which needs to be quantified. Also, influence on wheat rhizosphere chemistry if any, and microbial dynamics needs to be understood clearly.

Phosphate Solubilizers: The phosphate solubilizing bacteria have a role to play in the soils utilized for rice-wheat sequence. Mainly, by solubilizing the insoluble-P forms in soil. Both, rhizosphere and bulk soils are known to harbor P solubilizing microorganisms. For example, *Bacillus firme* is known to improve P recovery by rice crop treated with rock phosphate (Lata *et al*. 2002). Similarly, inoculation of wheat with *Penicillium bilaji*, a fungus, improves P uptake and yield. Several phosphatase releasing strains of Aspergillus, when tested on wheat, improved P recovery and biomass accumulation (Tarafdar and Rao, 1996). Similarly, other bacterial strains such as *Bacillus polymixa, B. firme* and *Pseudomonas striata* enhanced P availability in soils.

4. Sodicity versus Nutrient Dynamics

Sodic or salt-affected soils are encountered within the rice-wheat agroecosystem of the Indo-Gangetic plains. However, they are rampant in northwestern India, and adjoining areas in Pakistan. They constitute nearly 7 m ha in the Gangetic plains in India (Bajwa *et al.* 1997), and upto 0.8 m ha in the Indus agrobelt in Pakistan, mainly within Punjab and Sind provinces (Mohammed, 1987; Bajwa, 1992; Twyford, 1993; Hussain, 1994). Such maladies or deteriorations affect soil chemical transformations, the extent of nutrient availability as well as nutrient recovery and productivity of cereals. Hussain *et al.* (1994) state that sodicity in rice-wheat belt in Indus plains reduces N, P and K recovery. Therefore, replenishments need to be at higher rates to offset nutrient recovery related problems (Ali *et al.* 1999). Considering the vastness of this soil nutrient related malady, Bajwa and Josan (1989) inferred that mere amendments may not overcome sodicity and salinity. However, it is to be noted that rice, to a certain extent, withstands salty conditions and salinity in soils, without any loss to nutrient recovery and productivity. This has been the prime reason for adopting rice-wheat sequence in many parts of Indo-Gangetic belt. Chauhan *et al.* (1996, 1999) opine that under prevailing salinity in northern plains, rice is the best suited crop, since it can tolerate sodicity and alkalinity up to pH 9.5. According to Mishra (1994), rice is most preferred in any cropping sequence formulated in sodic soils because it tolerates nearly 55 to 60% exchangeable sodium. Among crops, the order of susceptibility to excessive exchangeable sodium is rice > sasbania > barley > wheat > linseed > cowpea > chickpea, pigeonpea, lentil, and blackgram. Notably, legumes tolerate exchangeable sodium percent of only between 10 and 20 in soil solution. Reclamation of sodic soil through introduction of rice in the sequence has been possible (Mehta, 1994). Rice enhances pCO_2 of soils, it therefore mobilizes Ca, which can reduce sodicity, atleast on surface layers of soil. Mehta (1994) believes that introducing rice, atleast delays salinity effects. In India, amendments with gypsum, plus introduction of rice enhanced cereal grain yield from 3 to 4 t ha^{-1} in the initial few years. Qadir *et al.* (1998) state that horizontal flushing of salts through irrigation and gypsum application ameliorates saline and solic soils of Pakistan.

More specifically, the major traits of saline/alkaline soils of Indo-Gangetic plains, such as excessive soluble salts, alkalinity, higher exchangeable sodium percentage (ESP), inappropriate air-water retention and lowered organic matter, impedes availability and recovery of nutrients by crops. Nutrients applied as fertilizer may not produce targeted effects. For example, fertilizer N recovery and efficiency are adversely affected, and the major reasons suggested are: (a) ammonium volatilization, denitrification and leaching losses can be significant, (b) sodicity is detrimental to normal physiological activity of cereal roots, (c) normal physico-chemical and biochemical transformations of N in soil gets severely disturbed and impeded, and (d) poor crop stand leaves fraction of fertilizer N unutilized, hence vulnerable to loss (Bajwa *et al.* 1997).

Chauhan et al. (1999) express that recovery of N applied into sodic soils seldom exceeds 20%. Volatilization as NH_3 is the major cause, which may be as high as 32 to 52% of applied N, if soil pH and alkalinity are uncongenial for absorption by cereal roots. At pH 10.5, they recorded a 65 to 85% N loss through volatilization. Once again, alternate wet and drying condition accentuates salinity/solicity related problems. Together, interactive effects due to redox changes and salinity can drastically curb nutrient availability. For example, nitrate-based N inputs during rice-phase (wet conditions) have been inferior in terms of efficiency. Whereas, wheat crop on arable soils in the next season, recovered N from both, NH_4^+ or NO_3^- sources equally well, especially where reclamation procedures were adopted (Bajwa et al. 1997). Within this rice-wheat belt, urea hydrolysis is fairly delayed whenever low temperature and sodicity related problems appear together. If, in normal soil at 30°C, urea hydrolysis is complete in 3 to 7 days, then sodicity and low temperatures (15°C) retard urea hydrolysis requiring nearly 14 days (Gupta et al. 1999).

Accumulation, losses, chemical fixation and decreased availability of P are important problems in saline/alkaline soils of Indo-Gangetic belt. Under submerged conditions, higher pH increases $NaHCO_3$ extractable P. Inorganic P forms such as Ca–P, Al–P and Fe–P become dominant. Whenever reclamation procedures are adopted, initially available P levels could be higher, but it decreases with time, because soluble Na–P is converted to less available Ca–P. Dearth of available P in reclaimed sodic soils is usually felt only after 5 to 6 crop cycles. In the Indus plains K^+ and P availability both are impaired. Ali et al. (1999) suggest that K^+/Na ratio is severely affected in sodic soils, and this needs consistent correction through K fertilizer inputs.

Sodicity and saline/alkaline conditions in soils, reduces available form of micronutrients. Swarup (1994) suggests that in this context interactions between major and micronutrient at varying levels of salinity/alkalinity is crucial. In sodic soils nearly 80 to 90% of Zn inputs get fixed as carbonates or hydroxides. Despite, Fe and Mn contents being sufficient in saline soils, their chemical forms may not be congenial for recovery by both rice and wheat crops. Actually, water soluble, exchangeable and reducible forms of Fe and Mn are depleted because of uncongenial pH, ESP, $CaCO_3$ and oxidized state of soils during wheat phase. Submergence for longer durations reduces Mn first, then Fe, both of which move into deeper layers of soils. As a consequence, recovery of these micronutrients by rice/wheat is hampered, the general nutrient balance is also affected (Nayyar et al. 1990; Kumar et al. 1996).

Managing nutrient dynamics and productivity favorably is a priority research item in northern Indo-Gangetic plains, both in India and Pakistan. It is stated that successful reclamation of sodic soils and attainment of favorable nutrient dynamics largely depends on the quality of irrigation source (Minhas and Gupta, 1992; Gupta *et al*. 1994), its salt content, sodium percent, pH and crop genotype. Generally, irrigation with salt-affected source, rapid transpiration and depletions of soil water results in accumulation of salts unfavorably around cereal roots. It retards nutrient acquisition by roots. Frequent irrigation with good quality water, low in salts reduces such ill effects by leaching away salts, and by reducing their concentration (Minhas *et al*. 1995). Monsoon water and surface runoff have been effective in leaching excessive salts. Reclamation through drainage has been highly successful. For example, 8 years of consistent drainage almost restored normal nutrient recovery ratio in wheat (Sharma and Singh, 1998). Amending irrigation water with gypsum too has provided better nutrient recovery. However, such advantages are limited, if irrigation source itself has high sodicity (>2.5 me L^{-1}). Generally, nutrient recovery and productivity of wheat, that follows rice on sodic soils suffers to a greater extent. Tripathi (1998) suggests following methods, which may alleviate the situation; they are (a) providing good drainage, (b) careful tillage at appropriate moisture level, avoiding deep tillage so that sodicity does not seep into lower layers, (c) since these sodic soils are deficient in organic matter, N, Ca and Zn, replenishments should be managed appropriately, (d) selecting crops, and their genotypes which tolerate sodicity better can be useful, and (e) irrigation schedules should be carefully prepared, and good quality water without excessive dissolved salts needs to be used.

5. Integrated Nutrient Management and Productivity

Sustaining rice-wheat cropping sequence in the Indo-Gangetic plains, particularly, the soil nutrient balance, meeting nutrient demands, attaining optimum nutrient recovery, as well as biomass and grain productivity has been the major concern of agricultural researchers in this zone. During the past two decades, a variety of programs that optimize soil fertility versus grain harvest equations, and those which maintain the net nutrient dynamics at an appropriately beneficial status have been devised, tested and adopted. Among them, integrated plant nutrition system (IPNS) has been utilized widely by researchers and farming community in Indo-Gangetic plains. At the minimal, it envisages use of a range of soil fertility and agronomic measures such as nutrient inputs through chemical fertilizers, organic manures, crop residues, nitrogen fixing Azolla-BGA complex, legumes in sequence and a range of cultural procedures that enhance nutrient recovery and use efficiency. All these measures, essentially aim at better nutrient dynamics and resultant higher rice-wheat productivity.

Basically, nutrient requirements depend on yield goals. Quite often, maximum attainable yield in a farm, may not be economically the best. Under IPNS, farmers are generally advised to adopt techniques that answer both concerns related to soil nutrient dynamics and economics (Swaroop and Rao, 1999). To quote an example, average nitrogen demand by wheat was 146 kg ha^{-1} for maximum yield, but economically better proposition was 128 kg ha^{-1}. Similarly, 103 kg P_2O_5 ha^{-1} was the requirement for maximum yield, but economically efficient level was only 73 kg P_2O_5 ha^{-1} (Sharma *et al.* 1999).

To ensure high productivity, the high N requirements of rice and wheat have to be satisfied. However, high N inputs may result in proportionately greater N loss through volatilization, denitrification and leaching. To attain better nutrient-use efficiency, splitting N inputs are suggested under IPNS in Indo-Gangetic plains. On an average, the two cereals—, rice and wheat—, need 180 to 220 kg N ha^{-1} for a moderately good harvest (Rajput, 1997), which needs to be channeled carefully without much loss to efficiency.

Nutrient Ratios: Rice-wheat cropping system is highly exhaustive in terms of soil nutrient depletion. Rapid and uneven depletion of nutrients can result in nutrient imbalances. Such imbalances are known to induce farmers to dump larger quantities of major nutrients without much concern to proper ratios of other secondary and micronutrients. This practice accentuates nutrient imbalance and disturbs optimum nutrient dynamics. One of the prime aims of IPNS is to attain nutrient balance, appropriate ratios and best possible nutrient dynamics (Kumar *et al.* 1999). Ideal ratios of the major nutrients (N, P, K) recommended based on consumption levels of rice-wheat sequence is 4:2:1 (FAI, 1999). Compared with this recommendation under IPNS, actual practices vary enormously. Actual nutrient ratios in vogue, which may or may not provide well balanced nutrition are 170:44:1 (in Punjab), 58:15:1 (in Haryana); 28:7 1(in Uttar Pradesh); 14:4:1 (in Bihar) 80:7:1 (in Punjab, Pakistan) (Yadav *et al.* 1999a, 1999b; FAO STAT, 1999). Also, these ratios may not replenish exactly the amount of nutrients removed. Pasricha and Brar (1999) have remarked that nutrient ratios utilized in the Indo-Gangetic plains have also varied with time. At the beginning of the last decade it was 1:0.42:0.17 (N:P_2O_5:K), but currently it is 1:0.29:0.29 (N:P_2O_5:K). Clearly, proportion of K inputs have increased. Obviously, sooner or later, nutrient imbalances may reappear, and affect rice-wheat productivity, if IPNS recommendations are not practiced. According to Singh *et al.* (1999b) yield decline, nutrient imbalance, uneven depletion or build up can be overcome by utilizing appropriate mixtures of major and micronutrients, and FYM. Evaluation over a 10-year period indicates that high yields of both, rice (7.1 t ha^{-1}) and succeeding wheat (3.8 t ha^{-1}) can be attained, without deleterious effects to normal nutrient turnover and soil fertility using IPNS recommendations.

Judicious applications of inorganic nutrients, along with organic manures, Azolla-blue green algae is indeed a crucial aspect of nutrient management in the Indo-Gangetic plains. Rathore (1996) reported that inorganic N, P, K at 40 kg N, 30 kg P_2O_5 and 15 kg K_2O per ha, applied along with FYM plus BGA resulted in optimum nutrient recovery, productivity, and provided best benefit:cost ratios. Productivity gains fluctuated between 36 and 115% over control. Long-term evaluations indicate that application of FYM, or incorporation of green manure increased organic C contents (Swarup, 1991; Bhat et al. 1991; Singh and Swaroop, 2000; Aulakh et al. 2001).

Judicious use of fertilizer formulations helps in enhancing nutrient use efficiency. For example, since rice roots preferentially absorb NH_4^+–N, nitrogen supplied as NH_4^+ salts or urea may provide better efficiency. Similarly, submergence which enhances soil pH may have adverse effects on phosphate rocks and their solubility. In addition, rice roots are less equipped to absorb P from phosphate rocks (Sanyal and De Datta, 1991). For best results, in terms of fertilizer P efficiency, it is advisable to apply phosphate rocks during wheat phase, and utilize residual P effects during the succeeding rice season. Such an arrangement manages P dynamics better during rice-wheat cultivation. However, it is observed that submergence and higher soil temperatures during rice phase enhances available P levels. Hence, P fertilizer applied to rice, as first crop seems sufficient to even support a good crop of succeeding wheat in sequence (Saggar et al. 1985). Normally rice/wheat crops utilize only 8 to 20% of P applied, and on an average 20 to 23 kg P ha^{-1} is required to produce 4 to 6 t grain ha^{-1} (De Datta et al. 1990). Nutrient inputs are often calculated based on requirements of individual crops, without consideration of carry-over effects, retention of nutrients in soil and residual effects (Patel et al. 1997). Residual effects of P (applied to rice) on wheat can be significant. Including this effect while deciding P inputs to wheat can make it efficient. In some locations, P fertilizer efficiency is enhanced by introducing a legume green manure crops, say sesbania, which then scavenges 80 to 90% of the residual P, in addition to providing N credits through symbiotic N fixation.

Appropriate seeding date is crucial. Evaluation under IPNS in a north-western location in Indo-Gangetic belt indicates that early planting, that is before July 1st week, yielded 4 t ha^{-1} rice grains. It decreased proportionately and was least at 1.86 t ha^{-1} if planted during August 2nd week. Obviously, nutrient recovery, recycling, and efficiency will be affected drastically if planting time is incorrect. Similarly, wheat sown between November 1st to 3rd week provided best nutrient recovery and propor-tionately high yield (6 t ha^{-1}), but decreased due to delay and it was only 4.7 t ha^{-1} if planted in December (Gangwar and Sharma, 1998). Optimum plant population is another important factor affecting both nutrient recovery and productivity of rice-wheat agroecosystem (Soni and Bhatia, 1989).

Introducing legumes, and deriving N credits from them is an important aspect of IPNS in Indo-Gangetic plains. Long-term evaluations indicate that advantages in terms of nutrient recovery and production efficiencies can be significant if legumes are introduced into the sequence (Table 5.7). Examination of soil fertility changes indicated that organic, total and available, all the three crucial fractions of N dynamics were altered beneficially. Nitrogen gains to the system were 9 to 15 kg ha^{-1} if legumes were used. On a long-term basis, P availability in soil was also enhanced (Singh *et al.* 1996a). Gain in production efficiency due to rice-wheat-legume sequence could range between 0 and 3 kg grain ha^{-1} day^{-1}, over the rice-wheat sequence (Table 5.7).

Equally crucial is to assess influence of weeds on rice-wheat productivity. Significant quantities of N, P, K are usually diverted, if weed flora are not removed in time. According to Singh *et al.* (1996b), major weed species in central and eastern Gangetic plains, such as *Chenopodium album, Amagallis arvensis, Melilitus alba, M. indica, Fumeria parviflora, Vicia sativa, Cyperus rotundus* and *Cyanodon dactylon* influence nutrient dynamics in rice/wheat fields, if unattended.

Overall, the IPNS results in a marginal depletion of N and K, but allows a slight build-up of organic P. There was a build up of organic C due to IPNS, attributable to addition of FYM, green manure and crop residues. Clearly, IPNS maintains nutrient status and reduces imbalance. Productivity levels of both rice and wheat under IPNS were maintained close to that obtained via incessant inorganic fertilizers. However, it is to be noted that under IPNS only 50% of the generally recommended inorganic fertilizers are used, even then, it consistently provided higher productivity than conventional farmers, practices (Nambiar *et al.* 1992; Biswas and Narayanswamy, 1997; Hegde, 1998; Singh *et al.* 1999a; Prasad, 2000).

4. Concluding Remarks

During the past 5 to 6 millennia, the cropping patterns in the Indo-Gangetic plains have changed periodically, depending on a range of natural and socio-economic factors. The rice-wheat sequence adopted since the past three decades is again attributable to edaphic/climatic factors, human preference to these two cereals, and economic considerations. However, it may be futile to forecast a continued dominance of rice-wheat cropping

Table 5.7: A long term (5 years) assessment of nutrient recovery, production efficiency and soil fertility change due to rice-wheat rotation in comparison to others practiced in Indo-Gangetic plains

Crop sequence	Nutrient recovery (kg ha^{-1})				Production efficiency (kg ha^{-1} day^{-1})			Soil fertility change		
	N	P	K	Total	Grains	Protein	Carbohydrates	Organic C (%)	Total N (kg ha^{-1})	Available P (kg ha^{-1})
Rice-Wheat	186	30	178	394	24	1.9	15.9	-0004	-8.0	+1.4
Rice-Legume	170	24	132	326	23	1.5	9.2	+0.006	+10.0	+4.8
Legume-Wheat	166	24	129	319	25	2.4	14	+0.006	+9.0	+8.8
Rice-Wheat-Legume	215	36	213	464	27	22	18	+0.010	+15.0	+13.8

Source: Singh et al. 1996a

in the Indo-Gangetic zone, because, better options may gain ground. Rice is preferred over legumes/oilseeds because it tolerates salinity/alkalinity better, and wheat grown in sequence is well adapted to dry winters in these plains. Adoption of rice-wheat sequence causes upheavals in soil physico-chemical environment, and biological component due to alternation of wet (submerged) and dry conditions. We have a certain level of understanding on soil physical aspects, and chemical transformations of essential nutrients (N, P, K, Ca, Fe, Zn) during flooding/dry cycles. However, longer term implication to soil ecosystem, particularly nutrient availability, soil quality, microbial component and productivity needs to be ascertained and forecasted with greater accuracy.

Cropping in the Indo-Gangetic plains has been generally intense during recent decades. It has lead to rapid depletion of nutrients, caused imbalances, and reduced productivity. Nutrient deficiencies, first noticed in 1960s, have spread rapidly. Hence, nutrient inputs through chemical fertilizers are mandatory. According to soil specialists, inputs of N, P and K have been generally high at 120-180:20-50:60-80 kg ha^{-1} respectively. They suggest that it just matches the net recovery of N, P, K by a rice-wheat sequence. Presently, the rice-wheat sequence yields between 6 and 10 t cereal grains ha^{-1} plus proportionate biomass. It may reach 10 to 15 t cereal grains ha^{-1} in future. The tendency is clearly towards intensification, using greater fertilizer inputs along with super rice varieties and hybrids that promise 30% higher yield. Obviously, at first, nutrient dynamics during the rice phase of the sequence will get altered significantly. Judicious use of chemical amendments and fertilizers can reduce salt accumulation, thwart spread of salinity/alkalinity, if not, soil deterioration seems imminent.

Nutrient use efficiencies have been generally low, and provides scope for improvement. At present only 25 to 40% of N, and 8 to 20% P_2O_5 applied as fertilizer is recovered by crops. Large retention of nutrients in soil leads to accumulation or loss through emissions. A green manure crop introduced into the sequence after wheat harvest seems to augment nutrient dynamics excellently, also helps in maintaining soil quality. In addition to influence of soil/crop management options, it would be interesting, perhaps useful to ascertain the effects of pests, diseases, drought/flooding on nutrient dynamics during rice-wheat sequence. Nutrients retained in soil, fraction vulnerable for loss through leaching, volatilization or transformation; and that recycled via biomass or removed in grains may differ, depending on the intensity of malady that afflicts. This aspect has not been investigated in detail.

Soil microbial ecology too is affected due to continuous cereal sequence, high nutrient inputs, alternate wet and dry spells and seasonal fluctuations in redox potential, pH, Ec, etc. Effects on N-fixation, nitrification/

denitrification, gaseous emissions (CH_4, CO_2, N_2O) needs to be understood in greater detail. Mainly because, consequences on soil and general agroenvironment can be then predicted better.

Further intensification of rice-wheat ecosystem in the Indo-Gangetic plains seems imminent, which needs to be achieved with least disturbance to nutrient and ecological equilibrium. At the bottom line, careful experimentation and data accrual, simulations in field, and virtually through computer models, accurate forecasts, shrewd agronomic measures (e.g. precision farming), excellent extension services are required. Such measures may lead us to better harvests required by anticipated higher levels of population in future.

REFERENCES

AICARP (1986-87). Annual report—All India Coordinated Rice Project. University of Agricultural Science, Bangalore, India, 114 pp.

Agarwal, S., Singh, T.A. and Bharadwaj, B. (1987). Inorganic soil phosphorus fractions and available phosphorus as affected by long term fertilizer and cropping pattern in Nainital-Tarai. *J. Indian Soc. Soil Sci.*, **35:** 25-28.

Agarwal, S., Singh, T.A. and Bharadwaj, B. (1993). Soil potassium as affected by long-term fertilizers and cropping with rice-wheat-cowpea rotations on Mollisols. *J. Indian Soc. Soil. Sci.*, **41:** 387-388.

Aggarwal, P., Garg, R.N., Das, D.K. and Sharma, A.M. (1999). Puddling, soil physical environmental and rice growth on a typic Clorochept. *J. Indian Soc. Soil. Sci.*, **47:** 355-357.

Ali, K., Javed, M. and Javed, A. (1999). Growth promotion of wheat by potassium application in saline soils. *J. Indian Soc. Soil Sci.*, **47:** 510-513.

Aulakh, M.S., Khera, T.S., Doran, J.W. and Bronson, K.F. (2001). Managing crop residue with green manure, urea and tillage in a rice-wheat rotation. *Soil Sci. Soc. Am.J.*, **65:** 820-827.

Bahl, G.S., Vig, A.C., Yashpal and Singh, A. (1998). Effect of green manure and cropping as P sorption in some soils of Punjab and Himachal Pradesh. *J. Indian Soc. Soil. Sci.*, **46:** 574-579.

Bajwa, M.I. (1992). Soil fertility management for sustainable agriculture. *In:* Proceedings of National Congress of Soil Science. Soil Science Society of Pakistan, Islamabad, Pakistan, pp. 7-25.

Bajwa, M.S., Choudhary, O.P. and Josan, A.S. (1997). Rice-wheat productivity on sodic soils and in soils irrigated with sodic waters. *Bull. of Indian Soc. of Soil Sci.*, **18:** 48-57.

Bajwa, M.S. and Josan, A.S. (1989). Prediction of sustained sodic irrigation effects on soil sodium saturation and crop yields. *Agric. Wat. Mgmt.*, **16:** 217-228.

Bandhyopadyay, K.K., Jena, D., Misra, C. and Shapers, J.S. (1994). Transactions of 15th International Soil Science Congress, Mexico, Vol. 56, 332 pp.

Bandhyopadyay, K.K. and Sarkar, M.C. (1999). ^{15}N Balance in rice-wheat cropping system. *Fertilizer News*, **44:** 39-47.

Bangar, K.C., Kapoor, K.K. and Mishra, M.M. (1999). Soil microbial biomass: its measurement and nutrient source. *Indian J. Microbial* **30:** 263-265.

Bansal, R.L., Nayyar, V.K. and Takkar, P.N. (1991). Field screening of wheat cultivars for manganese efficiency. *Field Crops Research,* **29**: 107-12.

Bansal, R.L., Nayyar, V.K. and Takkar, P.N. (1994). Tolerance of wheat and triticale to manganese deficiency. *Indian J. Agric. Sci.,* **64**: 382-386.

Bhandari, A.L., Sood, A., Sharma, K.N. and Rana, D.S. (1992). Integrated nutrient management in rice-wheat system. *J. Indian Soc. Soil. Sci.,* **40**: 742-747.

Bharadwaj, V., Bansal, S.K., Maheswari, S.C. and Omanmar, P.K. (1994). Long term effects of continuous rotational cropping and fertilization on crop yields and soil properties 3. Changes in the fractions of N, P, and K of the soil. *J. Indian Soc. Soil. Sci.,* **42**: 392-397.

Bharadwaj, A.K. and Omanwar, P.K. (1994). Long term effects of continuous rotational cropping and fertilization on crop yields and soil properties. 2. Effects on Ec, pH, organic matter, and available nutrients of soil. *J. Indian Soc. Soil Sci.,* **42**: 387-392.

Bharadwaj, K.K.R., Rao, V.R. and Pareek, R.P. (1997). Soil biological environment for rice and wheat crops. *In:* Sustainable soil productivity under rice-wheat system. T.D. Biswas and G. Naryanaswamy (eds.). *Bulletin of Indian Soc. of Soil Sci.* **18**: 58-69.

Bhat, A.K., Beri, V. and Sindhu, B.S. (1991). Effect of long term recycling of crop residues on soil productivity. *J. Indian Soc. Soil. Sci.,* **39**: 380-382.

Biswas, and Narayanaswamy, G. (1997). Sustainable soil productivity under rice-wheat system. *Indian Soc. of Soil Sci.* New Delhi, pp. 1-83.

Bushan, L. and Sharma, P.K. (1999). Effect of depth, bulk density and aeration status of root zone on productivity of wheat. *J. Indian Soc. Soil. Sci.,* **47**: 29-34.

Chattopandhyay, R., Harit, R.C. and Kalra, N. (2001). Evaluation of water and nitrogen production functions for assessing yield and growth of wheat. *Fertilizer News,* **46**: 43-53.

Chauhan, R.P.S., Ram, S., Singh, B.B., Singh, V.P. and Sigh, R.K. (1996). Nutrient management of rice in sodic soils. Paper presented at a review meeting of rainfed low land rice research consortium, New Delhi.

Chauhan, R.P.S., Singh, B.B., Singh, R.K. and Singh, V.P. (1999). Strategic nutrient management for sustained rice production in sodic soil. *Fertilizer News,* **44**: 13-26.

Chenkul, V. and Acharya, C.L. (1990). Effect of rice-wheat and maize-wheat rotations on soil physical properties including soil water behavior in acidic Alfisol. *J. Indian Soc. Soil. Sci.,* **38**: 574-582.

Chhonkar, P.K. and Tarafdar, J.C. (1981). Characteristics and location of phosphates in soil plant system., *J. Indian Soc. Soil. Sci.,* **29**: 215-219.

Dash, R.N., Mohanty, S.K. and Patnaik, S. (1981). Efficiency of HCl and H_2SO_4 acidulated rock phosphates for a rice leased cropping system. *Fertilizer Research,* **2**: 109-118.

Dash, R.N., Mohanty, S.K. and Patnaik, S. (1982). Transformations and availability of soil and fertilizer P under flooded conditions. *J. Indian Soc. Soil. Sci.,* **30**: 387-389.

De Datta, S.K., Biswas, T.K. and Charenchamatcheep, C. (1990). Phosphorus requirement and management for lowland rice. *In:* Phosphorus requirements for sustainable agriculture in Asia and Oceenia. International Rice Research Institute, Manila, Philippines, pp. 207-323.

F.A.I (1999). Fertilizer Association of India—Annual Review (1988-99). New Delhi.

FAO STAT (1999). Food and Agricultural Organization of the United Nation, *http://FAO.org* Rome, Italy.

Gangwar, K.S. and Sharma, S.K. (1996). On-farm assessment of management practices on the productivity of rice (*Oryza sativa*)—wheat (*Triticum aestivum*) system. *Indian J. of Agronomy,* **41**: 9-11.

Ghildyal, B.P. (1978). Soil and Rice. International Rice Research Institute, Los Baños, Philippines, pp. 317-322.

Ghildyal, B.P. (1982). Vertisols and rice soils of the tropics. Trans 12th Int. Congress Soil Science, *Indian Soc. of Soil Sci.* New Delhi, pp 121.

Gill, M.S., Rana, D.S and Narang, R.S. (1993). Response of maize (*Zea mays*), wheat (*Triticum aestivum*) and Gobi sarson (*Brassica rapus*) to balanced fertilization and Azotobacter inoculation in sub-humid Punjab. *Indian J. of Agronomy*, **38**: 463-485.

Gill, M.S., Khind, C.S., Bijay Singh and Singh, Y. (1998). Ammonia volatilization under flooded conditions as affected by urease activity of soils amended in the crop residue on long term basis. *J. Indian Soc. Soil. Sci.*, **46**: 448 - 458.

Goswami, J., Baroova, S.R. and Thakuria K. (1996). Direct and residual effects of phosphorus in rice (*Oryza Sativa*), wheat (*Triticum aestivum*) rotation. *Indian J. of Agron.* **41**: 144-146.

Goswami, N.N., Prasad, R., Sarkar, M.C. and Singh, S. (1988). Studies on the effect of green manuring in nitrogen economy in a rice-wheat rotation using C ^{15}N technique. *J. Agri. Sci. Cambridge*, **111**: 413-417.

Gupta, S.K., Chaudhary, M.L. and Das, S.N. (1999). Urea transformation: A comparative study in salt-affected soils. *J. Indian Soc. Soil. Sci.*, **49**: 546-548.

Gupta, R.K., Singh, N.T. and Sethi, M. (1994). Groundwater quality for irrigation in India. Central Soil Salinity Research Institute, Karnal, Haryana, India, **19**: 16.

Hegde, D.M. (1998). Effect of integrated nutrient supply on crop productivity and soil fertility in rice (*Oryza sativa*) and wheat (*Triticum aestivum*) variations in rice-wheat system. *Indian J. of Agron.*, **43**: 7-12.

Hegde, D.M. and Dwivedi, B.S. (1992). Nutrient management in rice-wheat cropping system in India. *Fertilizer News*, **37**: 27-42.

Hundal, H.S., Thind, S.S., Brar, J.S. and Arora, B.R. (1989). Efficiency of phosphatic fertilizers in rice-wheat rotation on an Alfisol of North India. *Intern. J. Trop. Agric.*, **7**: 65-68.

Hussain, K., Naseem, A.R., Ali, L. and Iqbal, J. (1994). *In:* Efficient use of plant nutrients. Proceedings of 4th National Soil Science Conference, Islamabad, Pakistan, pp. 43-47.

IRRI (1995). IRRI—Almanac (1995). International Rice Research Institute, Manila, Philippines.

Kar, S., Acharya, C.L. and Prihar, S.S. (1997). Soil management effects on physical edaphic environment and sustainability of rice-wheat system. *In:* sustainable soil productivity under rice-wheat system. T.D. Biswas and G. Naryanaswamy (eds.), *Indian Society of Soil Science Bull.*, **18**: 1-19.

Khan, M.A. (1997). Soil fertility education in teaching institution of dryland agriculture. *In:* Accomplishments and future challenges in dryland. Soil fertility research in the Mediterranean area. Ryan, J. (ed.) International Center for Agricultural Research in Dry Area, Syria, pp. 301-307.

Khan, Z.H., Mazumder, A.R., Mohiuddin, A.S.M., Hussain, M.S. and Saheed, S.M. (1998). Physical properties of some benchmark soils from the flood plains of Bangladesh. *J. Indian Soc. Soil Sci.*, **46**: 442-446.

Khan, A.R., Qayyum, A. and Chaudhary, G.A. (1989). A country paper on soil, water and crop management systems for dryland agriculture in Pakistan. Proceedings of soil, water and crop management systems in rainfed agriculture in Near East region, USDA, Washington, D.C.

Kumar, V., Govil, B.P. and Kasre, S.V. (1999). Nutrient management through IPNS in Farmers field—IFFCO experience. *Fertilizer News*, **44**: 89-106.

Kumar, D., Swarup, A. and Kumar V. (1996). Influence of levels and methods of N application on the yeild and nutrient of rice in a sodic soil. *J. Indian Soc. Soil. Sci.*, **44**: 259-263.

Kumar, A. and Yadav, D.S. (1993). Effect of long term fertilization on soil fertility and yield funder rice-wheat cropping system *J. Indian Soc. Soil. Sci.*, **41**: 178.

Kundu, B.S. and Gaur, A.C. (1984). Rice response to inoculation with N_2-fixing and P solubilizing microorganisms. *Plant and Soil,* **79:** 227-234 .

Ladha, J.K., Kundu, D.K., Angelo-Van Coppenoli, M.G., Peoples, M.B., Carangal, V.R. and Dart, P.J. (1996). Legume productivity and soil nitrogen dynamics in low land rice-based cropping systems. *Soil Sci. Soc. Am. J.,* **60:** 183-191 .

Lata, Saxena, A.K. and Tilak, K.V.B.R. (2002). Biofertilizers to augment soil fertility and crop production. *In:* Soil fertility and crop production. Krishna, K.R. (ed.) Science publishers Inc., New Hampshire, pp. 279-312.

Malik, D.M. and Hassan, G. (1992). Managing soil and water resources. *In:* Proceedings of 3rd National Congress of Soil Science. Soil Science Society of Pakistan, Islamabad, Pakistan, pp. 40-57.

Mandal, D.K. and Kar, S. (1995). Water and nitrogen use by rice-wheat on Ultic Haplustalf. *J. Indian Soc. Soil. Sci.,* **43:** 9-13 .

Meelu, O.P. and Yadvinder Singh (2000). Phosphorus management in rice and wheat in northern India. *Fertilizer News,* **45:** 31-38.

Mehla, D.S. (1999). Effect of frequency of zinc application in yield of rice (*Oryza sativa*) and wheat (*Tritium aestivum*) in rice-wheat sequence. *Indian J. of Agronomy,* **44:** 463-466.

Mehta, K.K. (1994). Chemistry of salt-affected soils. *In:* Salinity management for sustainable agriculture. D.L.N. Rao, N.T. Singh, R.K. Gupta and N.K. Tyagi (eds.) Central Soil Salinity Research Institute, Karnal, Haryana, India, 311 pp.

Minhas, P.S. and Gupta, R.K. (1992). Quality of irrigation of water-assessment and management. Indian Council of Agricultural Research, New Delhi.

Minhas, P.S., Sharma, D.R. and Singh, Y.P. (1995). Response of rice and wheat to applied gypsum and farm yard manure on an alkali water irrigated soil. *J. Indian Soc. Soil. Sci.,* **43:** 452-454.

Mishra, B. (1994). Breeding for Salt Tolesance in Rice. *In:* Salinity management for sustainable agriculture. D.L.N. Rao, N.T. Singh, R.K. Gupta and N.K. Tyagi (eds.), Central Soil Salinity Research Institute, Karnal, India, 266 pp.

Modgal, S.C., Singh, V. and Gupta, P.C. (1995). Nutrient management in rice-wheat cropping systems. *Fertilizer News,* **40:** 49-54.

Mohanty, S.K. and Mosier, A.R. (1990). Transactions of International Congress in Soil Science Commission IV, Kyoto, Japan.

Mohanty, S.K. and Patnaik, S. (1977). Effect of submergence on the chemical changes in different rice soils. III Kinetics of K, Ca, and Mg. *Acta. Agron. Acad. Sci. (Hungary),* **26:** 187-190.

Mohanty, S.K. and Singh, T.A. (1997). Nutrient changes and dynamics under rice-wheat system. *In:* Sustainable soil productivity under rice-wheat system. *India Soc. of Soil Sci.,* New Delhi, pp. 20-21.

Mongia, A.D., Singh, N.T., Mandal, L.N. and Guha, A. (1998). Effect of lime and phosphorus application on nutrient transformations in acid and acid sulphate soils under submergence *J. Indian Soc. Soil. Sci.,* **46:** 18-22.

Mohammed S. (1987). Presidential address. Section of Agriculture and Forestry. The 29th Pakistan Science Conference. University of Karachi, Karachi, pp. 122-28.

Nambiar, K.K.M. (1994). Soil fertility and crop productivity under long-term fertilizer use. Indian Council of Agricultural Research, New Delhi, 78 pp.

Nambiar, K.K.M., Soni, P.N., Vats, M.R., Sehgal, K. and Mehta, D.K. (1992). Annual report (1987/88). All India coordinate project in long term fertilizer experiences. Indian Council of Agricultural Research, New Delhi.

Nayyar, V.K., Takkar, P.N., Bansal, R.L., Singh, S.P., Kaur, N.P and Sadana, U.S. (1990). Micronutrient in soils and crops of Punjab. Research Bulletin. Department of Soils, Punjab Agricultural University, Ludhiana, India.

Ourry, A., Gordon, A.J. and Macduff, J.H. (1997). Nitrogen uptake and root nodule. *In:* Molecular approach to primary metabolism in higher plants. C.H. Foyer and W.P. Quich (eds.). Taylor and Francis, London, 237-253 pp.

Pabbi, S and Singh, P.K. (1998). Conservation and exploitation of Cyanobacteria. *In:* Trends in microbial exploitation. B. Rai, R.S. Upadhyaya and N.K., Dubey (eds.). International Society for Conservation of Natural Resources, Varanasi, India, pp. 27-37.

Pannu, B.S., Sangwan, B.S., Gayal, V.P. and Panwar, B.S. (1999). Comparative study of the morphology and characteristics of soils used for rice and non-rice based cropping sequence in Haryana. *J. Indian Soc. of Soil Sci.,* **47:** 105-109.

Pasricha, N.S. and Brar, M.S. (1999). Role of mineral fertilizer to increase wheat production. *Fertilizer News,* **44:** 39-43.

Patel, N.M., Ahlawat, R.P.S and Ardeshna, R.B. (1997). Direct and residual effect of phosphorus in rice (*Oryza sativa*), wheat (*Triticum aestivum*) sequence under South Gujarat conditions. *Indian J. of Agron.,* **42:** 18-21.

Patil, K.P and Jain, J.M. (1994). Effect of some organic chemicals on soil organic nitrogen fractions and their availability in a rice-wheat sequence *J. Indian Soc. of Soil Sci.,* **42:** 50-54.

Patil, R.G. and Sarkar, M.C. (1993). Mineralization and immobilization of nitrogen in soils mixed with wheat straw *J. Indian Soc. Soil. Sci.,* **41:** 33-37.

Ponnamperuma, F.N. (1972). The chemistry of submerged soils. *Advance in Agronomy* **24:** 29.

Prasad, R. (2000). Nutrient management strategies for the next decade: Challenges ahead. *Fertilizer News,* **45:** 21-28.

Prasad, R. and Power, J.F. (1995). Nitrification inhibitors for agriculture, health and the environment. *Advances in Agronomy,* **54:** 233-285.

Prasad, B. and Rokima, J. (1991). Integrated nutrient management—1. Nitrogen fractions and their availability in calcarous soils. *J. Indian Soc. Soil. Sci.,* **39:** 693-698.

Prasad, B. and Singh, K.N. (1992). Cumulative volatilization loss of ammonia from rice field under integrated nutrient management. *J. Indian Soc. Soil. Sci.,* **40:** 209- 210.

Prasad, R., Singh, D.K., Singh, R.K. and Rani, A. (1999). Ammonia volatilization loss in rice-wheat cropping system and ways to minimize it. *Fertilizer News,* **44:** 53-56.

Prihar, S.S., Ghildhyal, B.P., Painuli, D.K. and Sur, H.S. (1985). Soil physics and rice. International Rice Research Institute, Los Baños, Philippines, 57.

Prihar, S.S., Pandey, D., Shukla, R.K., Verma, V.K., Chaure, N.K., Chaudhary, K.K. and Pandya, K.S. (1999). Energetics, yield, water use and economics of rice-based cropping system. *Indian J. Agronomy,* **44:** 205-209.

Qadir, M., Qureshi, R.H. and Ahmad, N. (1998). Horizontal flushing: a promising ameliorative technology for hard saline-sodic and sodic soils. *Soil and Tillage Research,* **45:** 119-131.

Rajput, A.L. (1997). Effect of nitrogen and zinc split application on wheat (*Triticum aestivum*) and their residual effect on rice. *Indian J. of Agron.,* **42:** 22-35.

Rathore, A.L. (1996). Economics of bio, organic and inorganic sources of nutrients in rice (*Oryza sativa*), wheat (*Triticum aestivum*) cropping system. *Indian J. of Agronomy,* **41:** 502-505.

Rokima, J. and Prasad, B. (1991). Integrated Nutrient management—1 Nitrogen fractions and their availability in calcarous soils. Transformations of applied P into inorganic P fractions in relation to its availability and uptake in calcareous soils. *J. Indian Soc. Soil. Sci.,* **39:** 703-709.

Sadana, V.S., Nayyar, V.K. and Kaur, N.P. (1989). Micronutrients in rice-wheat cropping system. *Micronutrient News,* **2:** 1.

Sagger, S., Meelu, O.P and Dev, G. (1985). Effect of phosphorus applied in different phases in rice-wheat rotation. *Indian J. Agronomy,* **30:** 199-206.
Sahrawat, K.L. (1979). Ammonia fixation in some rice soils. *Commun. Soil Science and Plant Nutrition,* **10:** 1015-023.
Sakal, R., Nayyar, V.K. and Singh, M.V. (1997). Micronutrient status under rice-wheat cropping system for sustainable soil productivity. *In:* Sustainable soil productivity under rice-wheat system. T.D. Biswas and G. Narayanaswamy (eds.). *Bulletin of the Indian Society of Soil Science,* **18:** 39-47.
Sakal, R., Singh, A.P and Verma, M.K. (1991). Different reactions of some rice varieties to zinc deficiency. *Oryza, 28:* 55-58.
Sakal, R., Singh, A.P., Sinha, R.B. and Bhogal, N.S. (1993). Twenty-five years of research as micro- and secondary nutrient in soils and crops of Bihar. Research bulletin, Department of Soil Science, Rajendra Agricultural University, Pusa, Samastipur, Bihar.
Sanyal, S.K. and Datta, S.K. (1991). Chemistry of phosphorus transformations in soil. *Advances in Soil Science,* **16:** 1-119.
Sarkar, S., Rathore, T.R. and Sachan, R.S. (1991). Influence of wheat straw on yield of rice and ammonium nitrate contents of a soil. *J. Indian Soc. Soil Sci.,* **39:** 377-379.
Sarkar, S., Rathore, T.R., Sachan, R.S. and Ghildyal, B.P. (1989). Effect of wheat straw management on cation status of Tarai soils. *J. Indian Soc. Soil Sci.,* **37:** 402-404.
Sawhney J.S. and Sehgal, J.L. (1989). Effect of rice-wheat and maize-wheat crop rotations on aggregations, bulk density and infiltration characteristics of alluvium derived soils. *J. Indian Soc. Soil Sci.,* **37:** 235-239.
Saxena, A. and Tilak, K.V.R. (1998). Free-living nitrogen fixes their role in crop productivity. *In:* Microbe for health, wealth and sustainable environment A. Verma (ed.), Malhotra Publishing House, New Delhi, India, pp. 25-64.
Sharma, P.K. and De Datta, S.K. (1985). Soil Physics and Rice. International Rice Research Institute, Los Baños, Philippines, 217 pp.
Sharma, J.K. and Gangwar, K.S. (1995). Nitrogen and crop production. *In:* Fertilizer and integrated nutrient recommendations for balance and efficiency. H.L.S. Tandon (ed.) Fertilizer Development and Consultation Organization, New Delhi, pp. 79-93.
Sharma, D.P. and Singh, K. (1998). Effect of subsurface drainage system on some physico-chemical properties and wheat yield in waterlogged saline soil. *J. Indian Soc. of Soil Sci.,* **46:** 284-288.
Sharma, R.K., Chauhan, D.S. and Nagarajan, S. (1999). Current status of crop response to fertilizers in different agro-climatic zones experiences of All India Coordinated Wheat Improvement Project. *Fertilizer News,* **44:** 39-47.
Sharma, P.K., Gupta, B.A. and Ghosh, D. (1999). Fertilizer requirements of wheat for maximum and economic yield on an Inceptisol of Uttar Pradesh. *J. Indian Soc. Soil Sci.,* **47:** 164-165.
Sidhu, B.S. and Beri, V. (1985). Recycling of crop residue in agriculture. *In:* Soil biology Mishra, M.M. and Kapoor, K.K. (eds.) Haryana Agricultural University, Hissar, India, pp. 49-54.
Singh, M.V. (1999). Current status of micro and secondary nutrient deficiencies and crop response in different agro-ecological regions. *Fertilizer News,* **44:** 63-82.
Singh, M.V. (2001). Evaluation of current micronutrient sticks in different agro-ecological zones for sustainable crop production. *Fertilizer News,* **46:** 25-42.
Singh, P.K. and Bisoyi, R.S. (1989). Blue-green algae in rice fields. *Plant and Soil,* **28:** 181-195.
Singh, Y., Chaudhary, D.C., Singh, S.P., Bhardwaj, A.K. and Dheer Singh, (1996a). Sustainability of rice (*Oryza sativa*), wheat (*Triticum aestivum*) sequential cropping through introduction of legume crops of legume crops and green manure crops. *Indian J. of Agron,* **41:** 510-514.

Singh. Y., Dobermann, A., Bijay-Singh, Bronson, K.F. and Khind, C.S. (2000). Optimal phosphorus management strategies for wheat-rice cropping on a loamy sand. *Soil. Soc. Am. J.,* **64:** 1413-1422.

Singh, S.S., Ehsanullah, M., S. Singh and Mishra, S.S. (1996b). Weed control in wheat (*Triticum aestivum*) under rice (*Oryza sativa*) wheat system of North Bihar. *Indian J. of Agron.* **41:** 243-246.

Singh, A.K., Khan, S.K. and Nongkynrih (1999b). Transformation of Zn in wetland rice soils in relation to nutrition of rice crop. *J. Indian Soc. Soil Sci.,* **47:** 248-253.

Singh, D., Leelavathi, C.R., Krishnan, K.S. and Sarup, S. (1999c). Monograph on crop responses to micronutrients. IARI, New Delhi, 136 pp.

Singh, P.K., Panigrahi, S.C. and Satpaty, K.B. (1981). Comparative efficiency of Azolla, BGA and other organic manures in relation to N and P availability in a flooded rice soil. *Plant and Soil,* **62:** 35-44.

Singh, A.P., Sakal, R., Sinha, R.B. and Bhogal, N.S. (1998). Use efficiency of applied zinc alone and mixed with biogas slurry in rice-wheat cropping system *J. Indian Soc. Soil Sci.,* **46:** 75-80.

Singh, N.P., Sachan, R.S., Pandey, P.C. and Bisht, P.S. (1999a). Effect of a decade long fertilizer and manure application on soil fertility and productivity of rice-wheat system in a Mollisol *J. Indian Soc. Soil Sci.,* **47:** 72-80.

Singh A.L. and Singh, P.K. (1989). A comparison of use of Azolla and blue green algae biofertilizers with green manuring, organic manuring and urea in a transplanted and direct seeded rice. *Experimental Agriculture,* **25:** 485-491.

Singh, A.L. and Singh, P.K (1987c). Nitrogen fixation and balance studies of rice. *Biology and Fertility of Soils,* **4:** 15-19.

Singh, A.L. and Singh, P.K. (1987b). Influence of Azolla management on the growth and yield of rice and soil fertility. Z, N and P contents. *Plant and Soil,* **102:** 49-54.

Singh, A.L. and Singh, P.K. (1987a). Influence of Azolla management on growth, N_2 fixation, and growth and yield of rice. *Plant and Soil,* **102:** 41-47.

Singh, B., Singh, Y., Sadana, V.S. and Meelu, O.P. (1992). Effect of green manure, wheat straw and organic manures on DTPA extractable Fe, Mn, Zn, and Cu in a calcareous sandy loam soil, at field capacity and under waterlogged conditions. *J. Indian Soc. Soil Sci.,* **40:** 114-118.

Singh, G.B. and Swarup A. (2000). Lessons from long term fertility experiments. *Fertilizer News,* **45:** 13-24.

Singhania, R.A. and Goswami, N.N. (1978a). Transformation of applied phosphorus under simulated conditions of growing rice and wheat in a sequence *J. Indian Soc. Soil Sci.,* **26:** 193-197.

Singhania, R.A. and Goswami, N.N. (1978b). Transformation of applied phosphorus in soils of rice-wheat cropping sequence. *Plant and Soil,* **50:** 527-535.

Soni, P.N. and Bhatia, A.K. (1989). Input factors for high productivity in rice-wheat sequence under resource constraints. *Indian J. of Agron.* **34:** 200-204.

Srivastava, L.L., Mishra, B. and Srivatava, N.C. (1988). Recycling of organic waste in relation to yield of wheat and rice, and soil fertility. *J. Indian Soc. Soil Sci.,* **36:** 693-697.

Sur, H.S., Prihar, S.S. and Jalota, S.K. (1981). Effect of rice-wheat and maize-wheat rotations on water transmission and wheat root development in a sandy loam of Punjab, India. *Soil Tillage Research,* **1:** 361-371.

Swarup, A. (1991). Long-term effect of green manuring (*Sesbania aculeate*) on soil properties and sustainability of rice and wheat yield on a sodic soil. *J. Indian Soc. of Soil. Sci.,* **39:** 777-780.

Swarup, A. (1994). *In:* Salinity management for sustainable agriculture. D.L.N. Rao, N.T. Singh (eds.) Central Soil Salinity Research Institute, Karnal, Haryana, India pp 18.

Swarup, A. and Srinivasa Rao, C.L. (1999). Current status of crop responses to fertilizers in different agro-climatic zones. *Fertilizer News,* **44:** 33-43.

Tarafdar, J.C. and Rao, A.V. (1996). Contribution of Aspergillus strains to acquisition of phosphorus by wheat (*Triticum aestivum*) and chickpea (*Cicer aerietinum*) growth in a loamy sandy soil. *Applied Soil Ecology,* **3:** 109-114.

Tiwana, U.S. and Narang, R.S. (1997). Crop productivity and soil nitrate—nitrogen changes in rice (*Oryza sativa*), wheat (*Triticum aestivum*) cropping sequence grown under maximum yield fertilization. *Indian J. of Agron.,* **42:** 39-56.

Tripathi, B.R. (1998). Managing sodic soils for sustained crop production in Indo-Gangetic plains. *J. Indian Soc. Soil. Sci.,* **16:** 543-550.

Tripathi, S.C., Chauhan, D.S., Sharma, R.K. and Dhillon, O.P. (1999). Productivity and economics of different wheat-based cropping sequences. *Indian J. Agronomy,* **44:** 237-241.

Twyford, I.T. (1993). Fertilizer use at Farm level in Pakistan. *In:* Proc 4th National Congress on Soil Science. Islamabad, Pakistan, pp 47-71.

Velayuthum, M. (1997). Sustainable productivity under rice-wheat cropping system—Issues and imperatives for research. *In:* Sustainable soil-productivity under rice-wheat system. T.D. Biswas and G. Narayanaswamy (eds.). *Indian Soc. of Soil Sci.,* **18:** 1-6.

Venkatraman, G.S. (1979). Algal inoculation of rice fields. *In:* Nitrogen and rice. International Rice Research Institute, Los Baños, Philippines, pp. 312-332.

Venkatraman, G.S. and Tilak, K.V.B.R (1990). Biofertilizers in sustainable agriculture. *In:* Soil fertility and fertilizer use nutrient management and supply system for sustaining agriculture in 1990's. V. Kumar, G.L. Shrotniya and S.V. Kaore (eds.), pp. 357-371.

Vig, A.C., Bahl, G.S., and Milap Chand (1999). Phosphorus—its transformations and management under rice-wheat system. *Fertilizer News,* **44:** 33-48.

Vig, A.C. and Milap Chand (1993). Transformation of labile P in two alkaline soils amended with *Sesbania aculenta* at two moisture regimes. *Trop. Agric. Trinidad,* **70:** 305-308.

Vig, A.C., Singh, D., Milap Chand and Soroa, G.S. (1997). Release of phosphorus from added *Sesbania aculeate*. *J. Indian Soc. Soil Sci.,* **45:** 449-455.

Wiren, N., Gazzerine, S. and Frommer, W.B. (1997). Regulation of mineral nutrient uptake. *Plant and Soil,* **196:** 191-199.

Woodhead, T., Huke, E.R. and Balababa, L. (1994). Rice-wheat atlas of India. Indian Council of Agricultural Research, New Delhi, 147 pp.

Yadav, R.L., Dwivedi, B.S. and Prasad, K. (1999a). Current status of crop response to fertilizers in different agro-climatic zones. *Fertilizer News,* **44:** 45-60.

Yadav, R.L., Pal, S.S., Prasad, K. and Dwivedi, B.S. (1999b). Role of fertilizers in cereal production for food security and balanced diet. *Fertilizer News,* **44:** 75-883.

CHAPTER 6

The Dryland Agroecosystem of West Asia, North Africa and South Asia
Nutrients, Water and Productivity

CONTENTS

1. The Dryland Agro-ecosystem, in General
   A. Agro-environment Soils
   B. Water deficits, soil nutrient transformations and Microbial activity in dryland agroecosystems
2. Drylands of West Asia and North Africa (WANA)
   A. Expanse, agro-climate, soil resources and cropping pattern
   B. Nutrient dynamics
      - Nitrogen
      - Phosphorus
      - Potassium
   C. Productivity in the Dryland Ecosystems of WANA
3. Nutrients in Drylands of South Asia
   A. Expanse, agro-climate, soil, and cropping systems
   B. Nutrients, water and productivity
4. Concluding Remarks

1. The Dryland Agro-environment, in General

A. Agro-environment and Soils

The dryland agro-climate is encountered in several parts of the world. Such a climate is predominant in West Asia, North Africa, India, China and US Plains in the Northern hemisphere. In the Southern hemisphere, Australia, Southern African nations, Argentina and Brazil possess drylands. However, the intensity and expanse of drylands utilized for agricultural cropping varies widely. It is generally accepted that in dryland agroecosystems, firstly limiting levels of soil moisture dictates cropping patterns, then net nutrient availability in soil as well as its recovery and crop productivity. Added to it, are the frequently observed soil nutrient dearth's and droughts. In certain other ecosystems, such as wetlands, water is neither limiting nor the driving factor. Unlike wetlands in the dryland agroecosystems water and its interactive effects along with nutrients determines productivity. Liebig's 'law of minimum', if applied selectively to nutrients, may confound our inferences, because quite often soil moisture is intricately linked to nutrient dynamics and its effects on crop productivity (Power, 1990; Jones, 1997; Ryan, 1997). Again, unlike irrigated ecosystems, farmers in dryland belts cannot regulate or alter soil moisture levels, or the timing of water inputs. Precipitations received in whatever pattern or intensity has to be efficiently utilized. The capacity of dryland soils to supply nutrients also varies depending on large number of other natural factors, such as soil parent material, extent of weathering, cropping patterns adopted etc.

In the dryland belts of West Asia and North Africa (WANA), Xerosols, Cambisols and Lithosols predominate. These are calciferous, hence classified as Calciorthids or Calcixerolls (Matar *et al.* 1992). Alfisols, Inceptisols and Vertic inceptisols are frequent in South Asian dryland zones, Northern Australia, Africa and South America. These are less fertile zones, with N, P, Zn and Fe deficiencies. Mollisols, with a deep organic matter enriched A-horizon predominate the drylands of North America, North and East Asia. Since, N and P deficiencies are common, their application results in increased crop productivity. Whatever be the variations in soils and their ability to supply plant nutrients, the sustenance and productivity of dryland agroecosystems is strongly influenced by moisture. According to Power (1990), the productivity of dryland agroecosystem is almost always lower than other wetter agrozones. This is attributable to stress resulting from water and nutrients which often act additively. Greenwood (1965) has suggested long ago, a method to decipher such additive effects because of two stress factors acting simultaneously which is as follows:

$$S = (\log M_2 - \log W_2)/(\log M_1 - \log W_1) \times 100$$

where S = % stress, M_2 and M_1 = dry weight of unstressed crop and W_2 and W_1-dry weight of stressed crop at times t_2 and t_1, respectively.

Overall, a conglomerate of factors related to soil, atmosphere and hydrosphere influence the dryland, its ecosystematic functions and crop productivity (Fig. 6.1). However, this chapter mainly deals with nutrient dynamics of two major dryland belts, one occurring in West Asia and North Africa (WANA) and the other in the Indian subcontinent (Table 6.1).

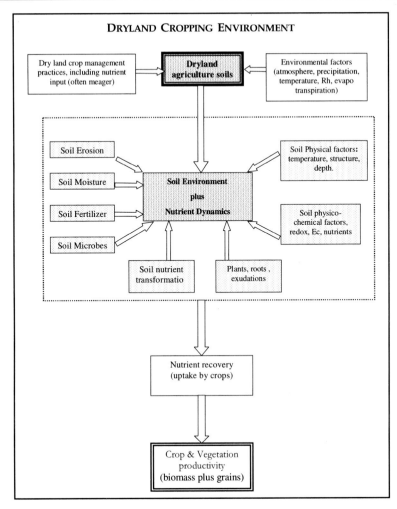

Fig. 6.1: Depicts the overall flow chart of factors, and their interactions in nature, under dryland conditions, that have consequences on nutrient recovery and productivity of agricultural crops or general vegetation.
Source: modified after Power, 1990

Table 6.1: The dryland agro-environments, salient features

West Asia and North Africa	
Expanse	This dryland zone extends between Morocco in the west, to Turkey in the north, Central Asian Republics in the east and to Yemen/Iran in the south. It includes the Nile valley in Egypt and Sudan, and Euphrates and Tigris valleys in Syria and Iraq. It occupies 128 m ha equivalent to 8% of 1.7 b ha of the total land area in WANA.

(Contd.)

Table 6.1: *(Contd.)*

Agro-climate	Precipitation ranges between 250-500 mm annually. Average temperature reaches <5°C in winter and >25°C in summer. In Mediterranean zone, cropping season temperatures average between +5° and 25°C.
Soils	Major soil types are Xerosols, Cambisols, Lithosols, Vertic Xerosols. Calciferous soils are common in West Asia.
Fertility	Moderate to severely deficient in N and P contents (<10 ppm Olsens P), needing yearly replenishments. Potassium levels are satisfactory. Micronutrient deficiencies (Zn, B) are also common. Boron toxicities are also perceived in West Asia. Organic C contents are low between 0.11% and 1.0%.
Major crops	Wheat (Durum and Bread) is preferred in zones receiving comparatively higher precipitation between 300 and 500 mm. Barley is cultivated in drier tracts that receive < 250 to 300 mm precipitation annually.
Production	Wheat yields at subsistence levels, and ranges between 0.5 t ha^{-1} and 2 t ha^{-1}, depending on precipitation/irrigation and nutrient inputs. Potentially, higher wheat yield at 5.0 t grain ha^{-1} are possible through high nutrient inputs and irrigation. Major legumes, lentil and Vetch yield between 0.5 and 1.0 t ha^{-1}.
Indian Subcontinent	
Expanse	Dry farming occupies larger tracts in northwest India, and on the fringes of the Thar desert, in India and Pakistan. Drylands are also frequent in southern India.
Agro-climate	Arid areas in northwest India and Pakistan receive an average precipitation between 250 and 450 mm during rainy season, and between 120 and 160 mm in post-rainy season. Whereas, in Peninsular India, the dry farming belt receives slightly higher rainfall between 500 and 700 mm. Sometimes pockets of these dry zones, overlap with semi-arid tropics that receive slightly higher precipitation at 700 mm annually. Winter temperatures in northwest India can be –5°C, but cropping season temperatures range from 17 to 30°C. During summer, temperature reaches >35°C. in peninsular India, cropping season temperatures range between 20 and 30°C.
Soils	Sandy Oxisols, Aridisols, Entsols, Inceptisols and Vertic Inceptisols are encountered in northwest of Indian subcontinent. Whereas, Alfisols, Vertisols and Laterittic soils are frequent in Southern Indian dry tracts.
Fertility	Soils are generally deficient in N and P, but sufficient in K. Vertisols in peninsular India possess higher CEC, therefore better nutrient buffering. Organic C in arid soils of northest India is low (0.35 to 0.25%), and it varies between 0.1 and 0.6% in the Inceptisols, Alfisols and Vertisols of South India. Soil pH may be 6.0 to 8.0. Zinc deficiency is common.
Major crops	Pearl millet, wheat, cowpea, cluster bean, cucurbits and agroforestry species are common in northwest India, and in eastern tracts of Pakistan. In southern India, Sorghum, Finger millet, Legumes (Cowpea, pigeonpea, and groundnut) or sunflower are common.
Productivity	Under rainfed dry farming conditions grain yields are low. Wheat yields 0.7 t ha^{-1}, pearl millet 0.5 t ha^{-1}, sorghum 0.8 t ha^{-1}, finger millet 0.65 t ha^{-1}. Legumes yield low at 0.5 to 0.7 t ha. For example chickpea 0.6 t ha^{-1}, oilseeds such as groundnut yield 0.8 t ha^{-1}, and sunflower gives 0.5-0.7 t ha^{-1}.

Source: Several. Mainly, Reports and Annual publications by International Center for Agricultural Research in Dry Areas (ICARDA) in Syria; International Crops Research Institute of the Semi-Arid Tropics (ICRISAT), Central Research Institute for Dryland Agriculture (CRIDA) Hyderabad, India and Central Arid Zone Research Institute (CAZRI), Jodhpur, India.

B. Water Deficits, Soil Nutrient Transformations and Microbial Activity in Dryland Agroecosystems

Scarce precipitation and intermittent droughts are common to drylands in any part of the world. Water deficits and resultant drier soils means increased concentration of salts in solution. Physico-chemical transformations such as phase changes—chemical precipitation versus solubility of salts, or even alterations in chemical forms may occur as soil dries. Normally, at salt concentrations beyond 5 d Sm^{-1} (2 to 3 d Sm^{-1} on oven dry soil basis), plant roots become sensitive and water relations get altered adversely (Power, 1990). Under such dry soil conditions, otherwise normally soluble anions such as bicarbonates (HCO_3), sulphates (SO_4^-), chlorides (Cl^-), and cations Na^+, Ca^{++}, Mg^{++}, K^+ are precipitated. In calcareous soils, Ca^{++} precipitates first, mainly as sulphate salts (e.g., gypsum). In gypsiferous soils of West Asia and northwest India, where both Ca^{++} and SO_4^{++} are predominant, electrical conductivity rarely exceeds 2 to 3 d Sm^{-1} due to gypsum precipitation. This phenomenon may, therefore, avoid adverse effects of salts on dryland crops. Water deficits may also induce Ca^{++} to precipitate as octo-calcium phosphates. It then affects P availability to plant roots. Further, soil P may also be precipitated as Fe and Al phosphates. Water deficits also delay or slow down weathering processes on soil minerals such as feldspars and K^+-rich micas. Hence, it affects chemical dynamics related to K^+ in soils. Root growth and its activities related to nutrient acquisition are vulnerable to water deficits. This is in addition to decreased diffusivity of nutrients caused by soil moisture deficits. Sometimes, root exudations alter pH, so that nutrients become more available, but this activity too may be affected due to water deficits.

Restriction to optimum biological activity due to suboptimal moisture regimes, results in impeded transformations and cycling of nutrients. It leads to decreased availability of major and micronutrients (Kissel et al. 1992). Soil microbial activity related to nutrient transformations immensely depends on moisture levels and its interaction with aeration (oxygen tension). Firstly, chemical and microbial processes such as organic matter mineralization, rates of N transformations, and nitrification reach their optimum when soil moisture is at field capacity. Actually, mineralization of organic N and nitrification effected by soil microbes such as *Nitrosomonas, Nitrococcus* and *Nitrobacter* is a two-step process. These microbes are highly sensitive to soil moisture deficits. Lengthier dry periods can practically limit N mineralization reactions. However, intermittent drought stress situations are common in dryland ecosystems. Such alternations in moisture, that is saturated and deficit conditions affect microbial population, its survival and N mineralization activity. It may cause upheavals and changes in soil microbial flora, which may either be beneficial or detrimental to soil nutrient dynamics (Power, 1990). For similar reasons of alternation in soil moisture and aeration levels, organic P mineralization and certain other reactions that enhance P solubility in drylands may get curtailed. However, intermittent soil moisture may also enhance P solubility (Matar *et al.* 1992). Rates of weathering and release of 'structural K' may also be altered by soil moisture deficits. The release of organic K from FYM through microbial activity is hastened by optimum soil moisture, but gets impaired by intermittent drought (Krauss, 1997). Soil moisture levels also play a key role in availability and defusibility of solution K in soil. The dynamic equilibrium between solution K and exchangeable K is also affected by soil moisture fluctuations. Alternate wetting and drying cycles induce corresponding changes in oxidation-reduction reactions, thus affecting availability of certain micronutrients, namely Fe or Mn.

Soil microbes also influence N dynamics in dryland agroecosystems through symbiotic and asymbiotic N fixation processes. In many dryland locations, N inputs from atmosphere via biological N fixation (BNF) perhaps sustains N requirements of crops. Again, soil moisture deficits and temperature affect "N credits" derived from BNF. If cold temperature affects BNF in West Asian drylands, excessively high soil temperature in the sandy soils of northwest India too lessens N credits derived through BNF. However, Rhizobium isolates that tolerate such vagaries in soil temperature are available (Peoples *et al.* 1996; Afandi *et al.* 1997; Venkateshwarulu and Mani, 1999; Lata *et al.* 2002). Phosphorus mobilization caused by arbuscular mycorrhizas (AM) may also be impeded by moisture deficits and extreme soil temperature. However, native AM isolates that grow and function in these dryland soils have been reported (Krishna and Lee, 1987; Sylvia and Williams, 1992). It would be useful to, firstly quantify P mobilization effects of AM isolates which are accustomed to drylands, examine their role in P dynamics, and then study the influence of extremes of temperature and moisture on this aspect.

2. Drylands of West Asia and North Africa (WANA)

A. Expanse, Agro-climate, Soil Resources and Cropping Pattern

Expanse: West Asia and North Africa (WANA) are vast, but agriculturally harsh regions to cultivate. This dry region extends from Morocco in the West, Turkey in the North to as far as Ethiopia and Yemen in the South. It includes central Asian republics. Excepting the irrigated Nile Valley in Egypt, and those around Euphrates and Tigris in Iraq and Syria, most regions practice dryland, subsistence agriculture. Vast expanses of Mediterranean dryland zones in the near east, comprising in all 26 nations, possess approximately 128 m ha arable cropping area. It constitutes only 8.0% of the total land area (1.7 b ha). This dryland belt caters to approximately 600 million people, which means 0.23 ha of arable land is available per capita, compared with a world average of 0.28 ha (Hamadallah, 1997; Van Schoonhoven, 1989). Obviously, thrust towards food self-sufficiency requires both higher nutrient inputs and area expansions. As such, nutrients applied into this agricultural zone through chemical fertilizers has increased from 0.5 mt in 1961 to 8.5 mt in 2000 (Ryan, 2001).

Agro-climate: The general agro-climate in West Asia and North Africa is highly variable. For example, Ryan (2001) states that variability in rainfall occurs even across small transects in the ground, and ranges between 150 and 600 mm in Syria and elsewhere in WANA. Such vagaries in precipitation can immensely influence soil nutrient dynamics, farming systems adopted and crop productivity. Such sharp variations in precipitation are also common in Morocco in North Africa, which receives < 600 mm rainfall annually. Overall, the average annual rainfall in WANA may range between 100 and 600 mm. Most conspicuously, intermittent drought sprees are common and they reduce crop productivity. Iran and Iraq, in central WANA region, receive 250 to 500 mm precipitation annually, which is distributed mainly during winter and spring period lasting from November to next May (Haghihatti, 1997). In southeastern WANA, the agricultural zones in Yemen receive between 100 and 800 mm annually. Average summer rainfall is 60 to 450 mm in northern highlands of Yemen, and spring rains range from 40 to 260 mm (Haidar, 1997). In many areas of WANA precipitation levels and crops' demand for moisture may tally in season, but temporal and spatial variations reduces water-use efficiency. Wheat yield in WANA is directly affected by precipitation levels, yielding a mere 0.5 t ha^{-1} in 250 mm zones, but over 4 t ha^{-1} at 530 mm rainfall (Pala, 1997). Supplemental irrigation needed to achieve the potential yields, actually decrease from about 400 mm in 250 mm rainfall zones, to 100 mm if rainfall is beyond 500 mm. The WANA region enjoys a cool Mediterranean climate, with temperature ranging between 5 and 18°C on an average. Cold winter temperatures <5°C and hot summers with >25°C are other features.

Soil Resources: Major soil types represented in dryland agrozones of the WANA are Entisols, Inceptisols, Alfisols, Mollisols and Vertisols. Aridisols are frequent in drier zones (Matar *et al.* 1992). It is estimated that nearly 70% of cropped area in WANA is occupied together by Xerosols (28 mha), Lithosols (26 mha) and Cambisols (17 mha), Regosols (10 mha) and Vertisols (5 mha). In West Asia, shallow Lithosols, Mollisols and Ardisols with relatively high organic matter are common, whereas Vertisols occur in Morocco (Monem *et al.* 1990a; Matar *et al.*, 1992; Ryan 2001). In the Nile

Valley and Delta, soils are recent and alluvial flood deposits transported from Ethiopian highlands. Montmorrillonite, Kaolinite, Feldspars, Quartz and some Illite are the dominant minerals (Sheta *et al.* 1981).

In general, soils of this region are predominantly calcareous, but still diverse. Fertility-wise, they are low in organic C, N and P, but supplied sufficiently with potassium. Low organic C contents in dryland soils (0.1 to 1%) limits proper expression of soil fertility. Hence, organic fractions of soil are said to influence N and P dynamics. Illite is the most abundant clay mineral in Mediterranean soils. Smectite occurs in many soils, but is dominant in Vertisols and 'Tirs' found in Morocco. Kaolinite is universal in these drylands, but occurs in small quantities. Matar *et al.* (1992) suggest that high pH and Ca contents literally dictates soil P availability and transformation. As such management of soil P dynamics in aridic and xeric soils is a major concern, because low soil moisture limits P availability. Above all, most soils in Mediterranean agro-belt are low in P (Orphanos, 1992). Since root systems confine to shallow layers, they scavenge limited quantities of P (Matar and Brown, 1989; Mohammed and Rawajfih, 1997). Potassium reserves are quite high ranging between 1.62 and 1.88% depending on location. The smectites with high CEC, and mica are primary K-bearing minerals (Badraoui *et al.* 1997). Calcite, dolomite and Mg-calcite which are common, also influence soil P dynamics. Among micronutrients, both deficiency and toxicity of boron are encountered (Orphanos, 1992). The ratio of Fe (oxidized) to Fe reduced is usually less than 0.2, and is attributed to crystalline Fe oxides in these xeric soils.

Cropping Pattern: Depending on the location and rainfall pattern, the cropping season in WANA spans between 5 and 7 months. Under Mediterranean climate it lasts from October to next May. In WANA, the principal crops are cereals, such as wheat (Fig. 6.2) and barley. In drier areas where barley predominates, wheat is produced for subsistence (Pala, 1997). Major legumes grown in rotation are lentil, chickpea and faba bean. The most common wheat-based cropping sequences, which confine to areas with slightly better precipitation are wheat-fallow; wheat-lentil; wheat-chickpea; wheat-water melon (summer crops); wheat-vicia or wheat-medic (pasture). Barley based cropping patterns common in WANA are barley-fallow; barley-barley; barley-chickpea; barley-lentil; barley-vicia (for hay); barley-vicia (for seed and hay), barley-medic (pasture) (see Christiensen *et al.* 1997; Ryan *et al.* 1997a; Ryan 2001). In most areas of WANA, it is the paucity of precipitation that regulates cropping pattern and choice of genotypes (Haidar, 1997). Pastures, after a major cereal is a very useful cropping sequence in WANA. The Mediterranean dryland pastures are actually characterized based on edaphic and climatic considerations. Badia or Steppe-type pastures are developed on flat lands

Fig. 6.2 Wheat farm in the West Asian Dry land Agroecosystem (Photo courtesy: International Center for Agricultural Research in Dry Areas, ICARDA, Aleppo, Syria)

with soils, which are also suitable for barley in the main season. During the pasture phase rainfall is generally erratic. Marginal pastures are developed on non-arable zones, demarcated based on edaphic considerations, mainly low soil fertility. Grazing is confined to spring season. Such marginal pastures occupy vast areas in West Asia (Osman, 1997).

B. Nutrient Dynamics

Nitrogen Dynamics: During the past 3 decades, N inputs into WANA agroecosystem increased six-fold. It was most conspicuous in 1980's and 1990's. Similar trends were not noticed for P and K. The need for K inputs continue to be neglected, because of the notion that sufficient levels of it occurs in WANA soil (Krauss, 1997; Ryan, 2001). However, enhanced crop productivity levels stimulated by greater N and P inputs causes increased K removal by crop. It then results in excessive K mining. Cropping systems in Mediterrenean region is dynamic, and the general trend has been towards intensification depending on availability of irrigation. According to Amar (1997), countries in WANA have traversed through the commonly known three stages of nutrient inputs into the agroecosystem. Firstly, the awareness stage which involves demonstration of benefits of chemical fertilizer inputs. The second stage involves extensive use of fertilizer technology, higher nutrient inputs and adopting methods that provide higher nutrient-use efficiency. The third stage is called the 'mature stage' when fertilizer inputs and recoveries are stabilized and the agroecosystem reaches equilibrium in terms of nutrient dynamics (Table 6.2).

Table 6.2: Fertilizer nitrogen inputs, recovery by crops and balance achieved in WANA agro-ecosystem during past 4 decades (×1000 t)

Period	Fertilizer N inputs	Crop N removal	Balance
1969-71	1059	2052	–993
1979-81	2951	2635	+316
1989-91	4982	3824	+1158
1998-2000	5260	3909	+1351

Source: Krauss, 1997; FAOSTAT 1999

Nitrogen Mineralization: Nitrogen supply from soil to plant is obviously dependent on two major factors, namely: a) the mineral N, that is soil NO_3 available to plant roots, and b) the fraction of N contributed by mineralization of organic matter during the crop season. Of course, N losses due to leaching and volatilization need due attention. The mineralization potential of WANA soils is considered substantial, but varies widely depending on soil, crop and environment related factors. For example, N minerlaized per cropping season ranged between 30 and 70 kg N ha^{-1} (Soudi et al. 1990), or 50 to 90 kg N ha^{-1} (Lamsalek, 1992), or 62 ot 273 kg N ha^{-1} in three different locations in Morocco (El-Gharous et al. 1990; El-Gharous, 1993). Avcin and Acvi (1992) reported that 100 to 150 kg N ha^{-1} could be mineralized in a wheat season in Turkish soils. Similarly, it was 130 to 170 kg N ha^{-1} at Telhadya in Syria (Asfary and Charanek, 1993). The nature of soil organic component is indeed a crucial factor because it influences mineralization rates. In certain soils from Syria, the average N mineralization rates varied between 17 and 340 mg N kg^{-1} soil, depending on the crop and its interaction with soil moisture and temperature. Mineralization of N peaked at 24°C with wheat (183 mg N kg^{-1} soil), but soils under legume crops reached peak N mineralization rates at 30°C (105 to 240 mg N kg^{-1} soil). Optimum soil moisture levels for N mineralization was 100% of field capacity. Hasbany et al. (1997) have pointed that intermittent drought stress common to WANA, and moderately low diurnal temperatures during January to May generally impeded N mineralization. Incidentally, this period coincides with maximum N demand by wheat/barley, hence it could be detrimental to optimum responses by crop.

Nitrogen Recovery and Losses: Fertilizer N recovery in different wheat based rotations could be low, between 8.7 and 26.4%, if immobilization and/or loss due to leaching, denitrification or volatilization are significant (Wood et al. 1997). Knowledge regarding distribution of N between the crop at harvest, and in soil layers as the wheat-fallow rotations proceeds is crucial while deciding on fertilizer N schedules. Certain evaluations indicate that nearly 26% of applied N remained in soil, and total recovery into crop was 38 to 49% of applied N. On those calcarous soils in WANA, which are high in pH, loss through volatilization together with leaching and denitrification can reach crop to 50% of applied N (Wood et al. 1997). However, ^{15}N measurements revealed that N loss through leaching/percolation could be comparatively small, < 5% at 20 to 30 cm depth. It could be attributed to low rainfall which does not cause percolation beyond 1.5 m depth. Fertilizer N recovery by wheat or other crops, is also dependent on the fertilizer formulations preferred by the WANA farmer. A series of field evaluations in WANA, which was co-ordinated by IAEA, Vienna revealed that out of the 120 kg N ha^{-1} broadcasted, N recovery ranged between 25 and 28% for urea, 24 and 65% for ammonium nitrate and 20 and 60% for ammonium sulphate (IAEA, 1974; Hera, 1997). Yet another evaluation of fertilizer formulations in different WANA countries revealed that 49 to 50% N was recovered from urea, and between 46 and 48% recovery was possible with ammonium nitrate or ammonium sulphate (Monem and Ryan, 1997). Appreciable levels of ^{15}N loss occurred whenever urea was applied to soil compared with ammonium sulphate (Wood et al. 1997). Obviously, soils in WANA are prone to volatilization losses after hydrolysis of urea. ^{15}N loss from nitrate based fertilizers too are high. Immobilization and denitrification are said to be other major routes of N loss. Actually, only 46% of nitrate-N applied was traced in plants. More recently in Syria, ^{15}N studies indicated that seasonal rainfall was an important factor affecting fertilizer N recovery. Nitrogen recovery ranged from 20 to 49%. Such variation was clearly attributable to rainfall pattern. Limited water supply often impeded recovery, and retention in soil as residual N (Monem and Ryan, 1997). Immediately after a rainfall event, percolation losses to layers measured beyond 20 cm can be significant. Since, a major fraction of fertilizer N applied to cereal crops in WANA is recovered through grain-N, it is not recycled but lost. Most often, stover serves as long term animal feed, which is not incorporated or recycled. Similarly, N contents in legume grains were higher at 22 to 71 kg N ha^{-1} compared with 12 to 30% in stover that may be recycled. Nearly, 18 to 48% N fixed in soil is via biological nitrogen fixation (Wood et al. 1997).

Cropping Systems and N Dynamics: Cereal-based rotations with legumes, forage, or fallow are common in WANA. However the productivity of such wheat or barley based rotations are immensely influenced by precipitation, and its interactive effects along with nitrogen. For example, in a dry year, wheat grain yield may be as low as 0.83 t ha^{-1} (235 mm), but a wet year may allow 3.62 t ha (486 mm). With sufficient N and optimum precipitation, potential wheat grain harvests can reach 4.5 to 5.0 t ha^{-1}. The underlying fact is that, N recovery fluctuates as yield levels vary depending on the cropping sequence. Ryan (1997) observed that among different cropping sequences, quantity of N recovered ranked as follows: Wheat ≤ Medic ≤ Chickpea ≤ Vetch ≈ Lentil < Melon ≈ Fallow. Further, long-term evaluation of these cropping sequences indicate that, during the past decades total soil nitrogen, mineral N in soil and organic C levels were generally of higher order with wheat medic sequence compared to wheat-fallow or wheat-wheat (ICARDA, 1993; Harris, 1995; Ryan *et al.* 1997a). Grazing the wheat stubble is a common practice in Syria and elsewhere in WANA. Grazing removes organic residues and limits carbon that could otherwise immobilize N into soil organic matter. Like other agroecosystems, WANA drylands too receive N inputs via symbiotic N fixation whenever cereal-legume rotations are practised. Wheat-legume rotations involving chickpea, Lathyrus, faba beans, peas, vicia are said to add between 20 and 130 kg N ha^{-1} as nitrogen derived from fixation (Ndff). Consistently higher Ndff, around 100 kg N ha^{-1} were obtained with *Vicia* in different locations of West Asia (Afandi *et al.* 1997). Christiensen *et al.* (1997) reported that long-term (7 years), wheat based rotations that include medic or vetch, enhanced total soil N concentrations by 15 to 20 times equivalent to 550 to 820 pp mm total N. Whereas, wheat rotations with lentils or melons added N to soil marginally.

The increased total soil N in medic or vetch sequences with wheat is attributed to larger organic N inputs into soil, through roots and stover incorporation. The extent of legume stover recycled obviously influences N recycled into soil. In fact, herbage incorporation into calcareous (Calci Xerollic Xerochrepts) soils of WANA adds to rapidity of N turnover in the agroecosystem. Further, Christiensen *et al.* (1997) observed that grazing animals may return appreciable quantities of N to soil, equivalent to nearly 20% of total N measured. On occasions, sheep grazing is said to have returned nearly 55% of N via urine and defecations.

Nitrogen-use Efficiency: In the dryland agroecosystem of WANA, recovery of N from fertilizers applied to soil is low. Hence, both soil researchers and farmers in this zone are interested in fertilizer management techniques that enhance N-use efficiency (El-Gharous, 1997). The fertilizer N-use efficiency (NFUE) values vary widely, depending on soil factors, crop, season and method of estimation. For example, fertilizer N recovery values were between 44 and 63% for cereals such as wheat and barley (Bloom *et al.* 1998; Nielsen, 1990; Asfary and Charanek, 1997). In Morocco, fertilizer N recovery was 25 to 35% depending on climate and cropping pattern (El-Mujahid, 1993). Whereas, for cereals cultivated in Greece N recovery was comparatively less at 17 to 32% (wheat); 25 to 40% (barley) and 17 to 37% (bread wheat) (Simmons, 1988). Precipitation levels or irrigations immensely influence fertilizer N-use efficiency (Table 6.3). Generally, wet

Table 6.3: Influence of soil moisture and N inputs on grain yield, N recovery from fertilizer, soil and N mineralized during a wheat crop season

Treatment (precipitation/irrigation) and N input	Grain yield kg ha^{-1}	Total N Input	N derived from Fertilizer (kg N ha^{-1})	N derived from Soil	Net N mineralized
Rainfed	2800	55.1	—	55.1	36
Rainfed 100 kg N ha^{-1}	3200	106.0	11.8	61.2	15.2
Irrigated	3600	66.4	—	66.4	30.3
Irrigated 100 kg N ha^{-1}	5200	125.5	49.1	76.4	-2.0

Source: Excerpted from Garabet *et al.* 1997
Note: N mineralized = (Total N uptake + soil inorganic – N at harvest + N loss) – (initial soil inorganic N + N fertilizer applied + N added through rainfall)

soils allow higher NFUE. Methods of fertilizer N application, its timing and crop growth pattern also influence N-use efficiency (Monem *et al.* 1990b). Variations in soil moisture and temperature may impose restrictions on NFUE (Garabet *et al.* 1997). Residual N caused by incessant N inputs is a factor to be considered while estimating N-use efficiency. Considerable amounts of NO_3-N may accumulate in the soil profile (Orphanos, 1992). For example, in Moroccan soils, nearly 70 to 100 kg residual N was traced in top 60 cm of soil profile (Soltanpour *et al.* 1989; Monem *et al.* 1990a). Effects of this residual N in subsequent crops can be considered. In general, higher crop yield responses was possible whenever residual soil NO_3-N due to fertilizer inputs reached above 12 mg N kg soil^{-1} (Orphanos, 1992).

Considering crop phenology, NFUE was maximum, when wheat reached anthesis in the field, and it ranged between 65 and 75%. It decreased thereafter during grain filling stage. At harvest, the cumulative NFUE ranged between 42 and 63% depending on the technique used to measure it (Garabet *et al.* 1997).

Such differences in NFUE values are attributable to soil mineralization/immobilization turnover (MIT) that occurs due to high microbial activity mediated transformations in warm climatic conditions. In fact, MIT is said to be a dominant factor that controls N dynamics in WANA agroecosystems (Garabet *et al.* 1997). Further, soil N dynamics and resultant NFUE attained are also dependent on seasonal changes in WANA. Since, N uptake occurs at lower rates during the early growth phase of the wheat crop, NFUE measured during February and March are generally low. Maximum NFUE was possible during April, which coincides with rapid growth phase and high N demand by the crop. Thereafter, NFUE either stabilizes or declines through ripening and at harvest (Garabet *et al.* 1997). Overall, after due considerations to a range of factors, Ryan (2001) states that NFUE varies between 20 and 70% in WANA dryland farming zones.

Phosphorus Dynamics: In general, phosphorus occurs in comparatively smaller quantities in dryland soils of WANA. Phosphorus in soil can severely affect crop productivity. It undergoes complex physico-chemical trans-formations that affect its chemical behavior and dynamics in soil. Ryan (2001) prefers to term phosphorus as the 'intractable element' in dryland agricultural systems of WANA.

Forms of Phosphorus: Total phosphorus content of Mediterranean dryland soils ranges widely depending on several factors that include parent material, soil genesis and cropping pattern. Mattar *et al.* (1992) report that total P values may range between 100 mg P g^{-1} soil and 2500 mg P g^{-1} soil. Flour appetite ($Ca_5(PO_4)F$) which is the predominant P bearing mineral, weathers easily in the xeric regimes encountered in WANA drylands. The released P is often re-precipitated as secondary phosphate, or assimilated into organic phase, or adsorbed on to clay surface or occluded. Inorganic P forms constitute the major fraction of total P, but organic P accounts for only 0.5% to 1%. It is attributable to inherently low levels of organic matter in Xerosols of WANA, and low recycling of crop residue. Quite often, organic P/inorganic P ratio could be as low as 0.05 to 0.15 (Ryan and Zhgard, 1980; Torrent *et al.* 1989). Chemical fractionation of Xerosols, Lithosols and Cambisols of WANA drylands indicate that Al, Fe, Ca-bound and reductant soluble Fe are the predominant inorganic P forms. Soil parent material also influences the extent of different P fractions encountered. Since, soils are calcareous and neutral or alkaline, Ca-P predominates over Al-P and Fe-P.

Availability of Phosphorus: According to Janat *et al.* (1997), there is much to investigate and learn regarding the complex nature of P chemistry in soils in WANA drylands, particularly its dynamic conversion into insoluble fractions, fixation into soil minerals, availability and mobility. It is felt so because, despite incessant research efforts to improve fertilizer P efficiency, only 10 to 15% of actual quantities are recovered into crops. In these predominantly calcareous soils of WANA, acidification affects P solubility and consequent recovery into crops (Ryan and Strochlein, 1979). Recently, Janat *et al.* (1997) too have reported that acidification or use of fertilizers that enhance acidity (e.g., urea), provided dominant chemical effect in enhancing availability and mobility of P. However, they summarize that despite relatively better mobility, and higher levels of available P accumulations at a depth of 10 to 20 cm, it may still not be sufficient to enhance net P recovery and fertilizer P efficiency. Phosphorus inputs into these drylands are often regulated depending on soil tests. Generally, accepted critical P limit is 6 mg Olsen's P kg $soil^{-1}$ (Ryan *et al.* 1997b). Using Cate-Nelson graphical method, critical levels for optimum wheat and barley production were standardized at 5 and 10 ppm P respectively.

However, soil researchers at ICARDA, Syria opine that it is P availability versus soil moisture interactions that needs greater attention. During wet years, despite only 4 or 5 ppm Olsen's P in soil, nutrient recovery may not be affected. Whereas, in dry years, critical P levels needed may be as high as 8 to 9 ppm. Obviously, soil moisture paucity hinders normal P recovery and hence P dynamics.

P Balance/Residual Effects: Information on trends in P balance that ensue after each crop season is useful while scheduling P inputs, gauging recovery into crops and the residual P retained in soil. There are indeed several reports on this aspect (see Matar *et al.* 1992; ICARDA, 1994; Ryan, 1997). Continuous cropping without replenishment, mines soil P resources. However, in many locations on these calcarous soils, even modest P applications at 30 or 60 kg P_2O_5 ha^{-1} results in positive P balance (Table 6.4). Persistence of residual P varies depending on soil type, location, climate and cropping systems. Residual effects of a single P application event can be seen up to 3 to 4 years (Ryan *et al.* 1997b). It is possible that P fertilizer inputs into WANA dryland soils is fixed irreversibly into unavailable forms. Despite immobilization reactions that occur between soil constituents and soluble P fertilizers, a large fraction of residual P is still left after each harvest. Highly significant residual P effects have been discerned in many locations (Ryan, 2001). Therefore, thus reducing fractionally the amount of repeated P inputs needed. To quote an example, yearly inputs of 60 kg P ha^{-1} on a sandy loam in Cyprus resulted in a build up of 4 to 36 ppm P in soil in 5 years. Similarly, yearly inputs of 8 kg P ha^{-1} on wheat-lentil sequence raised Olsen's P from 2 to 8 ppm within 4 years. Clearly, such a build up of residual P, if utilized judiciously, can enhance P fertilizer-use efficiency.

Table 6.4: Phosphorus balance in soil after 8 years of fertilizer P inputs to calcareous soils (Xerolchrepts) in WANA dry lands

Location	P application (kg P_2O_5 ha^{-1})		
	0	30	60
Jindress, Syria	−107	+27	+165
Tel Hadya, Syria	−89	+81	+228
Breda, Syria	−78	+87	+261

Source: ICARDA, 1995

Mycorrhizal Fungi: Mycorrhizal fungi, both native and inoculated isolates have been examined for their role in P dynamics in WANA drylands. They definitely add to the mobility of P towards roots. However, the exact levels of P benefits to the crop in terms of recovery needs to be investigated. Arbuscular mycorrhizal fungi may also help crops to withstand lowered levels of critical P in soils, because of their better affinity and scavenging ability. Their role in the P nutrition of WANA pastures needs to be ascertained. Naturally, a sizeable fraction of P is mobilized into cereal-legume sequences, and pastures through mycorrhizal fungi. Overall, understanding relevance of mycorrhizal fungi to P dynamics is important, equaly so, their role in soil microbial ecology and their interactions with symbiotic N fixers that occur on legumes and leguminous pastures needs emphasis (Kothari, 1991; Mohammed, 1993; Mohammed and Rawajfih, 1997).

Potassium in WANA Drylands: WANA soils possess much higher reserves of K, and equally higher capacity to release available K forms required by crops. The rainfall pattern in Mediterranean environment causes low leaching. Therefore, it allows sufficient available K in the root zone. In fact, potassium rarely leaches beyond 0.5 to 1.0 m even in normal rainfall tracts. Potassium leaching is supposed to be least in heavier Vertisols and 'Tirs'. Above all, the comparatively low K demand created by marginal productivity of cereal/legume or pasture in rainfed drylands, really limits interest and priority bestowed on investigating K dynamics in WANA drylands (Ryan, 2001). In view of this situation, Krauss (1997) terms potassium as the "Forgotten nutrient", whose dynamics in WANA drylands needs greater attention. In WANA soils, K is encountered in three major fractions namely, solution K, exchangeable K and structural K. At any time, a dynamic equilibrium is established between solution K, absorbed K and fixed K. This equilibrium is controlled by the clay mineralogy (Badraoui *et al.* 1997; Bouabid *et al.* 1991). In vertic and calciferous soils, K–Ca equilibrium and selectivity coefficients for K versus Ca may need consideration (Moujahid, 1993). In most soil types, levels of exchangeable (NH_4OAC) plus water soluble K is closely related to plant uptake. Wherever values are < 0.5 C mol C kg^{-1}, a K deficiency in terms of crop productivity is likely. However, in some areas of WANA, < 80 mg K kg^{-1} is considered the critical value for optimum crop productivity (Ryan, 2001). Recovery of K by crops often corresponds to K release characteristics, i.e., clay contents. Higher clay contents allow better K buffering and higher amounts exchangeable K (Krauss, 1997). In the absence of sufficient K reserves and buffering, K replenishments are needed to maintain crop productivity. For example, in Pakistan, K reserves depleted by 4 to 7 ppm K per annum and by 12 to 20 ppm K, if the crop is irrigated, but without K inputs (Krauss and Malik, 1995). Such drastic loss of K in soil has also been observed in several other locations in Iran, Syria and Morocco (Krauss, 1997). It is not just the exchangeable K that gets depleted, even general K reserves decreased by 40%, and it required 6 to 9 years consecutive K replenishments.

Potassium Fixation: Fertilizer K, upon application into WANA soils undergoes chemical transformation. It includes incorporation into mineral structures of soils, which then becomes inaccessible to plants. Relatively higher percentage of K fixing clay minerals such as vermiculite or smectites in dryland soils results in greater amounts of K lost through fixation reactions. Extended K fixation process may result in reduction of exchangeable K, and with continuous cropping the situation may get aggravated (Tributh *et al.* 1987). Intense K fixation reaction may obscure soil test based K recommendation. Therefore, Krauss (1997) suggests that in such a case, K inputs should first overcome fixation effects, and still be in excess for crops to recover K and respond. In fact, if we ignored K fixation, our perception about K deficiency or sufficiency may be erroneous to that extent.

Potassium Recovery by Crops: Physiologically, K demand and absorption by dryland crops are not uniform. Instead, a major fraction of K required is acquisitioned within a short span of time. In cereals, it coincides with growth period, between stem elongation to spike emergence. During this period K absorption rates are expected to fluctuate between 6 and 8 kg K_2O ha^{-1} day^{-1}. A legume may even draw up K at 12 kg K_2O ha^{-1} day^{-1}. Seasonal pattern of K recovery by biannual and perennial pastures too vary. After such rapid phase of K acquisition retranslocation, and or even leaching from leaves are to be expected. Such variations will influence K dynamics in soil and the dryland ecosystem. Obviously, K replenishment if any, or K management procedures adopted should aim at providing for sufficient level of available K. Krauss (1997) concludes that to meet K demand by cereals/legumes in WANA drylands, K^+ should occupy at least 3 to 5% of cation exchange capacity. Sometimes, split inputs of K can overcome such deficiency or restricted K release. In summary, it is evident that greater insights into K dynamics in WANA drylands are needed, and methods to sustain demand-supply equation at optimum levels are a necessity.

C. Productivity in the Dryland Ecosystems of WANA

Productivity in the dryland ecosystems of WANA is immensely influenced by soil moisture (rainfall) and nutrient input interactions. Congenial rainfall distributions totaling approximately 485 mm, and optimum N inputs (e.g. 60 kg N ha^{-1}) quadruples wheat biomass yield to 12,300 kg ha^{-1} from a mere 3700 kg ha^{-1} biomass attained at 238 mm rainfall (Wood *et al.* 1997). However, Orphanos (1992) opines that, although it is readily evident that rainfall limits crop yields in the drylands, indeed soil fertility and nutrient dynamics related factors may more often be the major constraints to crop productivity. As a corollary, if one wants to exploit a wet year better, then, appropriate nutrient inputs and their management can fetch the desired result. Seasonal variability in precipitation, soil N fertility status and other environment related factors, often make yield prediction difficult or may become erroneous. However, certain efforts to simulate wheat yield at various water and N regimes, using various models have been useful (Pala, 1997). In simple terms, wheat grain yield simulations indicate that it increases linearly from 0.5 t ha^{-1} at 250 mm rainfall to about 5.5 t ha^{-1} at 530 mm, and then plateaus off. Predictions on N requirements indicate that for optimum yield wheat crop demands 50, 80, 100 and 120 kg N ha^{-1} at moisture regimes 250, 350, 450 are 570 mm respectively.

Jones (1997) suggests that soil nutrient status together with inherent dynamics is one of the 4 major factors that influences Mediterranean agroecosystem. Some specialists believe that soil nutrients, as a component is more variable and dynamic than crop physiological factors. Nutritional factors that influence productivity of this ecosystem vary widely spatially. Fortunately, such factors can be managed through fertilizer inputs. In view of this, Jones (1997) has suggested following broad and integrated approaches that aim at more congenial nutrient dynamics in the Mediterranean dryland agroecosystem, namely:

(a) Appropriate nutrient cycling: returning residues, manuring, selecting deep rooted crops to scavenge nutrients in lower horizons.
(b) Crop rotations with legumes and due weightage for N fixation potential.
(c) Matching nutrient needs and replenishments, after considering nutrients removed, annual additions and net nutrient balance achieved.
(d) Matching actual crop needs with fertilizer rates; top dressing, banding etc.
(e) Using soil test-crop response based recommendations etc.

Indeed there has been marked awareness regarding need to replenish nutrients to the agroecosystem. Firstly, significant chemical fertilizer based nutrient inputs began in 1985, which quadrupled by 2000 A.D., Clearly, the intention has been to intensify this cropping ecosystem through nutrient inputs. Given that irrigation potential will also increase, further intensification of this Mediterranean ecosystem seems imminent. Whatever be the level of productivity aimed at through intensification, the bottom line is use nutrients judiciously, maintain nutrient dynamics congenially, and achieve greater yield stability. For this to happen, there is clear need to work out a synchrony between crop stages, with water, N and P supply to achieve greatest possible water and nutrient-use efficiency. This aspect actually gains importance as we intensify cropping in WANA. Such efforts have been successful in other agroecosystems. Some of these suggestions are easier said than accomplished.

3. Nutrients in Drylands of South Asia

A. Expanse, Agroclimate, Soil and Cropping Systems

Expanse: The rainfed dryland agroecosystem is an important component of agriculture in the Indian subcontinent. Such dryland ecosystem constitutes 66% of the total 141 m ha arable cropping ecozone in India, and contributes 40% total food requirements of the country. These rainfed dry farming tracts experience the twin problems of 'water thirst' and 'plant nutrient hunger' simultaneously (Singh et al. 1999a, 1999b).

Agroclimate: Within the Indian subcontinent, dryland belts occupy three different precipitation zones. They are arid (150 to 400 mm); low rainfall dry farming zone (500 to 750 mm); and semi-arid areas (750 to 1150 mm). The semi-arid zones are dealt separately in Chapter 7. The arid and low rainfall dry farming areas occupy 23.4 m ha and 32.5 m ha respectively in India alone (Singh et al. 1999b). A sizeable patch of arid and dry farming zone exists in Pakistan along the fringes of Thar Desert. The mean rainfall in the arid zones of Rajasthan in northwest India varies between 185 mm and 400 mm (mean 300 mm). The cropping season, that begins in July with monsoon rains extends for 8 to 12 weeks. Such short durations of precipitations and cropping season mean that only early maturing, short duration crops would be preferred. Maximum ambient temperatures range between 36 and 39°C, while that in soil may reach even 42°C during September/October. Such extremes in temperature may be critical in terms of microbial transformations of nutrients, root activity and nutrient recovery by crops. The amount of rainfall and its pattern, both are crucial in dryland ecosystems, because they affect a range of soil and crop related functions. Variations in annual rainfall for example at (792 mm), Hyderabad in South India is 20%. Whereas, in northwest India, Jodhpur receives only 382 mm but at a high 43% variation. This clearly affects agricultural operations, such as nutrient inputs, if any, and cultural operations. In a nutshell, the agroclimatic parameters of drylands in Indian subcontinent immensely influence nutrient dynamics, cropping sequence and productivity.

Soils: In the northwest of Indian subcontinent, the Arenosols possess low moisture holding capacity. High seepage losses and evaporation are other characteristics. Soil erosion due to high wind velocity and run off, both cause significant loss to arable soils and their nutrients. Hardpan farming, $CaCO_3$ ('kankar') layers are frequent in some parts. This limits root growth and infiltration of water. For pearl millet cultivated on sand dunes, the dune stabilization, moisture and nutrient availability are major soil related constraints to productivity.

The subgroups of gypsiferous soils, namely petrogypsifers are common in dune zones, and haplogypsifers are frequent in sandy ridges and low dunes. If not planted to pearl millet or moth bean, vegetation is generally scanty (Joshi, 2000). In peninsular, central and northern India, Alfisols and Vertisols are two major soil types encountered in dryland eco-zones. Alfisols tend to be less fertile. Other soil types encountered are Inceptisols, Entisols, Vertic Inceptisol, Lateritic and loamy soils.

According to Singh (1991), the alluvial soils encountered are less prone to severe nutrient deficiency, but major nutrient problems exist with Alfisols (52 m ha) and Vertisols (72 m ha). Poor infiltration of moisture and nutrients, sheet erosion and leaching reduces productivity of Vertisols in dryland zones. Moisture holding capacity of Alfisols is generally low, and nitrogen inputs need greater attention. In all soil types of Indian drylands, N and P deficiencies are rampant. Surveys have indicated that nearly 63% of soils in dryland belts are low in N status and 32% are medium in N status. Phosphorus deficiency is a major constraint in all locations (Swindale, 1982). Soil erosion and runoff related losses may reach up to 8.4 mt N, P, K annually from the dry farming zones of India (Singh, 1991). Measures to check erosional losses seem mandatory. Soils utilized incessantly with only N, or N and P inputs, obviously show up micronutrient imbalance.

Cropping Pattern: The main cropping seasons encountered in dryland ecosystems of India and Pakistan are 'kharif' which is the rainy season, extending from June or July till October. Whereas, 'Rabi' is the post-rainy season. During this period cultivation depends on the availability of stored moisture. Crops and cropping systems are dictated immensely by soil moisture, precipitation levels and nutrient availability. In northwest and peninsular India, shallow soils with limited moisture (approx 100-150 mm) in a season allows only a single crop. Usually cereals such as pearl millet or sorghum are cultivated. Medium soils at 150 mm moisture may support longer duration crops or sometimes intercropping. While in deep soils at 200 mm moisture double cropping is a possibility. At soil moisture >300 mm, this ecosystem supports cultivation of wheat, pea, or chickpea. Whenever, precipitation is suspected to be less than 300 mm, 'desi' wheat, barley, lentil and chickpea are possibilities. Stored moisture levels at <100—150 mm supports cultivation of rape seed or tara mira (*Eruca sative*).

Therefore, it is glaring that in the dryland farming zones, irrespective of nutrient levels in soil, cropping patterns are matched with precipitation levels. It is a common practice, especially in northwest India and Pakistan to grow perennial shrubs/trees interspersed with field crops. They are useful in creating congenial micro-climate under the canopy, and add organic matter. Leguminous trees fix atmospheric nitrogen and support grazing. Most common species are *Prosopis junifer, P. nilotica, Acacia tortilis,*

A. senegal in scanty precipitation zones (150-300 mm). *Acacia, Dalbergia, Zizyphus, Cenchrus, Albizia* are preferred in zones with >400 mm precipitation.

B. Nutrients, Water and Productivity

As in other dryland belts, water is an important factor affecting nutrient demand, its recovery and consequent productivity. Therefore, major focus while devising any improved agricultural practice has been to minimize water loss through erosion, percolation and evapotranspiration. This is because, nutrients and water are tightly linked in terms of absorption and use efficiency. Long-term assessments in peninsular India, indicate that proper management of soil nutrients and water can enhance grain yield from a mere 0.8 to 1.0 t ha^{-1} to 4.0 t ha^{-1} (Table 6.5). Clearly, soil moisture and fertilizer interactions significantly influence nutrient dynamics and productivity of Indian drylands (Singh and Das, 1986; 1990; 1995). In fact, fertilizer inputs are strictly guided by precipitation and stored moisture levels. Nutrient inputs can also be regulated using plant morphgenetic index (leaf area, roots), evapotranspiration, etc. These methods can help in attaining greater nutrient-use efficiency. Conversely, Prihar and Gajri (1998) believe that water use efficiency (WUE) could be enhanced through appropriate nutrient input schedules. Fertilizer inputs, firstly enhance root growth and its spread leading to better exploration of soil moisture. However, much depends an extent of soil moisture stored and available, and extent to which roots need to penetrate to harness it.

Table 6.5: Management of nutrients and water in drylands of peninsular India

Farming systems	Annual precipitation (mm)	Water used by crop (mm)	Water losses		Soil loss (t ha^{-1})	Productivity (kg ha^{-1})
			Surface run off (mm)	Evapo-transpiration (mm)		
Improved practices	904	602	130	172	1.5	4000
Traditional system	904	271	227	406	6.4	800

Note: Single crop, on flat beds, meager nutrient inputs (20:0:0, N:P:K), double cropping, broad bed, furrow, nutrient input at 80:40:20, N:P:K
Source: Virmani, 1991

With regard to attaining better nutrient efficiency, once again, timing and amount of nutrient inputs are important. Nutrients applied early at seedling stage induces better rooting, hence may help in tolerating end season drought/nutrient deficiency better. Top dressing of N avoids nutrient dearth that may occur at later stages of the crop, therefore increases nutrient recovery and water harvest. Since precipitation levels strongly influence nutrient dynamics, suitable modeling and simulating their interactions can be beneficial. It may help in devising approximate nutrient input schedules. Although not discussed here, there are several crop simulation studies that deal with soil moisture and nutrient dynamics in dryland ecosystems (see Jones, 1990). Generally recovery of N is low, hence to avoid risks of mismanagement, farmers tend to curtail higher N inputs during years when drought or sub-normal precipitation occurs.

Residual Nutrients: Nutrient credits, especially N from previous legume species have been investigated extensively. To quote a few examples, maize cultivated after groundnut received 15 kg N as residual N in soil from the previous crop. On an average, a crop of cowpea which is a common species in drylands of South India provides 20 to 25 kg N as residual nutrient in soil to the succeeding crop. A greengram plus pigeonpea intercrop, provided as much as 97 kg N to succeeding cereal. In order to achieve better nutrient efficiency, sometimes shorter duration legume such as green gram, horse gram or cluster beans are preferred over longer duration pigeonpea. Such a selection aims at achieving better nutrient efficiency in space and time. Like N credits, benefits in terms of P can also be discerned whenever stubble reincorporation takes place. Since, only 15 to 20% of P fertilizer applied to drylands is utilized by the first crop, a sizeable portion is generally available as residual P for the next crop.

Organic Matter, Green Manuring and Productivity: The organic matter content of sandy soils encountred in the drylands of northwest India is often low, and the situation is somewhat similar with soils in peninsular Indian drylands. Additionally, organic matter decomposition, measured as CO_2 evolution is generally rapid. Actually, higher temperatures, that range above >35°C results in rapid decomposition of organic matter, hence depletes organic C levels. This, in turn deteriorates soil structure and fertility, also disturbs soil microbe mediated nutrient transformations (Singh *et al.* 1999b). In general, nutrient inputs achieved through organic matter, stover, and crop residue varies depending on the source, its biochemical composition and ensuing biochemical transformations. In the drylands, decomposition of any source of organic residue is rapid during the first week. Later on, immobilization of N sets in, but by 14 to 16th week mineralization process during a crop season is complete. FYM or green manuring procedures adopted in dryland ecosystem improve the overall nutrient status in dryland ecosystem including micronutrients. It is believed that 'hidden hunger' for micronutrients could be efficiently managed through organic manures (Katyal and Sharma, 1991; Katyal and Reddy, 1997). Thus, green manuring is a time-tested nutrient recycling procedure for farmers in Indian drylands (Katyal *et al.*1994). A traditional green manure crop may add between 30 to 40 kg N ha^{-1} after 40 to 60 days growth. An improved green manure cultivar, on an average adds 60 kg N and 5 to 8 kg P ha^{-1} upon mineralization.

Selection of green manures depends on their ability to supply nutrients in the dryland ecosystem, but sometimes, short rainy season too dictates the selection of green manure species. *Leucana* and *Glyricidia* are currently popular green manure species preferred both in northwestern and peninsular Indian drylands. Their succulent tissue contains 3% N on dry weight basis. In the drylands characterized by sparse vegetation, crop residues (stalks, straw and by products) are highly valued sources of organic C and other nutrients. Efficient use of off-season precipitation by cultivating legumes such as horse gram, cowpea or green gram may provide up to 1 to 1.5 t biomass ha^{-1} which is equivalent to 20 to 30 kg N. Within the integrated nutrient management procedures, FYM and other organic manures are significant components. On a macro-scale, FYM and animal wastes generated in Indian drylands can contribute nutrients equivalent to 87.2 mt NPK. However, Katyal *et al.* (1999) believe that only 30% of this potential is utilized. They also suggested that using green manures with low C:N ratio, and adopting good composting procedure using N and P supplements can enhance efficiency of green manuring. Green manures are also excellent soil conditioning agents, and can raise productivity of drylands, particularly when used in conjunction with chemical fertilizers (Table 6.6).

Table 6.6: Organic manures and chemical fertilizer inputs and their influence on productivity of finger millet grown on Alfisols of peninsular India (Bangalore, South India)

Annual nutrient inputs	Productivity (kg ha^{-1})
Nil (control)	1510
Organic manures (FYM 10 t ha^{-1})	2550
NPK 50:50:50 (kg ha^{-1})	2940
Organic manure (FYM 10 t ha^{-1}) plus NPK 50:50:52 (kg ha^{-1})	3570

Source: Hegde *et al.* (1988)

There is yet another useful dimension in growing green manure species in drylands. High desiccating winds common in northwestern India result in significant soil erosion, and with it loss in nutrients, both inherent and that applied as FYM or chemical fertilizers occur. Shelter belts planted using green manure species such as Khejri (*Prosopis grandiflora*) thwarts soil loss by over 200 t ha^{-1}. Clearly, efficient green manuring procedures help in sustaining nutrient dynamics and crop productivity in drylands.

4. Concluding Remarks

The cropping activity in dryland agroecosystems are excessively regulated first by water resources, then via nutrient dynamics. Harsher temperature and other environmental factors too limit the productivity. Productivity is generally low at 0.5-1.0 t cereal grain ha^{-1}. However, in pockets where water resources are augmented, and nutrient inputs are increased, the grain yield enhances many folds, reaching sometimes 2.5 t to 4 t ha^{-1}. Analyses indicate that in WANA, nutrient inputs into the agroecosystem have consistently risen during the past 4 decades, causing positive N and P balances in the soil. Incidentally, N and P are the key elements that regulate crop yields in drylands. Intensification of cropping through nutrient inputs seems to continue wherever water resources permit and/or precipitations are congenial. However, in due course nutrient imbalances may creep in if replenishments of all essential nutrients do not occur in correct proportions. Such nutrient imbalances should be forecasted and avoided carefully.

Generally, nutrient recycling through organic residue incorporation or stubble recycling is least in drylands. High soil temperatures may deplete soil organic fraction quickly. This aspect needs due attention, so that organic component, and soil quality are maintained. Integrated plant nutrient supply systems, that envisage use of both organic and inorganic fertilizers may enhance fertilizer-use efficiency, delay soil deterioration, and enhance crop productivity (CAZRI, 2000).

Many of these observations and suggestions made here could be equally applicable to dryland ecosytems in different continents, such as in northwest India, southern India, China, Southern Africa etc.

REFERENCES

Afandi, F., Trabulsis, N. and Saxena, M.C. (1997). Biological nitrogen fixation by cool season legume. Impact on wheat productivity in Syria and Lebanon. *In:* Accomplishments and future challenges in dryland soil fertility research in the Mediterranean area J. Ryan (ed.), International Center for Agricultural Research in Dry Areas, Aleppo, Syria, pp. 88-94.

Amar, B. (1997). Phosphate use potential in Mediterranean countries. *In:* Accomplishments and future challenges in dryland soil fertility research in the Mediterranean areas. J. Ryan (ed.), International Center for Agricultural Research in Dry Areas, Syria, pp. 3-8.

Asfary, F. and Charanek, A. (1997). Nitrogen fertilzer-use efficiency studies by Syrian Atomic Energy commission using ^{15}N-labeled fertilizers. *In:* Accomplishments and future challenges in dryland soil fertility research in the Mediterranian area. J. Ryan (ed.), International Centre for Agricultural Research in Dry Areas, Syria, pp. 64-70.

Avcin and Acvi, M. (1992). Soil, water and inorganic nitrogen accumulation at sowing time of wheat in a two-year rotation as influenced by previous crops under Central Anatolian conditions. *In:* Fertilizer use efficiency under rainfed agriculture in WANA. Proceedings of Fourth Regional Workshop, Agadir, Morocco, ICARDA, Aleppo Syria, pp. 163-174.

Badraoui, M., Soudi, B., Moujahid, Y., Bennani, F., Bouhlassa and Mikou, M. (1997). Mineralogical considerations in soil fertility management in Morocco. *In:* Accomplishments and future challenges in dryland soil fertility research in the

Mediterranean areas. J. Ryan (ed.), International Center for Agricultural Research in Dry Areas, Syria, pp. 3-8.

Bloom, T.M., Sylvester-Bradley, R., Vaidhyanathan, L.V. and Murray, A.W.A (1988). Apparent recovery of fertilizer nitrogen by winter wheat. *In:* Nitrogen efficiency in agricultural soils. D.S. Tenkinson and K.A. Smith (eds.). Elsevier Applied Science, Amsterdam, pp. 27-45.

Bouabid, R., Badaoui, M. and Bloom, P.R. (1991). Potassium fixation and charge characteristics of soil clays. *Soil Sci. Soc. Am. J.,* **55:** 1493-1498.

CAZRI (2000). Integrated farming systems research—Annual report. Central Arid Zone Research Institute, Jodhpur, India, pp. 35-45.

Christiensen, S., Nersoyan, N., Goodchild, A., Nordblum, T., Shome, F., Thomson, E., Bahnady, F., Gintzburger, G., Osman, A. and Singer, M. (1997). Evolution of ICARDA's longterm cereal/pasture and forage legume rotation trials. *In:* Accomplishments and future challenges in dryland soil fertility research in the Mediterranean areas. J. Ryan (ed.), International Center for Agricultural Research in Dry areas, Syria, pp. 118-127.

El-Gharous, M. (1993). Long term N fertilization effect on N mineralization potential under different cropping systems and deficient environments. Ph.D. Dissertation, Agronomy Department, Oklahoma State University, Stillwater, OK, USA.

El-Gharous, M. (1997). Nitrogen challenges in drylands. *In:* Accomplishments and future challenges in dryland soil fertility research in the Mediterranean area. J. Ryan (ed.), International Center for Research in Dry Areas. Syria, pp. 47-56.

El-Gharous, M., Westerman, R. and Soltanpour, P.N. (1990). Nitrogen mineralization potential of arid and semi-arid soils of Morocco. *Soil. Sci. Soc. Am. J.,* **54:** 438-443.

El-Mujahid, K. (1993). Effect of N on yield, N uptake and water use efficiency of wheat in rotation systems under semi-arid conditions of Morocco. Ph.D. Dissertation, University of Nebraska, Lincoln, Nebraska, USA.

FAOSTAT (1999). Food and Agricultural Organization of the United Nations Statistical Databases. Online at *http://apps.fao.org.*

Garabet, S., Wood, M. and Ryan, J. (1997). Field estimates of nitrogen-use efficiency by irrigated and rainfed wheat in a Mediterrenean type climate. *In:* Accomplishments and future challenges in dryland soil fertility research in Mediterranean area J. Ryan (ed.), International Center for Agriculture Research in Dry Areas, Syria, pp. 95-106.

Greenwood, E.A.N., Goodale, D.W. and Titmanis, Z.V. (1965). The measurement of nitrogen deficiency in grass swards. *Plant and Soil,* **23:** 97-103.

Haghihatti, A. (1997). Synthesis of dryland fertilizer trials in western Iran. *In:* Accomplishments and future challenges in dryland soil fertility research in the Mediterranean area J. Ryan (ed.). International Center for Agricultural Research in Dry Areas, Syria, pp. 322-323.

Haidar, A.A. (1997). Soil fertility research in Yemen: Nitrogen phosphorus. *In:* Accomplishments and further challenges in dryland soil fertility research in the Mediterranean area J. Ryan (ed.), International Center for Agricultural Research in Dry Areas, Syria, pp. 169-174.

Hamadullah, G. (1997). Fertilizer use in the Near East: changing patterns in Egypt and Saudi Arabia. *In:* Accomplishments and further challenges in dryland soil fertility research in Mediterranean area J. Ryan (ed.), International Center for Agricultural Research in Dry Areas, Syria, pp. 35-43.

Harris, H.C. (1995). Long-term trails on soil and corp management at ICARDA. *Advances in Soil Science,* **19:** 447-467.

Hasbany, R., Atallah, T., Garabet, S. and Ryan, J. (1997). Laboratory assessment of nitrogen in a wheat based rotation trail. *In:* Accomplishments and future challenges in dryland soil fertility research in Mediterranean area. J. Ryan (ed.). International Center for Agricultural Research in Dry Areas, Syria, pp. 107-117.

Hegde, B.R., Krishne Gowda, L.T. and Parvathappa, H.C. (1988). Organic residue management in red soils under dryland conditions. *In:* National symposium on recent advances in dryland agriculture, CRIDA, Hyderabad.

Hera, C. (1997). Use of isotopics in agriculture with special reference to soil fertility and plant nutrients. *In:* Accomplishments and future challenges in dryland soil fertility research in Mediterranean area. J. Ryan (ed.), International Center for Agricultural Research in Dry Areas, Syria, pp. 21-34.

IAEA (1974). Isotope studies in wheat fertilization. Technical report series No. 157. International Atomic Energy Agency, Vienna.

ICARDA. (1993). Productivity of crop rotations. *In:* Farm resource management program—Annual report—1992. International Center for Agricultural Research in Dry Areas, Syria, pp. 137-166.

ICARDA (1994). Annual report (1993). International Center for Agricultural Research in Dry Areas, Syria.

ICARDA (1995). Annual report—1994 International Center for Agricultural Research in Dry Areas, Syria.

Janat, M., Strochlein, J. and Ryan, J. (1997). Use of radioactive phosphorus in soil fertility resource; mobility in an acidified soil. *In:* Accomplishments and future challenges in dryland soil fertility research in Mediterranean area, J. Ryan (ed.), International Center for Agricultural Research in Dry Areas, Syria, pp. 181-188.

Jones, C.A. (1990). Use of crop simulation models in dryland agriculture. *Advances in Soil Science,* **13:** 332-345.

Jones, M. (1997). Soil and crop management research for sustainable improvement in agricultural productivity. *In:* Accomplishments and future challenges in dryland soil fertility research in Mediterranean area. J. Ryan (ed.). International Center for Agricultural Research in Dry Areas, Syria, pp. 123-134.

Joshi, D.C. (2000). Morphogenetic characterization of gypsiform soils of arid Rajasthan. *J. Indian Soc. Soil Sci.,* **48:** 134-139.

Katyal, J.C. and Reddy, K.C.K. (1997). Plant nutrient supply needs: Rainfed food crops. *In:* Plant nutrient needs, supply, efficiency and policy issues. 2000-2025. J.S. Kanwar and J.C. Katyal (eds.), National Academy of Agricultural Sciences, New Delhi, pp. 91-113.

Katyal, J.C. and Sharma, B.D. (1991) DTPA-extractable and total Zn, Cu, Mn and Fe in Indian soil and their association with soil properties. *Geoderma,* **49:** 165-79.

Katyal, J.C., Sharma, K.L., Narayana Reddy, M. and Neelaveni, K. (1999). Integrated nutrient management strategies for dryland farming systems. *In:* Fifty years of dryland agricultural research in India. Central Research Institute for Dryland Agriculture, Hyderabad, India, pp. 357-368.

Katyal, J.C., Venkateshwarulu, B. and Das, S.K. (1994). Biofertilizers for nutrient supplementation in dryland agriculture: Potentials and problems. *Fertilizer News,* **39:** 27-32.

Kissel, D.E., Ritchie, J.T. and Richardson, C. W. (1992). A stress day concept to improve nitrogen fertilizer utilization: Dryland sorghum in Texas Black land Prairie. Texas Agric. Exp. Stn. MP-201.

Kothari, S.K., Merschner, H. and Rimheld, V. (1991). Contribution of the VA mycorrhizal hyphae in the acquisition of phosphorus and zinc by maize and bean under green house conditions. *Plant and Soil,* **131:** 177-185.

Krauss, A. (1997). Potassium 'the forgotten element' in West Asia and North Africa. *In:* Accomplishments and future challenges in dryland soil fertility research in Mediterranean area. J. Ryan (ed.), International Center for Agricultural Research in Dry Areas, Syria pp. 9-20.

Krauss, A. and Malik, D. M. (1995). The effect of K application to genetically different soil series on K status of soils and plants and yield in Punjab/Pakistan. NFDC Technical Bulletin. Islamabad.

Krishna, K.R. and Lee, K.K. (1987). Management of vesicular arbuscular mycorrhiza in tropical cereals. *In:* Mycorrhizal in the next decade. D.M. Sylvia, L.L. Hang and J.H. Graham (eds.). University of Florida, Gainesville, Florida, USA, pp. 43-45.

Lamsalek, M. (1992). Contribution a l etude methode de raisonnement de la fertilizatuion azotee du ble en irrique (cas de Tadla). Memoire de 3 kme cycle. 1AV Hassan 11, Rabat, Morocco.

Lata, Saksena, A.K. and Tilak, K.V.B.R. (2002). Biofertilizers to augment soil fertility and crop production. *In:* Soil fertility and Crops production. Krishna, K.R. (ed.). Science Publishers Inc. New Hampshire, USA, pp. 279-312.

Matar, A.E. and Brown, S.C. (1989). Effect of rate and method of placement on productivity of durum wheat in Mediterranean climate II. Root distribution and P dynamics. *Fertilizer Research,* **20:** 83-88

Matar, A., Torrent, J. and Ryan, J. (1992). Soil and fertilizer phosphorus and crop responses the dryland Mediterranean zone. *Adv. in Soil Science,* **18:** 82-147.

Mohammed, M.J.A. (1993). Wheat growth and P uptake responses to mycorrhizal inoculation and deep P placement. Ph.D. Thesis. Washington State University, WA, USA.

Mohammed, M.J. and Rawajfih, Z. (1997). Optimizing phosphorus nutrition for dryland cereals. *In:* Accomplishments and future challenges in dryland soil fertility research in Mediterranean area. J. Ryan (ed.), International Center for Agricultural Research in Dry Areas, Syria, pp. 162-168.

Monem, M.A., Azzaoin, A., El-Gharous, M., Ryan, J. And Soltanpur, P. (1990a). Residual nitrogen and phosphorus for dryland wheat in central Morocco. Proceedings of third regional soil test calibration workshop, in Amman, Jordan. International Center for Agricultural Research in Dry Areas, Aleppo, Syria, pp. 163-174.

Monem, M.A., Azzaoni, A., El-Gharous, M., Ryan, J. Soltanpur, P.N. (1990b). Fertilizer placement of dryland wheat in central Morocco. Proceedings of III Regional workshop in Amman, Jordan. J. Ryan and Matar, A. (eds.) International Center for Agricultural Research in Dry Areas, Aleppo, Syria, pp. 149-162.

Monem, M.A. and Ryan, J. (1997). Nitrogen fertilizer use efficiency in WANA as determined by ^{15}N techniques. *In:* Accomplishments and future challenges in dryland soil fertility research in Mediterranean area. J. Ryan (ed.), International Center for Agricultural Research in Dry Areas, Syria, pp. 57-63.

Moujahid, Y. (1993). Mineralogic de argiles et exchange K-Ca dan les Vertisol du Morac: Incidences sur la fertilite potassique des sols. M.Sc., thesis, Universite Mohammed V, Faculty of Resources, Rabat, Morocco.

Nielsen, N.E., Schjorring, J.K. and Jenson, H.E. (1990). Efficiency of fertilizer nitrogen uptake by spring barley. *In:* Nitrogen efficiency in agricultural soils. D.S. Jenkinson and K.A. Smith (eds.), Elsevier Applied Science, Amsterdam, Holland, pp. 62-72.

Orphanos, P.I. (1992). Barley response to nitrogen fertilization under rainfed agriculture in West Asia and North Africa. Proc. Fourth Regional Workshop, Agadir, Morocco. International Center for Agricultural Research in Dry Areas, Aleppo, Syria, pp. 169-176.

Osman A.E. (1997). Potential benefits of phosphorus fertilizer on marginal lands. *In:* Accomplishments and future challenges in dryland soil fertility research in Mediterranean area. J. Ryan, (ed.), International Center for Agricultural Research in Dry Areas, Syria, pp. 154-161.

Pala, M. (1997). Use of models to enhance nitrogen use by wheat. *In:* Accomplishments and future challenges in dryland soil fertility research in Mediterranean area. J. Ryan (ed.), International Center for Agricultural Research in Dry Areas, Syria, pp. 135-144.

Peoples, M.B., Herridge, D.F. and Ladha, J.K. (1996). Biological nitrogen fixation: an efficient source of nitrogen for sustainable agricultural production. *Plant and Soil,* **174:** 3-28.

Power, J.F. (1990) Fertility management and nutrient cycling. *Advances in Soil Science,* **13:** 131-151.

Prihar, S.S. and Gajri, P.R. (1988). Recent advances in dryland agriculture. Central Research Institute for Dryland Agriculture, Hyderabad, India.

Ryan, J. (1997). Accomplishments and future challenges in dryland soil fertility research in Mediterranean area. International Centre for Agricultural Research in Dry Areas, Syria, pp. 31-70.

Ryan, J. (2001). A perspective on available soil nutrients and fertilizer use in relation to crop production in the Mediterranean area. *In:* Soil fertility and crop production. K.R. Krishna (ed.) (in press).

Ryan, J., Masri, S., Garabet, S. and Harris, H. (1997a). Changes in organic matter and nitrogen with a cereal legume rotational trial. *In:* Accomplishments and future challenges in dryland soil fertility research in Mediterranean area. J. Ryan (ed.), International Centre for Agricultural Research in Dry Areas, Syria, pp. 79-87.

Ryan, J., Mashi, S. and Pala, M. (1997b). Residual and current effects of phosphorus in rotational trials. *In:* Accomplishments and future challenges in dryland soil fertility research in Mediterranean area. J. Ryan (ed.), International Centre for Agricultural Research in Dry Areas, Syria, pp. 175-180.

Ryan, J. and Strochlien, J.L. (1979). Sulfuric acid treatment of calcareous soils: effects on phosphorus solubility, inorganic phosphorus solubility, inorganic phosphorus form and plant growth. *Soil Sci. Soc. Am. J.,* **43:** 731-735.

Ryan, J. and Zhgard, A. (1980). Phosphorus transformations with age in a calcareous soil chronosequence. *Soil. Sci. Soc. Am. J.,* **44:** 168-169.

Simmons, A.D. (1988). Studies on nitrogen use efficiency in cereals. *In:* Nitrogen efficiency in agricultural soils. D.S. Jenkinson and K.A. Smith (eds.), Elsevier Applied Science, Amsterdam, Holland, pp. 110-129.

Sheta, T.H., Gobran, G.R., Dufey, J.E. and Laudebert, H. (1981). Sodium-calcium exchange in Nile-Delta soils: Single value for Vaneslow and Gaines-Thomos selectivity co-efficients. *Soil. Sci. Soc. Am. J.,* **45:** 749-756.

Singh, R.P. (1991). Problems and prospects of dryland agriculture in India. Central Research Institute for Dryland Agriculture, Hyderabad, India, pp. 267-275.

Singh, R.P. and Das, S.K. (1986). Prospects of fertilizer use in drylands. *Fertilizer Industry,* **14:** 181-190.

Singh, R.P. and Das, S.K. (1990). Nutrient management in drylands. *In:* Soil fertility and fertilizer use. V. Kumar (ed.), IFFCO, New Delhi, pp. 207-219.

Singh, R.P. and Das, S.K. (1995). Soil fertility and fertilizer management *In:* Sustainable development of dryland agriculture in India. R.P. Singh (ed.), Science Publishers, Jodhpur, India, pp. 117-138.

Singh, H.P., Sharma, K.L., Venkateshwarulu, B. and Neelaveni, K. (1999b). Fertilizers in rainfed areas: problems and potentials. *Fertilizer News,* **44:** 27-38.

Singh, H.P., Sharma, K.L., Venkateshwarulu, T., Vishnumurthy and Neelaveni, K. (1999a). Prospects of Indian agriculture with special reference to nutrient management under rainfed systems. *In:* Long-term soil fertility management through integrated plant nutrient supply. A. Swarup, D.D. Reddy and R.N. Prasad (eds.), Indian Institute of Soil Science, Bhopal, India, pp. 34-50.

Soltanpur, P.V., El-Gharous, M., Azzaoni, A. and Monem M.A. (1989). Soil test based N recommendation for dryland wheat. *Communication in Soil Science and Plant Analysis*, **20:** 1053-1068.

Soudi, B., Chiang, C.N. and Zerouli, M. (1990). Variation saisonniere de l'azote mineral et effect combine de la temeperature it del'humidite du sol sur la mineralization. *Actes Inst. Agron. Vet.* **10:** 27-38.

Swindale, L.D. (1982). Review of soil research in India. Transactions of 12th Congress of International Soil Science Society Plenary Sessions, New Delhi, pp. 67-100.

Sylvia, D.M. and Williams, S.E. (1992). Vesicular-arbuscular mycorrhizae and environmental stress. *In:* Mycorrhizae in sustainable agriculture. *American Soc. of Agron.*, **54:** 101-124.

Torrent, J., Barron, V. and Schoonhoven, V. (1989). Phosphate adsorption and desorption by goethitics differing in crystal morphology. *Soil Sci. Soc Am. J. pp.* 100-102.

Tributh, H., Baguslavshi, E.V., Lieres, A.V. and Mengel, K. (1987). Effects of potassium removal by crops on transformation of Illitic clay minerals. *Soil Sci.,* **143:** 404-409.

Van Schoonhoven, A. (1989). Soil and crop management for improved water use efficiency in dry areas: The challenge. *In:* Proceedings of soil and crop management for improved water use efficiency in rain fed areas. H. Harris (ed.). International Centre for Agricultural Research in Dry Areas. Ankara, Turkey, pp. 3-8.

Venkateshwarulu, B. and Mani, S.P. (1999). Biofertilizers: An important component of integrated plant nutrient supply (IPNS) in drylands. *In:* Fifty years of dryland agricultural research in India. H.P Singh, Y.S. Ramakrishna, K.L. Sharma, and B. Venkateshwarulu. (eds.), Central Research Institute for Dryland Agriculture, Hyderabad, India, pp. 379-394.

Virmani, S.M. (1991). Agricultural climate of semi-arid India: Some issues, problems and solutions. *In:* Dryland agriculture in India. Central Research Institute for Dryland Agriculture, Hyderabad, India, pp. 259-265.

Wood, M., Pilbeam, C., McNeil, A. and Harris, H. (1997). Nitrogen cycling in dryland cereals-legume rotation system. *In:* Accomplishments and future challenges in dryland soil fertility research in Mediterranean area. J. Ryan (ed.), International Centre for Agricultural Research in Dry Areas, Syria, pp. 71-78.

CHAPTER 7

The Semi-arid Tropical Agroecosystem
Nutrient Dynamics

CONTENTS

1. Introduction
2. Pearl Millet Agroecosystem in Sahelian West Africa
   A. Pearl millet agro-environment
   B. Nutrient dynamics
   C. Soil organic matter
   D. Nutrients versus water
   E. Cropping system, physiology and nutrient dynamics
3. Sorghum-based Cropping System in Semi-Arid Tropics
   A. Expanse, agro-climate and soils
   B. Nutrient dynamics.
   C. Integrated nutrient management and productivity
   D. Soil microbes and nutrient dynamics
4. The Legumes Ecosystem in Semi-Arid Tropics
   A. SAT legumes and their agro-environment
   B. Nutrient dynamics
5. Concluding Remarks

1. Introduction

The definitions and classifications of ecogeographical regions termed as semi-arid tropics (SAT) have varied depending on the criteria and purpose. Broadly, considering the agricultural purpose, Vleck and Mokwunye (1990) include zones with 2 to 7 wet months, wherein precipitation exceeds evapotranspiration as SAT. Drier SAT are characterized by mono-model rainfall distribution, with rainy season lasting a mere 3 to 4.5 months, and averaging between 250 mm and 700 mm annually (Troll, 1966). In this book, drier SAT are discussed in chapter 6 titled 'Dryland Agroecosystem'. Hence, the main focus here is on wetter SAT, wherein precipitation ranges between 400 and 1200 mm annually, and rainfall is bi-model allowing two cropping seasons. Variations in precipitation levels, both in time and space, as well as extended or intermittent droughts are a common feature in SAT. Although not strictly, it is accepted that soil fertility, particularly depreciated levels of N, P and organic C (in sandy soils), and interaction with moisture dictates productivity of this agroecosystem. Of late, continuous cropping without appropriate replenishment of all essential nutrients has caused micronutrient imbalances. The SAT agroecosystem encompasses vast cropping zones in different continents. In Africa, it covers about 485 m ha (Sanchez, 1976), which includes the sub-Saharan belt in West Africa extending from Senegal to Chad. It includes parts of Ethiopia, Kenya in the east, and the fringes of Kalahari in southern Africa. In Asia, SAT cropping systems are conspicuous in western, central and peninsular, India and large tracts in Pakistan. The Brazilian 'Cerrado' is a vast expanse of SAT in Latin America. Larger tracts of SAT also exist in northern Australia. Indeed, SAT is a vast expanse utilized for agricultural activity, wherein nutrient dynamics, ecology and productivity aspects have been studied and reported in great detail. However, within this chapter, only the pearl millet agroecosystem in the Sahelian West Africa, sorghum-based ecosystem in peninsular India, and the legume belt in central and peninsular India are discussed in detail.

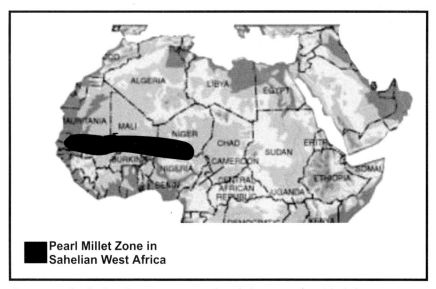

Fig. 7.1 Pearl Millet-based cropping zone in the Sahelian West Africa (Shaded area represents cropping belt where pearl millet is predominent, with cowpea or groundnut as intercrop. Demarcation is not to scale).

2. Pearl Millet Agroecosystem in Sahelian West Africa

> *A perpetuating cropping ecosystem, evolved and mended through the ingenuity of Sahelian population. It is a conspicuous example of crop adaptation and mass food production under harsher climatic, and marginal soil fertility conditions. Since, five millennia, this pearl millet agroecosystem, with its low input based nutrient dynamics has nourished and sustained Sahelian population. This, despite periods of drought, sand storms and poor nutrient status of soils. With greater infusion of improved agro-technologies, moderate expansion and intensification, higher grain productivity and better economic returns seem imminent.*

A. Pearl Millet Agro-environment

The pearl millet based agroecosystem of Sahelian West Africa is a stable, well equilibrated ecosystem lasting since 3rd millennium B.C. The seeds for this pearl millet agroecosystem were sown through the domestication of pearl millet land-races, along the river Niger. Today, pearl millet literally dominates the cropping pattern of almost all West African nations, spreading into an area over 14 m ha (Fig. 7.1; 7.2 A, B). It is the staple carbohydrate source for population residing in the West African Sahel. Pearl millet, in this agrozone, exhibits remarkable tolerance to harsher climatic conditions and poor soil fertility situations. In fact, such physiological traits have allowed it to proliferate, and carve out a niche with a superimposing effect on the general cropping pattern in West Africa.

Unlike intensive cropping ecosystems elsewhere such as rice in Southeast Asia, or wheat in North America and Europe, this pearl millet ecosystem seems to have resisted change towards higher inputs and intensification. Intermittent drought stress and limited supply of fertile soils keeps nutrient turnover at sub-optimal levels, reduces productivity and confines pearl millet to zones with marginal soils. Under such subsistence cropping systems, efficient water use and nutrient cycling are key factors that regulate the ecosystem functions and productivity. However, periods of higher nutrient inputs, rapid fluxes and enhanced nutrient turnover ratios could be expected in near future in the Sahel. In the West African Sahel, pearl millet is cultivated on sandy Aridisols, Ustalfs, Udalfs, Ultisols and Entisols. Generally, nutrient inputs are feeble due to subsistance farming methods. Locations that practice comparatively intensive farming procedures may receive nutrient inputs (40 to 80 kg N and 20 to 40 kg P_2O_5), but it depends on water resources and economic viability. Further, pearl millet is confined to zones with < 600 mm annual rainfall. Greatest fraction of precipitation, nearly 300 to 400 mm is received during June to September. Continuous cropping, without semblance to

Fig. 7.2 (A) Pearl Millet (mono-crop) in the Semi-arid Tropics of Peninsular India (ICRISAT, Hyderabad, India). (B). Pearl Millet/Cowpea inter crop in the Sahelian West Africa (ICRISAT Sahelian Center, Niamey, Niger). Note the differences in the plant morphology, planting density, soil type, its texture, fertility and agronomic practices followed between the two locations. All of the above and several other factors affect the nutrient dynamics with in the two agroecosystems. (Photo courtesy: Dr. K.R. Krishna, ICRISAT, Hyderabad, India)

nutrient replenishment, particularly in the fragile sandy soils leads to deterioration in productivity. In some areas, crop yield has reduced from 300 kg ha^{-1} to 100 kg ha^{-1} (Vleck and Mokwunye, 1990). Average productivity ranges between 600 and 700 kg ha^{-1} in Sahel, which is partly due to neglect of nutrient inputs. Crop residue recycling may increase productivity. In the recent years, introduction of better genotypes and nutrients have increased productivity to around 1000 kg ha^{-1}. During the past two decades, excellent strides have been made in deciphering soil nutrient availability, nutrient dynamics and devising appropriate fertilizer schedules, mainly through national agricultural agencies, and international agricultural organizations such as Office de la Recherche Scientifique et Technique Ontre-Mer (ORSTOM), Paris, France; International Fertilizer Development Centre (IFDC), Muscle Shoals, Alabama, USA; and International Crops Research Institute for the Semi-Arid Tropics (ICRISAT), Patancheru, A.P., India.

B. Nutrient Dynamics

In general, fertilizer based nutrient inputs into this pearl millet agroecosystem is simply too low. When compared with Asian farmers, who utilize, on an average 15 kg inorganic nutrients ha^{-1} on arable fields, West African farmers averaged a paltry 1.6 kg nutrients ha^{-1}. Actually, subsistance farming and meager *in situ* nutrient cycling through stover, stubbles and organic residues are characteristic of this ecosystem. This situation is deemed precarious by scientists, because Sahelian farmers, in reality, will be mining away nutrients through this mono-cropping procedure (ICRISAT, 1998; Shapiro and Sanders, 1998).

Tillage and Soil Nutrient Management: Tillage versus nutrient interactions, their effects on root proliferation, in turn their ability to scavenge soil moisture and nutrients is worth noting (Ikpe *et al.* 1999). It has direct bearing on the nutrient dynamics in the ecosystem. Although soils are sandy, tillage reduces bulk density, conserves soil moisture, and enhances root proliferation (Chapart, 1983). Tillage aids in incorporating organic matter appropriately (Fussel *et al.* 1987), and can enhance water use efficiency. Actually, extensive rooting that ensues after tillage results in exploitation of entire depths of soil profile. Porosity increases due to tillage has ranged from 10 to 20%. While, bulk density decreased from 1.55 or 1.65 t m^{-3} before tillage to 1.3 t m^{-3} after tillage (ICRISAT, 1985). The synergistic effects of tillage and soil fertility management practices is known to induce higher nutrient recovery and crop productivity (Pieri, 1985a, b; Fussel *et al.* 1987; Kouyate *et al.* 2000). Nicol and Chareau (1985) believe that a late season tillage may be crucial to better nutrient and moisture recovery, plus it may help in removing weeds.

Nitrogen: Nitrogen balances reported so far indicate that N losses are significant, and could be due to leaching, denitrification and volatilization. However, the extent and mechanism of N loss may vary depending on location, soil type and environmental parameters encountered within the pearl millet zone. Soil profile analysis indicates that in wet years, N leaching towards lower horizons could be insignificant (Mughogho *et al.* 1986), but in locations where calcium ammonium nitrate was applied, nitrate leaching was severe. Soil nitrogen loss could also be caused by wind erosion (sand drifts). Ammonia volatilization is suspected to be a major cause of N loss in acid sandy soils of West Africa. Warmer temperatures can further enhance N volatilization as ammonia. Denitrification also causes N losses, particularly in drier belts (Westerman and Tucker, 1979). Since, knowledge regarding N dynamics in Sahelian zone is crucial to raise fertilizer N recovery, its use efficiency and productivity, these aspects have been consistently investigated for the past 3 decades (Christianson and Vleck, 1991). Obviously, ^{15}N based studies have been useful in deciphering N dynamics.

Nitrogen losses from soil, estimated using ^{15}N isotope method reached 50%, whereas a mere 25% was recovered in to pearl millet plant. Nearly 10 to 11% of total N recovered by plants was traceable in grains. In another set of study by researchers at ORSTOM in Senegal, ^{15}N recovery into pearl millet ranged between 30 and 32%. Nearly 17 to 21% N was partitioned into seeds and rest retained in stover. ^{15}N loss through volatilization and leaching reached 37 to 45% (Ganry and Guinard, 1979; Gigou and Dubernard, 1979; Table 7.1). Bationo et al. (1989) suggest that splitting fertilizer N dosage appropriately, can help reduce such heavy N losses in Sahelian soils. Band placement of N is known to provide better recovery, and therefore higher fertilizer N efficiency. Whereas, point placement of N fertilizer seems less efficient, perhaps due to greater leaching and volatilization. Regarding distribution of N applied into soil, it is generally accepted that nitrates leached into farther depths in soil. ^{15}N analysis indicates that at harvest, generally much of ^{15}N applied stays at surfaces layers (0 to 15 cm depth). Nitrogen recovery from such surface layers ranged between 15 and 30% depending on the formulation. However, least recovery, between 0 and 2% occurred at depths beyond 50 cm till 175 cm (Bationo et al. 1989). Two major factors, namely rapid percolation of water and dissolved nutrients in sandy soils of Sahel, and dryness that sets in quickly at surface horizons, limits N extraction by roots. Therefore, it reduces N recovery and fertilizer N-use efficiency. Vleck and Mokwunye (1990) suspect that in acidic, highly sandy soils of Sahel, substantial N loss occurred due to volatilization as NH_3. Therefore, appropriate placement of different N sources is crucial. They opine that deep placements may not be a big advantage in reducing N losses.

Table 7.1: Fertilizer-N recovery and partitioning, net N losses and productivity of pearl millet in Sahelian West Africa (Niamey, Niger)

Year	Productivity level (kg ha^{-1})	^{15}N Recovery (%)			Net loss (%)
		Grain	Plant	Soil	
1982	1070	19.0	31.0	37.3	31.7
1983	1040	9.8	22.8	39.2	38.0
1984	470	5.5	20.0	40.1	39.9
Mean	860	11.5	25	38	37

Source: Christianson et al. 1990
Note: Years 1982 and 1983 received good rainfall (> 500 mm), whereas 1984 is drought-stricken at mid-season. N source i.e. urea was banded in split dose.

Phosphorus: Phosphorus is a key element that partly regulates productivity of this pearl millet agroecosystem (ICRISAT, 1997). It is also one of the most studied and investigated nutrients in the Sahelian agro-belt. Its dearth in soil can be so severe that pearl millet seedlings cease to grow, tiller or flower (Pichot and Roche, 1972; Payne *et al.* 1996). The acidic sandy soils are generally deficient in P. Its deficiency is felt throughout the Savanna and Sahelian belt, and all through the pearl millet growing zones. Alfisols, Entisols and the widely distributed Aridisols in West Africa rarely exceed 100 mg P g^{-1} soil. Whereas available P levels (Olsen's or Bray-II P) are usually < 2 to 3 mg g^{-1} soil (Bationo, 1989). Hence, timely P inputs become mandatory. However, P application induces spectacular increases in biomass and grain yield (Bationo, 1985, Bationo *et al.* 1989; Vleck and Mokwunye, 1990; Christianson *et al.* 1991). 'Corrective fertilization' is a concept that may suit this agro-ecozone. It involves an initial large input of P into soil This results in rapid flux of P into the system and sustains crop productivity, although during later years only smaller P inputs take place (Bertrand *et al.* 1972; Vleck and Mokwunye, 1990). Obviously, residual P effects will dictate the extent of P recovery in later years. Low phosphate sorption capacity of these sandy soils may allow extended periods of residual P effects. Repeated P inputs as soluble or rock phosphates even though small can result in gradual accumulation of P. One such evaluation in Niger indicated that available P (Bray-II P) increases from 3.6 ppm to 7.2 ppm for 40 kg P_2O_5 inputs during 3 consecutive years (Bationo *et al.* 1986). Phosphorus inputs through untreated or partially acidulated versions of phosphate rocks are helpful in attaining better P balance. Phosphate rock inputs can be valuable in maintaining soil P fertility status, as well in providing P to pearl millet crop. Its benefits are known to accrue to both the current crop and those cultivated in the next 2 or 3 crop cycles (Bationo *et al.*, 1986; 1989). However, the dynamics of P release into soils, P availability to pearl millet roots and persistence may vary with phosphate rock source. Acidulation of phosphate rocks generally enhances solubility, and available P derived from them. Compared with soluble P fertilizers such as super phosphates, whose effectivity if equated as 100, then, West African phosphates rocks sourced from Parc-W, Kodjari, and Telmsi locations ranged between 60 and 93. (Hammond *et al.* 1986; Chien and Menon, 1995). Overall, on a long-term basis, P inputs either as phosphate rocks or soluble P fertilizers can alter soil P balance positively, and enhance pearl millet productivity. Mycorrhiza enhances P recovery by pearl millet, and it would be useful to understand their contribution to P dynamics in Sahelian zone (Krishna, 1985).

Sulphur inputs can enhance productivity of pearl millet in Sahel. However it is not practiced commonly. Sulphur inputs, quite often occur along with P whenever single super phosphate (SSP) is used. Isotopic analysis has shown that nearly 6 to 25% of S applied through SSP is traceable in stover and grains (IFDC, 1988). In the pearl millet zone, recovery of S from elemental S was 9%. It is much less compared with SSP. About 27% of elemental S input was held in soil, 24% as organic S and 3% retained as sulphate-S. However, fertilizer S recovered into crop could be as high as 78% of sulphate-S which was applied to those sandy soils. Wind erosion, sand drift and rain storms during cropping season accounts for up to 20% sulphate-S loss. Considering that nearly 40 to 50% sulphate-S still remains in soil, residual effects of S could be significant (Vleck and Mokwunye, 1990).

With regard to micronutrients, if sandy soils in Sahel are cultivated continuously, then Zn deficiencies may limit plant growth. Hence, recovery of other nutrients too will be affected. Ordinarily, to produce 1.0 t grain, pearl millet recovers 160 g Fe, 20 g Mn, 40 g Zn and 8 g Cu per hectare. Replenishment through chemicals or stubble recycling will generally sustain micronutrient balance.

C. Soil Organic Matter

Recycling crop residues, and with it the nutrients, as well as growth stimulating factors, if any, is a key procedure in Sahelian zone. Crop residue incorporations enhance soil organic C contents. Often, best levels of nutrient recovery and productivity have been possible by combining inorganic nutrient inputs with residue reincorporation. Such effects have also been spectacular, in many places in the Sahel. Grain and stover productivity increases have ranged from a mere 50 to 100 kg ha^{-1} without organic matter additions, to nearly 2000 kg ha^{-1} with organic matter plus inorganic fertilizer—a 20-fold increase (Bationo *et al.* 1989; Buerkert *et al.* 2000). When assessed at the end of cropping season, organic matter recycling did not significantly influence pH, secondary elements and Al contents. The extent of nutrients returned through stover incorporation may vary depending on the soil management system adopted, crop stage, stover condition (fresh or hay) etc. Approximations indicate that, if stover is incorporated without burning, then straw derived from one tonne biomass (pearl millet) enriches Sahelian soil by 20 kg N, 10 kg P_2O_5, 5 kg S, 10 kg K_2O, 22 kg CaO_2 and 30 kg MgO. This is not a small input by any standard, hence quite crucial to stabilize soil fertility and productivity. One of the striking observations regarding rapidity of nutrient depletion in sandy soils is that, if pearl millet is cultivated without organic matter replenishment for four consecutive years, then plant growth or grain formation practically ceases (Bationo *et al.* 1989).

The sandy soils encountered within the Sahelian pearl millet belt have feeble buffering strength. Therefore, consistently higher productivity may not be a possibility despite continuous nutrient inputs. Since organic matter is low, often <1.0%, nutrient availability, its buffering and dynamics in general are affected unfavorably. Hence, to obtain appropriately congenial soil conditions, favorable nutrient dynamics, consistent and repeated organic replenishments seem necessary. High soil temperature, accompanied with humidity during certain periods of year, supports rapid mineralization. Hence, if stover application is not timed properly, mineralizations are so high that benefits from organic matter may fade away (Vleck and Mokwunye, 1990). Overall, within the Sahelian agroecosystem, maintaining organic matter levels and carbon cycle intact is crucial, if pearl millet productivity is to be sustained. Let alone, pearl millet, even the productivity of general vegetation in Sahel has been immensely influenced by soil carbon dynamics, as well as carbon and other nutrient recycling procedures adopted (Bird and Cali, 1998). It is even suggested that first step in raising agricultural productivity in Sahelian West Africa is to improve fertility and quality (carbon content) of soils (IFPRI, 2002).

D. Nutrients versus Water

According to the law of limiting factor, whichever it is, either nutrients or precipitation will take precedence in influencing productivity of pearl millet agroecosystem. Precipitation levels seem to be the predominant factors whenever optimum soil nutrient levels are reported. Years with optimum or higher rainfall have generally fetched better nutrient recovery and grain productivity. However, the distribution of rainfall events, their timing and intensity also plays crucial role in determining nutrient dynamics in sandy soils. Bationo *et al.* (1989) have observed that mid or late season precipitation (nearly 280 mm) enormously benefited the crop, in terms of nutrient recovery and productivity. To quote an example, under optimum precipitation, 30 kg N input fetched best N-use efficiency of 12 kg grain kg N^{-1} input. Later, Christianson *et al.* (1990) and Vleck and Mokwunye (1989) observed that pearl millet nutrient recovery, stover and grain formation actually depended on 'critical rainfall level' over a 7- to 8-week period in mid-season beginning in July.

E. Cropping System, Physiology and Nutrient Dynamics

Pearl millet in Sahel is usually intercropped. Most common crop associations are millet/cowpea, millet/sorghum, millet/groundnut, millet/maize, millet/sorghum/cowpea. Such intercrops account for 87% of this predominantly pearl millet belt. Pearl millet is usually the first sown crop (Swinton *et al.* 1984; Sawadogo and Kobore, 1984). Obviously, nutrient dynamics encountered with a sole crop pearl millet compared with intercrops differ. Fussel and Serafini (1985) consider that intercrops are more stable in terms of nutrient recovery and grain harvest. Although pearl millet is the dominant crop, its productivity as well as extent of nutrients garnered from soil by it, could be lessened due to the presence of another crop (say cowpea). However, intercrops have often resulted in higher fertilizer use-efficiency and better productivity ranging between 10 and 100% over monocropped zones (Fussel and Serafini, 1985). This has been attributed to the combination wherein pearl millet matures early. Then, cowpea that matures later, allows extended periods of exploitation of nutrients from soil. Fussel (1985) states that such intercrops are also efficient in terms of water use. Better N dynamics are also attributable to biological nitrogen fixation by nodulating cowpea. In addition, variations in planting densities of both pearl millet and/or cowpea can alter nutrient loss from soil, nutrient recovery, nutrient-use efficiency and productivity (Fussel *et al.* 1987). They have suggested that planting density is a crucial aspect that influences intensification of pearl millet based farming systems. Combinations of high nutrient inputs (NPK—40:40:0) and high density planting, act synergistically and induce rapid nutrient turnover in the ecosystem (Fussell *et al.* 1987). However, during practical cropping, Sahelian farmers deliberately plant pearl millet seedlings widely at 1 × 1 m, mainly, to achieve better expression of tillering, tallness, large canopy and to exploit soil fertility and moisture efficiently. Even the recently released pearl millet composites (e.g. CITV) are spaced widely. Pearl millet roots grow exuberantly exploring large volume of soil under each hill. Intensification through, closely spaced, comparatively shorter statured genotypes and high nutrient inputs may have advantages, if conditions permit for its adoption (Bationo *et al.* 1992). As we know, rice or wheat based ecosystems in other continents underwent drastic changes in nutrient dynamics because of densely planted, semi-dwarfs, that utilized high levels of soil nutrients and partitioned it efficiently to seeds. However, such drastic changes in physiological genetics have not been preferred yet by Sahelian farmers. The African pearl millet genotypes are actually taller, possess longer vegetative phase until floral initiation, low in harvest index, hence ratio of nutrients partitioned into seeds will be comparatively low. However, stover could be recycled efficiently. If the aim is to introduce shorter genotypes, with better nutrient partitioning between stover and seeds, then dwarfing (D_2) genes are available. Dwarfing genes could be transferred and utilized, so that appropriate changes in nutrient dynamics could be effected in this pearl millet zone. Again dwarfed pearl millets, and close spacings are still not a preferred practice. Only semidrawfs, at 2 to 3 m tallness, and wide planting are in vogue in Sahel. Perhaps, several factors other than just availability of dwarfs, may influence intensification, e.g. dearth of water. Definitely, close spacing will require greater access to fertilizers, irrigation and other ingredients.

Albeit, there exists a clear possibility to enhance nutrient recovery, nutrient partitioning ratio, yield, and therefore better nutrient turnover ratios by effecting genetic changes in pearl millet genotypes presently adopted in the Sahelian West Africa. Perhaps, it is worthwhile to strive and emulate the kind of changes that occurred in nutrient inputs/recovery and quantum increases in productivity of wheat/rice, and replicate it in this pearl millet belt in Sahel. The genetic gains, if any, for loci related to nutrient recovery and utilization; especially N and P will be helpful. At this point of time, we have no exact idea regarding limits placed by environment and crop physiological genetics to the enhancement of nutrient recovery, use-efficiency and the concomitant raise in pearl millet grain harvests.

3. Sorghum-based Cropping System in SAT

A. Expanse, Agro-climate and Soils

Sorghum is cultivated extensively in the SAT regions of India and Africa. It is generally confined to zones with slightly higher rainfall, when compared with pearl millet agro-belts. In India, 90% of the sorghum based cropping zone is concentrated in peninsular and central zones, wherein temperatures range between 27 and 35°C during the two main cropping seasons, namely 'Kharif' (rainy) and 'Rabi' (post rainy). The West African sorghum ecosystem spans the higher rainfall zone (800-1200 mm per annum) of the sub-Sahara. The cropping season lasts from June to October, when the temperatures range between 25 to 35°C. Sorghum is also an important crop in the SAT regions of southern and eastern Africa. In southern Africa, it is a predominant cereal, sown in September and harvested in February. The sorghum ecosystem also extends into Latin America, mainly the Brazilian Cerrados. It has been gaining in area in central America and southern plains of the United States of America (Unger and Baumhardt, 1999). Within this chapter, the focus, however is confined to Indian sorghum belt with frequent references to West African and southern African sorghum zones.

There are different ways of zonating sorghum ecosystems in India. Recently, Mandal *et al.* (2000) have attempted to classify it based on 'growth index' values (GI). The GI considers water satisfaction index (WI) and solar radiation index (RI). Therefore, GI = WI × RI, which may be utilized in dividing sorghum, ecozones, into namely: a) sorghum humid —SH (GI < 0.15); b) sorghum sub-humid—SSH (GI 0.15 to 0.30); c) sorghum semi-arid—SSA (GI 0.30 to 0.45) and d) sorghum arid—SA (GI > 0.45). The SSH, SSA and SA constitute 60% of sorghum belt in India, and the rest 40% is SH. These growth indices in fact correlate excellently with productivity (Mandal *et al.* 2000), as well as reflect approximately the nutrient status and agro-climatic factors.

Based on such characterizations, it can be easily deciphered that deep soils (Vertisols) with moderate levels of nutrient reserves in humid or sub-humid tracts of Deccan plateau are best suited to sorghum productivity. Perhaps, more accurate demarcations will be possible, if nutrient availability indices of soils are superimposed with the above set of data. After all, soil nutrients are major determinants of sorghum crop sustenance and productivity. Major soil orders on which sorghum is preferred are Vertisols, Vertic soil types (Vertic Inceptisols) and to a certain extent Alfisols. By texture, these soils are predominantly clayey (50 to 70%). Silt contents range between 30 and 40% and sand between 2 and 20%. Vertisol and Vertic intergrades in Deccan plateau, which fall under well irrigated areas are often prone to salinity, sodicity and waterlogging. Such soils are frequent in the peninsular states of Karnataka and Maharashtra. They also possess high smectite clay minerals. These Vertisols are high in CEC ranging between 40 and 70 me 100 g^{-1} soil, which is attributable to montmorillonite clay. Low hydraulic conductivity, and high contents of exchangeable Mg^{+2} and Na^{+2} are other traits of Vertisols. Within soil orders of SAT (Andhra Pradesh), both strongly bonded and complexed forms of Zn (34% of Zn) and Cu (35% of total Cu) are high. Soil pH ranges between 8 and 9, water retention ranges between 23 and 46 kg $soil^{-1}$. Sodium contents, generally increase with depth and ranges from 3.6 to 57 C mol kg^{-1} (Singh, 2001). Nutrients-wise, N, P and Zn are generally deficient compared with crops' demand for it. Nutrient paucity may regulate production to low levels at 0.5 to 1.0 t ha^{-1} in Indian SAT. All other factors being equal, the differences in productivity due to shallowness of the soil could be around 1 t ha^{-1} sorghum grains. Infact, farmers in many locations in SAT India are actually advised regarding N and P inputs, only after judging soil depth and moisture holding capacity (Tandon and Kanwar, 1984). Soil moisture in profile plays a crucial role in regulating nutrient dynamics and productivity of sorghum, particularly in the post-rainy season. Often deeper soils, with better moisture holding capacity support better nutrient recovery compared with shallow soils.

B. Nutrient Dynamics

Overall the inherent nutrient status of soils encountered in sorghum regions in peninsular India is low or marginal with respect to N and P, and is sumptuous or marginal in K. Vertisols are commonly more fertile than Alfisols. Thiagarajan (1998) suggests that better fertility is due to fine texture, and better clay-organic complexes compared with red soils (Alfisols). Vertisols in this sorghum belt are considered productive, because of their better water holding capacity (Sahrawat, 2000). In terms of physio-chemical properties, a typical Vertisol suited best for sorghum cultivation in northern Karnataka in peninsular India exhibits a pH 7.3 to 8.3; Ec 0.21

to 0.72 dS m$^{2-1}$; organic C 4 to 5%; CaCO$_3$ 7 to 13%; particle sizes—coarse sand 8 to 8.5%, fine sand 3.5 to 5.0% and clay 54 to 70% (Venkatesh and Satyanarayana, 1999).

Nutrient Recovery: On an average, a hybrid sorghum crop that yields between 4 and 4.5 t ha^{-1} under well fertilized conditions, removes 80 kg N, 35 kg P$_2$O$_5$ and 120 kg K$_2$O into its stover and grains. Around 45 kg Ca, 30 kg Mg and 32 kg S, and micronutrients Fe (700 g), Mn (450 g), Zn (125 g) and Cu (40 g) are also accumulated into plant tissues. Yet another summarization of several field trails, mainly in Vertisols, Entisol and Alfisols of peninsular India suggests that to produce one ton of sorghum grains, 20 to 25 kg N, 9 kg P$_2$O$_5$ and 30 kg K$_2$O are recovered (Tandon and Kanwar, 1984). In southern Africa, Clegg (1996) reports that sorghum crop removes 30 kg N and K each, and 11 kg P$_2$O$_5$ for every 1000 kg grains produced. Similarly, Kanwar and Youngdahl (1985) reported that for a ton sorghum grain yield, micronutrients 705 g Fe, 53 g Mn, 72 g Zn, 5 g Cu and 2 g Mo are recovered from soil. Obviously, variations in nutrient recovery by sorghum crop should be expected, depending on the genotype, soil environmental parameters, and micronutrient inputs, if any.

Nutrient Balance: An understanding about the net nutrient accumulations, losses and recycling rates is crucial, if the aim is to manage nutrient dynamics within the sorghum ecosystem appropriately, and to achieve a sort of equilibrium. Excessive nutrient inputs may occur resulting in positive balance. Table 7.2 depicts one such situation in sorghum growing zones of Cameroon (West Africa). Vleck and Mokwunye (1990) reported that sizeable amounts of nutrients are recycled into the ecosystem through stover reincorporation. A good fraction, nearly 10 to 15% fertilizer N inputs could be leached out. Potassium recycled through stover can almost satisfy its requirement for the next season.

Table 7.2: An example of nutrient gains and losses from a sorghum based cropping system in sub-Saharan Africa (Cameroon)

C. Nutrients dynamics	N	P$_2$O$_5$	K$_2$O	CaO	MgO
Removal (−)					
Nutrient recovery by crop	−186	−112	−403	−102	−62
Leaching loss from soil	−22	−Tr	−7	−131	−32
Input (+)					
Fertilizer inputs	+ 200	+ 270	+ 300	+ 150	0
Sorghum straw recycling	+ 111	+ 45	+ 393	+ 74	+ 51
Balance	+ 103	+ 203	+ 283	−9	−13

Source: Gigou, 1982 as quoted by Vleck and Mokwunye, 1990
Treatments: Urea at the rate of 50 kg ha^{-1} + 10 t ha^{-1} sorghum straw every 2 years, 45 kg ha^{-1} TSP; and 60 kg ha^{-1} KCl.

Fertilizer N Recovery: Recovery of fertilizer N by sorghum crop may range between 20 and 45%, but nearly equal amount may be lost from soil through leaching, volatilization and denitrification (Ganry and Guinard, 1979; Gigou and Dubernard, 1979). Fertilizer-N use efficiency and recovery may actually depend on variety of factors such as fertilizer placement, formulation, timing etc. Tandon and Kanwar (1984) reported that N recovery by sorghum crop is immensely influenced by precipitation levels and pattern of rainfall distribution. Obviously, in wetter years, or higher rainfall areas (e.g. 1200 mm yr^{-1}), the crop recovered nearly 60 to 63% of N applied. Whereas at 570 mm yr^{-1} precipitation, N recovery dropped to 40%.

Irrespective of whether local landraces or high yielding genotypes were used, sorghum response to P application could be confirmed in 92% of field trials conducted in savanna zone (Mokwunya, 1979). Soil type strongly influences the net P recovery and the P-use efficiency achieved by the sorghum crop in Peninsular India. For example best P-use efficiency is possible in Alfisols (17 to 33 kg grain/kg P) followed by Entisols (11 to 34 kg grain/kg P) and least with Vertisol (7 to 27 kg grain/kg P). Partially acidulated phosphate rocks from local quarry (Kodjari) was agronomically 80% as effective as SSP, when tested on sorghum grown in Burkina-Faso. However, if untreated, its effect in terms of P recovery and grain productivity was equivalent to only 30% of soluble fertilizer (Vleck and Mokwunya, 1990). Under such conditions residual effects are to be expected, because P release is often slow from untreated phosphate rocks. Vertisols that support sorghum cultivation in peninsular India are considered low in P status (Tandon, 1987; Sahrawat, 1999). Residual effects of P are common, however, seasonal variation could be rampant. Sahrawat *et al.* (1995) suggest that simplest indicator to discern residual effects is to compare biomass productivity. Effectiveness of residual P decreases from 80% to 40% during first three years. Such information may be useful. Phosphorus inputs may also influence recovery of micronutrients by sorghum (Sahrawat *et al.* 1998).

Fixation of potassium added as fertilizer or otherwise is an important physico-chemical process in the Vertisols. It affects K dynamics. If surface horizons are considered, then smectite Vertisols and Vertic subgroups fix highest quantities of K, ranging between 26 and 32%. Higher quantity of smectites in clay fraction was attributable to basaltic parent material of southern Indian soils. Alfisols and Inceptisols in SAT fix between 23 and 29% of added K. Amongst sub-surface soils, Illitic groups fixed maximum K (36 to 52%) followed by smectitic group (26 to 36%), then kaolinitic group (17 to 29%). Clearly, mineralogical properties and texture affect K fixation by soils in the sorghum belt of peninsular India.

Sulphur levels in soil, its recovery and utilization are crucial to the productivity of sorghum grown in sub-Saharan Africa. Reports by IFDC scientists working in Niger and Burkina-Faso, suggest that only 6 to 25% of fertilizer based S, supplied through SSP can be traced in sorghum stover and grains. Even considering that a sizeable amount of S is stored in underground root system, which cannot be accounted, the recovery of applied S is indeed less. However, slightly higher recovery was possible by increasing plant population, provided sufficient moisture is available. Sulphur balance studies showed that recovery of S from SSP at 78% is better compared with that from elemental sulphur (37%). Largest fraction, nearly 53% S from SSP was traced in soil as organic-S (29.9%) and Sulphate-S (23%). Within sorghum plants, around 20% of applied 35 S was traceable in stover and 4% in grains. In contrast, only 9% of applied elemental S reached plants parts and 28% ^{35}S got retained in soils as organic-S (25%) or sulphate–S (3%). Obviously, mineralization-immobilization reactions involving S, effective root system to scavenge available-S, and long-term residual effect of S if any, contribute to the agronomic efficiency of S fertilizer application.

C. Integrated Nutrient Management and Productivity

Nutrient, Moisture and Productivity: Water and nutrients are crucial to sustenance and productivity, of any cropping system, in SAT. Clegg (1996) opines that efficient management of these two factors, which are interlinked ensures sorghum productivity in southern Africa. Efficient use of water is of utmost necessity in rainfed sorghum zones. Fertilizer N inputs and appropriate cropping sequences can alter water-use efficiency. In a year with expected precipitation at 720 mm, nitrogen inputs between 50 and 200 kg N ha^{-1} can enhance water use efficiency of sorghum sole crop from 3 to 12 kg grain mm^{-1}, and that of a sorghum in rotation from 9 to 13 kg grain mm^{-1}. In the southern Indian sorghum belt, again levels of precipitation and its distribution can immensely influence recovery of nutrients, particularly N, P and K. In higher rainfall zones (> 1000 to 2000 mm yr^{-1}), this crop recovers 60 to 63% of fertilizer N applied, equivalent 40 to 45 kg N ha. In the post-rainy season, stored moisture can be vital in deciding the nutrient inputs. Average rainfall during rabi, in interaction with N inputs affect net N recovery levels, fertilizer N-use efficiency and productivity. Literally, N dynamics seems to sway depending on moisture levels (Table 7.3). Generally, post-rainy season sorghum yields better in deep soils with higher moisture storage capacity (e.g. 110 mm), when compared with shallow soils (50 mm). In fact, N inputs are decided based on soil depth, because biomass and grain formation increases proportionately, only upto 25 kg N ha^{-1} in shallow soils, but responses continue upto 50 kg N ha^{-1} in deeper soils. Clearly, soil moisture storage capacity is the underlying factor. There are indeed several reports which confirm that moisture and N inputs are interrelated aspects in SAT sorghum zones (Singh *et al.* 2000). Recent reports from the peninsular

Indian sorghum belt indicates that N inputs up to 50 kg N ha^{-1} stimulated yield from 1.2 t to 1.7 t ha^{-1}, at an N-use efficiency of 8 to 8.5 kg grain kg N^{-1}. (Patil and Sheelavantar, 2000). Concomitantly, water-use efficiency increased from 47 kg ha^{-1} cm to 59 kg ha^{-1} cm.

Table 7.3: Nutrient input versus moisture (precipitation) interactions in southern Indian sorghum agroecosystem during post-rainy (rabi) season

Location	Average precipitation (mm)	Agronomic N-use efficiency (kg grain/kg N)	Productivity (t ha^{-1})
Bellary	500	39	1.58
Bijapur	680	50	1.56
Sholapur	722	48	1.45
Kovilpatti	730	70	2.12

Source: Singh et al. 2000

Further, compared with rainy or post-rainy season sorghum grown in peninsular India, the summer crop is well fertilized and irrigated. Mainly because, nutrient recovery and productivity of summer sorghum is much higher at 6 to 7 t grain ha^{-1}. Analyses indicate that irrigation supplements are crucial and it accounts for 1.0 t ha^{-1}, while soil fertility alone (restricted irrigation) provides for 2 to 3 t grain ha^{-1}. Soil nutrients and irrigation (yearly supplements) fetches best productivity at 7 t ha^{-1}. Nutrient recovery could be further enhanced by intercropping system that includes groundnut or any other legume.

Soil Management: Soil nutrient management procedures influence both the nutrient-use efficiency and productivity levels attained. The traditional methods of sorghum farming may result in lowered levels of fertilizer-use efficiency compared with improved methods. For example at Hyderabad and Bellary, which are in the Vertisol zones of peninsular India, if nutrient-use efficiency varied between 9 and 20 kg grain per kg nutrient (NPK) with traditional farming methods, then, it ranged between 21 and 31 kg grain per kg nutrients with improved methods. Proportionate changes in total grain productivity were clearly discernible, which jumped from 0.47 t ha^{-1} to 2.75 t ha^{-1} due to improved procedures (Yadav *et al.* 1998). An earlier summarization of over 30 years data revealed that productivity averaged only 800 kg ha^{-1}, if traditional soil management procedures were adopted. Whereas, mere low key improved procedures, even under low fertility condition enhanced nutrient use-efficiency to 42 kg grain ha^{-1} and raised grain harvest to 2.6 t ha^{-1}. Further, high fertility and well tuned improved soil management altered nutrient dynamics through better fertilizer-use efficiency (47 kg grain kg nutrient^{-1}), and grain harvests raised to 4.7 t ha^{-1} (ICRISAT, 1983; CRIDA, 1997; Singh *et al.* 2000). Yadav *et al.* (1998) have cautioned that despite several years of research, and improved procedures adopted by SAT farmers, farming systems that aim at better nutrient and water-use efficiency needs higher priority over several other aspects of sorghum farming.

In the SAT, sorghum is intercropped and sequenced in several different combinations, each with its unique effect on nutrient dynamics and productivity levels attained. Most commonly, sorghum is intercropped with legumes such as cowpea, horse gram, pigeon pea, groundnut, mung bean, soybean, urd bean, or field bean. Intercropping with legumes provides N credits to the system through symbiotic N fixation. Nitrogen credits from legume to sorghum may average between 30 and 70 kg N ha^{-1} depending on legume species (Lata et al. 2002). Legume stover with better C : N ratios upon recycling benefits soil nutrient storage (Mishra et al. 1996; Clegg, 1996). Under such farming systems, productivity of sorghum grown alone will be inferior to crop combinations. Productivity of sorghum based cropping system is then influenced by a combination of factors, such as intercrop, nutrient input combinations, soil and environment (Mahakulkar et al. 1998). Evaluation in the Vertisol belt (Bijapur, India) suggests that green manures such as subabul, glyricidia or sun hemp, in combination with inorganic fertilizers provided better infiltration rates (1.2 cm hr^{-1}), and water use efficiency (Hundekar et al. 1999). Conjoint use of inorganic N and coirpith also enhanced N recovery by sorghum (Duraisami et al. 2000).

In SAT India, over 200 t each of *Azospirillum* and *Azotobacter* based biofertilizers are used annually on cereals including sorghum (Biswas et al. 2001). We may therefore expect sizeable quantity of N inputs through asymbiotic N fixation in cereal rhizospheres. Nitrogen inputs through these microbes induced higher levels of sorghum forage yield, ranging between 40 and 60 q ha^{-1} more over uninoculated control. Nitrogen gain through such asymbiotic nitrogen fixation however will depend on effectivity of microbes, soil and environmental conditions.

Soil K reserves and its availability in SAT is so high, that despite 20 cycles of cropping, response to K inputs were not discernible. However, mining K from SAT soils has been rampant. According to Murugappan et al. (1999), intensive cultivation of land, with inappropriate K replenishments has disturbed the equilibrium. In particular, in Indian SAT yearly K removal through crops is 11.6 mt yr^{-1}, but fertilizer replenishment totaled only 1.2 mt yr^{-1}. It is equivalent to only 5% of K removals. Hence, in the long-run we can forecast K deficiency and nutrient imbalance.

Nutrient management within sorghum cropping ecosystem should aim at stabilizing, firstly the nutrient input/output relations, nutrient availability, nutrient recycling, and nutrient turnover rates, because these aspects influence the ecosystem productivity. Long-term sorghum based cropping may also influence soil properties, nutrient transformations and nutrient ratios, which could be congenial or detrimental. To quote an example, evaluation of soils (chromostols) from southern India, after 14 years of sorghum based sequences suggested that availability of major nutrients, namely NPK and organic C levels were enhanced, whenever soils were regularly fertilized with N and P. Such nutrient accumulations were greatest if major nutrients NPK were supplemented with organic manures (Table 7.4). Similar observations were made in other locations in this sorghum belt. Further, there is every possibility that ratios of various N fractions, such as NO_3–N, NH_4–N, organic–N, similarly total P and its fractions such as available P, Al–P, Fe–P and total available K, exchangeable and non–exchangeable K will increase due to continuous sorghum cropping (Santhy et al. 1999). Overall, inorganic nutrient inputs along with organic manures can sustain nutrient availability at appropriate levels.

Table 7.4: Influence of long-term sorghum cropping for 14 years (1982 to 1996) on soil nutrient status of a typic Chromostert in southern India (Tamil Nadu)

Treatment	Org C (g kg^{-1})	Available N (kg ha^{-1})	Available P (kg ha^{-1})	Available K (kg ha^{-1})	DTPA-Zn (mg kg^{-1})
Control (no Inputs)	4.2	107	8.2	433	0.71
Nitrogen (40 kg ha^{-1}) + Phosphorus (20 kg ha^{-1})	4.6	125	8.8	445	0.68
FYM + N (20 kg ha^{-1})	4.8	126	10.5	484	0.85
FYM (5 t ha^{-1})	5.9	115	9.7	483	0.07

Source: Data excerpted from Suresh et al. 1999 and is based on a long-term sorghum rotation trial in a Vertisol in SAT (731 mm rainfall).

D. Soil Microbes and Nutrient Dynamics

Sorghum rhizospheres harbor a dynamic population of microbes that mediate different types of nutrient transformations. Most importantly, they may influence nutrient availability to plant roots. The microbial loads in the rhizospheres are generally higher, and it may aid higher rates of nutrient transformations. Patil and Varade (1998) state that dominant microbial species in sorghum rhizosphere may vary depending on soil, crop genotype and environmental factors. Fertilizer based inorganic nutrient inputs and FYM definitely influence rhizosphere microflora and their functions. Asymbiotic nitrogen fixing microbes found in the rhizosphere, and in outer soil may provide N credits to soil ecosystem. Depending on genotype and soil parameters, inoculation with efficient strains of asymbiotic N fixer (*Azotobacter, Azospirillum*) may add between 30 and 60 kg N ha^{-1} (Lata et al. 2002).

Soil microbes found in the sorghum rihzospheres are involved in yet another aspect of nutrient dynamics, namely phosphate solubilization.

Such a biochemical activity enhances P availability to sorghum roots. Inoculation of sorghum seedlings grown on Vertisols or Alfisols with efficient strains of phosphate solubilizers (e.g. *Aspergillus awamore, Pseudomonas striata, Bacillus polymixa*) can solubilize low grade phosphates and enhance plant growth proportionately (Quereshi and Narayanaswamy, 1999a, b). Definitely, a portion of soil P transformation is affected by P solubilizers, but significance of this microbial function to the P dynamics, in general, within sorghum ecosystem needs to be ascertained and highlighted if significant.

Arbuscular mycorrhizal (AM) fungi too contribute to soil P dynamics of the sorghum ecosystem. These symbiotic fungi influence P transport positively at soil-plant root interphase. On an average, in a P deficient Alfisol or Vertisol, AM fungi can enhance P extraction from soil by 5 to 10 kg P ha^{-1}, with proportionate higher plant growth rates (Krishna and Bagyaraj, 1981; Seetharama *et al.* 1987; ICRISAT, 1983; Raju *et al.* 1990). Quantification of such P mobilization effects suggests that, it may vary depending on AM fungal isolates, soil, its background P status and other environmental factors. These AM fungi scavenge, only the available P and sparingly soluble P fraction from soil, and are not known to extract it from chemically bound or fixed P portions. If extrapolated to sorghum cropping belt as a whole, natural AM flora should be mediating transfer of a sizeable amount of P traced in the crop. There are clear indications that, a change in sorghum genotype/cultivar alters AM activity, particularly its influence on P transport and acquisition by crop (Krishna and Lee, 1987). The AM mediated P absorption by sorghum is also influenced by soil factors such as temperature and moisture. Series of field trails in low-P containing Alfisols and Vertisols at ICRISAT (southern India), revealed that AM fungi enhance P recovery by sorghum. Also, natural AM flora may often be effective in mobilizing P (personal observations). Clearly, in nature, AM flora are important soil microbial components that influence P dynamics.

4. The Legumes Ecosystem in SAT

A. SAT Legumes and Their Agro-environment

The emphasis under this section is on SAT legumes predominantly cultivated in India, namely pigeonpea, groundnut, chickpea, green gram, black gram, and cowpea. On a smaller scale, other legumes such as moth bean, cluster bean, horse gram, and field bean also form this SAT legume agroecosystem. Soybean is an import legume in central India, because of its wide adaptability to this SAT zone and better market value (Dubey *et al.* 1996). These legumes are most frequently sequenced or intercropped with cereals. In India, large mono-cropping belts of legumes are not common, excepting the sizeable patches of groundnut area in western India (Gujarat) and peninsular India, and larger pigeonpea expanses in central and peninsular India. This legume belt in SAT is however the largest covering 33% of global legume area, but its production has not matched the requirements of the populace, mainly because productivity of rainfed SAT legumes is fairly low at 500 to 600 kg ha^{-1}. The need for legume grains in SAT India is high because of large population, and it equates to 22% of global legume needs. Obviously, it has caused decline in per capita availability of protienaceous grains from 70 g in 1956 to 34 g in 2000 A.D. It could be deleterious to human health, unless supplemented via other sources. Further, Masood Ali and Mishra (2000) state that projected consumption of legumes in India are 22.3 mt in 2001, 25.7 mt in 2011, and 30.2 by 2021 A.D. Currently, legume grain production through this belt is short by 2.6 mt, if it has to catch up with stipulated requirement of 55 g legume per capita per day. Hence, intensification through enhanced nutrient inputs become a necessity. Area expansions over the present 24 m ha, perhaps will be achieved by rotating or intercropping legumes more frequently. Whatever be the agricultural strategy, nutrients and their dynamics will play the vital role in this legume-based cropping agroecosystem in future years.

Agroclimate and Soils: Pigeonpea cultivated in India is a warm season crop (kharif). The cropping season precipitation ranges between 300 and 520 mm, where annual rainfall may be 700 to 800 mm. Temperatures fluctuate from 20 to 40°C, but optimum is 25 to 32°C. The water-use efficiency (WUE) of SAT legumes is approximately 500 kg water kg dm^{-1}, compared with cereals at 300 to 350 kg water kg dm^{-1}. Pigeonpeas are cultivated mostly under rain-fed conditions, and with high density planting, it demands 55 to 60 cm. It consumes nearly 20 to 25 cm water to form 1 t grain. Pigeonpea is grown in wide range of soil types in India, including Vertisols, Entisols, Inceptisols, Ultisols and Oxisols. However, this crop thrives best in Vertisols of southern and central India, where it is a predominant legume (Hunsigi and Krishna, 1998). Among SAT crops, pigeonpea tolerates salinity and alkalinity only to a certain extent.

Groundnut is an important protein and fat source in Asia, Africa and America. Large patches of groundnut cropping zones occur in West African nations namely Senegal, Mali, Burkina-Faso and Nigeria. The West African groundnut season lasts from June till October. Annual precipitation ranges from 600 to 1000 mm, and the essential ET is 550 mm. In Senegal, groundnut is cultivated on sandy Oxisols, Ultisols, Alfisols and Entisols. In West Africa, groundnut-pearl millet intercrops are frequent. It is a dominant oilseed crop is SAT India, covering the western plains zone (Gujarat) and peninsular India (Karnataka, Maharashtra,

Andhra Pradesh and Tamil Nadu) (Fig. 7.3). Temperature in this groundnut belt varies between 20 and 30°C, and annual precipitation ranges between 700 and 1400 mm. Average rainfall during crop season in some parts of southern Indian groundnut belt varies between 350 and 450 mm. Mean pod yield ranges between 0.5 and 3.0 t ha^{-1} depending on frequency of drought and precipitation pattern. In India, groundnut is a major dryland crop, grown on Alfisols, Usterts and Oxisols. Loamy soils (Ustifluvents) in western Indian groundnut belt are moderately prone to erosional losses (with it nutrients) because of erratic rainfall distribution, poor soil structure and low organic matter (Sutaria and Patil, 1999). Cropping season includes both the kharif (rainy season, June to October) and rabi (post-rainy season, November to February), as well as in summer (March to June). Salinity in soil and/or irrigation water is encountered in western India, and in Vertisol belt of peninsular India. It is deleterious to legume productivity. It reduces seedling emergence, supports only inferior crop stand, and alters nutrient dynamics in soil (Tripathi and Praveen Kumar, 1999).

Fig. 7.3 Groundnut mono-crop in the Semi-arid Tropics of Peninsular India. (Photo courtesy; Dr KR Krishna, ICRISAT, Hyderabad, India)

B. Nutrient Dynamics

Within reasonable limits of variation, the extent of nutrients recovered by SAT legumes are dependent on species, geographic location and soils, fertilizer-based nutrient inputs etc. There are indeed several reports on this aspect. To quote one, Ahlawat and Masood Ali (1993) reported that to produce one ton biomass, legume crops remove 30 to 50 kg N, 2 to 7 kg P_2O_5, 12 to 30 kg K_2O, 3 to 10 kg Ca, 1 to 5 kg Mg, 1-3 kg S, 200 to 500 g Mn, 5 g B and 1 to 5 g Cu. Another recent estimate indicates that on an average SAT legumes produce 0.9 to 2.0 t grain ha^{-1}. To produce it, they recover 60 to 80 kg N t grain 6 to 14 kg P t grain^{-1}, and 20 to 50 kg K t grain^{-1} (Masood Ali and Mishra, 2000; Hegde, 2000). However, individual legume species or its genotypes may actually remove different levels of nutrients, depending once again, on yield goals, farming practices and fertilizer inputs. Recovery of P by SAT legumes is generally poor in Vertisols than in Alfisols, which is often attributed to excessive sorption of phosphate. However, certain observations indicate Vertisols in SAT India have comparatively low phosphate sorption capacity, than Oxisols or Ultisols, in the same agroclimate (Sahrawat, 1999). Recently, feeble responses of SAT legumes to added P in Vertisols were attributed to higher P buffering capacity. Since, this inference pertains to surface horizons of Vertisols, behavior of P in deeper horizons of Vertisols needs examination (Sahrawat, 1999). While it is agreed that potassium and its forms vary widely in SAT soils utilized for legume production, K distribution depends largely on the homogeneity of K bearing material. Analysis indicate that fractions such as water soluble K, exchangeable K, non-exchangeable and total K indicate existence of a dynamic equilibrium between these forms (Tewatia *et al.* 1989; see Krishna, 2001). While water soluble K fraction correlates well with exchangeable K levels, the non-exchangeable and total K were dependant on clay and silt contents. Knowledge regarding forms of K, their distribution and chemical dynamics is perhaps crucial to legume K nutrition and productivity (Yadav *et al.* 1999). Grain legumes, in general, recover larger quantities of K. Quite often, productivity enhancements are attributable to large K recovery into grains via retranslocation from vegetative tissues (Yahiya *et al.* 1996).

The need to study and understand sulphur nutrition of legumes, and its dynamics in soils is gaining priority. Trivedi *et al.* (1998) remind us that knowledge about sulphur status and dynamics in soil, particularly availability of SO_4–S and its chemical transformations is essential in this legume belt. Evaluation of major soil types in the pulse belt, namely Vertisols, Alfisols and Vertic intergrades indicate that all sulphur forms decrease with depth, and is related to organic carbon content. Water soluble S correlated with clay content and its nature, whereas SO_4–S contents were related to total S, organic S and heat soluble S.

Intensive cultivation using high yielding varieties has rampantly depleted S levels. According to Venkatesh and Satyanarayana (1999), if 10 ppm S is considered critical limit, then soil profiles with S deficiency are common. The C : N : S ratios crucial to crop growth and productivity were 100 : 8.7 : 3.2, and the ratios depended on soil parent material, cropping pattern, its intensity and extent of replenishments. In general, legumes may demand higher quantity of S compared with cereals. In SAT India, groundnut which is both fat and protein source, demands much higher quantities of S. Sulphur recovery is dependent on genotypes, and their ability to accumulate it in seeds. Groundnuts grown in sandy loams (Aeric Haplustalfs) recovered on an average 3 to 3.5 kg S ha^{-1} to produce 1 t pod ha^{-1}. The S-use efficiency of grain legumes varies. For example, it ranged between 114 and 290 g grain kg S with green gram. Among, S sources, phosphogypsum is known to allow better S recovery, hence higher fertilizer-S use efficiency (Singh *et al.* 1998).

Cropping Systems and Nutrient Dynamics: In SAT India, legume-based cropping sequences often include a cereal. Legume-cereal or legume-legume intercrops are most frequent. To quote a few examples, Pigeonpea-sorghum, groundnut-sorghum, groundnut-rice sequences etc. Such intercropping systems aim at enhancing land equivalent ratios, C fixation, nutrient recovery and efficiency (Verma and Warsi, 1997; Sidhu *et al.* 1997). Legume intercrops, for example, pigeonpea with black gram, soybean or cluster bean are known to stabilize productivity, provide better nutrient input : output ratio, and help in sustaining minimum essential nutrient dynamics (Yadav *et al.* 1997).

There are innumerable examples, wherein, legume crop provides residual 'N credits' to the succeeding cereal crop. Extent of N credits derived may vary depending on the legume and cereal grown in sequence, the soil and environmental factors. It is interesting to note that N credits derived are also dependent on the genotype of the legume. For example, Ghosh and Dayal (1997) verified this aspect in the major groundnut belt in Gujarat, India, and found that yields of wheat crop that succeeds legume ranged from 2.5 t ha^{-1} to 3.7 t ha^{-1}. Actually, proportionate variations in $NH_4^+- N$, NO_3-N and total N ascertained that, N credits derived from groundnuts were responsible for variation in productivity of wheat. Intercropping is also beneficial because it enhances water use efficiency. Basically, moisture versus nutrient interactions are important in SAT. Water-use efficiency index (kg m^{-3}) was generally higher, whenever groundnut was intercropped with cereals (3.05 kg m^{-3}) compared with groundnut mono-cropping (0.52 kg m^{-3}) (Harsono and Karsono, 1997). Mishra and Rout (1996) generalize that introducing legume in sequence, often thwarts adverse effects of cultivating cereal continuously. If jute or a cereal occurs in sequence with groundnut then depletion in organic C, total N, available P and K were appreciable. Whereas, including black gram instead of a cereal improved organic C, N contents improved by 22 to 46 kg ha^{-1}, available P by 15 to 16 kg ha^{-1} and available K by 13 kg ha^{-1} (Mishra *et al.* 1997).

Planting Density: In addition to variations in legume-based cropping patterns, modifications in planting density, geometry, and soil nutrient management procedures also affect nutrient dynamics. Firstly, phostosynthetic efficiencies and nutrient recovery rates are influenced, then the nutrient-use efficiency. For example, Sharma and Rajput (1996) found that in a pigeonpea-groundnut intercrop, planting density/geometry enhanced land equivalent ratio, nutrient recovery and pigeonpea yield equivalent. It is also possible that higher planting density improves soil physio-chemical properties such as structure, stability index, porosity and hydraulic conductivity, in turn enhancing rooting and nutrient availability (Nagaraju and Singa Rao, 2000). Actually, paired planting geometry at 30/90 cms and intercrop ratio at 1 : 2 (pigeonpea : groundnut) was best in terms of nutrient recovery. Nitrogen credits through rhizobial symbiosis was also higher. In case of legumes sown in central India combinations of agronomic procedures, mainly spacing, planting density and inoculation with N-fixing rhizobium can be crucial to nutrient dynamics. For example, row spacing at 30 cm seems optimum to attain good root spread and better nutrient scavenging ability. It improved N recovery by 10 kg ha^{-1} and P uptake by 1 kg ha^{-1} (Shukla and Dixit, 1996; Singh *et al.* 1996b).

Legume N fixation: Rhizobium species and their role in atmospheric N fixation, extent of N inputs into legume cropping ecosystems in SAT, the residual effects that ensue, and amount of N credits to the succeeding crops have been amply reported since past two decades. Indeed, a vast body of knowledge has been accrued on this aspect, but detailed discussions are not within purview of this chapter. In India alone, over 102 t of Rhizobium based bio-inoculants are utilized during legume cultivation (Biswas *et al.* 2001). According to them, at nominal N fixation rates possible through Rhizobium, effect of such bio-inoculants should be equivalent to 75% total N utilized by all the legume species in SAT India. Obviously, Rhizobium plays a key role in N dynamics of SAT legume belt, and it becomes more relevant in soils generally deficient in N. The extent of N inputs derived through Rhizobium is controlled by its activity, the soil, plant, environmental factors. Rhizobium isolates used, and the agronomic procedures adopted are equally crucial. For example, recent evaluations on chickpea-rhizobium symbiosis suggests that, N increase in the soil, as well as total nutrient (N, P, K) recovery by the crop are highest, if an efficient strain (Rhizobium H 45) and P supplements are practiced (Singh and Mishra, 1998; see Tables 7.5 and 7.6). Yet another soil microbe-legume crop symbiosis is formed with arbuscular–mycorrhizal (AM) fungi. Hence, in nature, fixation of C (photosynthesis), N fixation and phosphate transport are interlinked through a tripartite relation. Again, there are several reports regarding influence of dual inoculation (Rhizobium plus Arbuscular Mycorrhizal fungi) in different legume crops cultivated in SAT (Tilak, 1987; Subba Rao and Krishna, 1988; Mahadevan *et al.* 1989; Lata *et al.* 2002). Both, N and P dynamics are influenced. Particularly, their recovery and balance obtained in soil. While

Table 7.5: Rhizobium and phosphate interactions on chickpeas cultivated in post-rainy season, SAT, India

Treatment	N Credits (kg ha^{-1})	Nutrient uptake by chickpea (kg ha^{-1})		
		N	P	K
Control	137	105	6.7	20.9
Rhizobium BGI	308	141	8.4	28.6
Rhizobium H 45	339	145	9.0	28.8
Rhizobium BGI + 22 kg P ha^{-1}	478	169	9.6	27.2
Rhizobium H 45 + 22 kg P ha^{-1}	478	167	11.9	34.8
SE±	40.3	16.4	1.13	3.52

Source: Singh and Mishra, 1998
Note: N credits refer to amount of N increase in soil because of N-fixing activity by Rhizobium.

Table 7.6: Influence of tripartite symbiotic relationship with Rhizobium (Rh) plus Arbuscular Mycorrhiza (AM), and P fertilizers on N balance

Treatment	N recovery (mg kg^{-1} soil)	Total soil N after harvest (mg kg^{-1} soil)	N balance (mg kg^{-1})	Soil residual P (kg ha^{-1})	Available P utilized by crop (kg ha^{-1})	P balance (kg ha^{-1})
Control	0.4	280	13.7	13.6	0.6	0.6
Mussori Rock PO$_4$ + Rh + AM	1.6	287	21.5	30.4	3.1	0.1
Single super PO$_4$ + Rh + AM	1.5	291	25.6	30.2	2.6	0.8

Source: Das et al. 1999
Note: N balance = [Total soil N after harvest + N uptake]—[initial soil total N + Fertilizer N added]; P balance = [Available P in soil after harvest + P utilized by plant]—[initial available P in soil + P applied]

there are methods to quantify N credits derived from Rhizobium, direct evaluation of AM benefits, in terms of phosphate mobilization has been difficult. However, measuring increase in P contents and biomass has been useful. It is generally accepted that a legume crop, because of tripartite symbiosis expends higher amounts of photosynthetic energy. However, legumes may not suffer any deficiency in photosynthetic rates due to agro-climate. Symbiosis stimulates slightly higher levels of photosynthesis. Hence, it is believed that extra needs for carbohydrates are easily met (Kucey and Paul, 1981).

The legume rhizosphere harbors a diverse population of microbes that mediate several types of soil nutrient transformations. Plant growth promoting rhizobacteria (PGPR) are known to mediate chemical transformations that augment nutrient availability. They secrete siderophores, organic acids and enzymes that enhance nutrient availability for plant roots to absorb. For example, Pal *et al.* (1999) report that certain

Pseudomonads and *Bacillus* traced in Alfisols in western Indian groundnut belt, produced siderophores that preferentially chelates Zn and Fe, enhances P solubilization through organic acids and induces rooting. Hence, it may lead to higher nutrient recovery. In several groundnut fields, PGPR induced higher levels of nodulation and N-fixation, and it ranged from 86 to 108 mg plant^{-1} (Day *et al.* 2000).

Weeds Affect Legume Nutrition: Weeds are an important component during legume production because they divert nutrients and compete for radiation interception. Weeds, if unchecked, definitely alter nutrient dynamics of the legume ecosystem and reduce grain productivity. For example, in a green gram field left unweeded for first 30 days, the grain yield reduced by 35 to 68% (Singh, 1996a; Rana *et al.* 1999). Indeed the legume canopy development versus weed build-up is crucial. In certain places, it is common to intercrop pigeonpea with yet another legume (e.g. green gram) or cereal, mainly because it effectively thwarts weeds. Evaluations in central India indicate that pigeonpea sole crop allowed 16.8 kg ha^{-1} weed biomass to be generated and 42, 7.5, 40 kg ha^{-1} N, P, K to be removed from the soil. Whereas an intercrop with cowpea or greengram allowed only 14.3 kg ha^{-1} weed biomass, and 36, 6, 35 kg ha^{-1} N, P, K respectively (Patil and Pandey, 1996). A good hand weeding in legume field can avoid loss of 50 kg N, 9 kg P and 45 kg K from the system. However, it must be remembered that, if done timely, weed reincorporation can be effective in recycling nutrients. In off-season, weed reincorporation can be a good technique, if we use them as 'catch crops'. In the groundnut belt of western India (Gujarat), weeds can cause severe nutrient imbalance, and remove as much as 26 kg N and 8 kg P (Dev Kumar and Gajendra Kumar, 1999). According to them, two weeding exercises in groundnut fields are equivalent to 11 q ha^{-1} pods, and 80 kg N, 12 kg P, 60 kg K, and 10 kg S per ha^{-1}, which otherwise would have been diverted to weeds.

5. Concluding Remarks

The SAT agroecosystem supports an assortment of crops including cereals, legumes, oilseeds and agroforestry species. Agricultural productivity has been low or moderate owing mainly to soil nutrient and water resource related constraints. Farming is generally subsistent, consequently nutrient inputs, recycling and turnover rates are of lower order. Nutrient dearth, particularly N and P, is greatest in SAT soils. Hence, their replenishments should be consistent, if rapid depletion and imbalances are to be thwarted. As such, crop residue recycling and organic manure inputs are minimal in several locations of SAT. Agronomic procedures that minimize nutrient loss, enhance fertilizer recovery and use efficiency should be devised and utilized on a priority. At present, N recovery by pearl millet/sorghum is low at 30 to 35%, whereas, loss through volatilization/leaching could reach beyond 37 to 45%. Recovery of P into crops is very low at 8 to 20%. Obviously, measures that exploit residual N and P will be useful. For example, a carefully weighed out cropping pattern that scavenges more of fertilizer based nutrients, appropriate fertilizer schedule and placement methods can together enhance fertilizer use efficiency. In this regard, adoption of integrated plant nutrition systems should be considered on a priority. Efforts to maximize 'N credits' derived from legumes will be helpful in terms of nitrogen dynamics.

Low input based nutrient dynamics, subsistence farming and low productivity seems perpetual in Sahelian zone, parts of Indian SAT and elsewhere. Is it true for ever? Perhaps not, because potentially higher crop productivity is a clear possibility. Intensification through enhanced nutrient inputs, well augmented water supply, and an appropriate cropping pattern (genotypes) can fetch the desired levels of productivity, be it in Sahel, SAT India or Brazilian Cerrado. To a good extent intensification of SAT cropping system also depends on governmental policy and economic viability of individual farmers. For such fragile agro-environments, Swaminathan (1998) suggests restoration and intensification of crop productivity through well focused ecoregional agrotechnology and precision farming methods.

Intensification, no doubt, causes upheavals in nutrient dynamics and ecological equilibria that existed since ages. It affects soil biota, and transformations they bring about. Some of these aspects could be tested experimentally. We may realize that, as plant breeders generate new variations with significantly higher yields and introduce them into SAT farming zones, they definitely affect nutrient dynamics. For example, a change in harvest index or root : shoot ratio will proportionately alter nutrient demand, the extent recovered and partitioned into stover and seeds, and that recycled. Accurate computer-based simulations on farm unit or agroecoregion basis, are needed to forecast such effects on nutrient dynamics better.

REFERENCES

Ahlawat, I.P.S. and Masood Ali (1993). Fertilizer management in pulse crops. *In:* Fertilizer management in food crops. H.L.S. Tandon (ed.) FDCO, New Delhi, pp. 114-138.

Bationo, A., Christianson, C.B. and Mokwunye, U. (1989). Soil fertility management of the pearl millet producing sandy soils of Sahelian West Africa: The Niger experience. *In:* Soil, crop and water management systems for rainfed agriculture in the Sudano-Sahelian zone. ICRISAT-Sahelian Center, Niamey, Niger, pp. 159-169.

Bationo, A., Christianson, C.B., Baethgen, W.A. and Mokwunye, A.V. (1992). A farm level evaluation of nitrogen and phosphorus fertilizer use and planting density for pearl millet production in Niger. *Fertilizer Research,* **312:** 175-184.

Bationo, A. (1985). Agronomic evaluation of phosphate fertilizer alternatives in sub-Saharan Africa—An internal report. ICRISAT Sahelian Center, Niamey, Niger.

Bationo, A., Mughogo, S.K. and Mokwunye, U. (1986). Agronomic evaluation of phosphate fertilizers in Tropical Africa. *In:* Management of nitrogen and phosphorus fertilizers in sub-Saharan Africa. A.V. Mokwunye and P.L.G. Vleck (eds.) Martinus Nijhoff, Dordrecht, Netherlands, pp. 213-218.

Bertrand, R., Natios, J. and Vicaire, R. (1972). Exportations minerals par le Mil it. I'Arachide consequences sur la definition. d'une Fumure, de'Entration d'un Sol Ferrgineaux Tropical Developpe sur Materialax soliens and Tarna. (Niger). *Agronomic Tropical,* **27:** 1287-1303.

Bird, M.I. and Cali, J.A. (1998). A million years record of fire in sub-Saharan Africa. *Nature,* **394:** 767-768.

Biswas, B.C., Soumitra Das and Kalwe, S.P. (2001). Crop response to bio-fertilizer. *Fertilizer News,* **46:** 15-24.

Burkert, A., Bationo, A. and Dossa, K. (2000). Mechanism of residue mulch-induced cereal growth increases in West Africa. *Soil Sci. Soc. Am. J.,* **64:** 346-358.

Chien, S.H. and Menon, R.G. (1995). Agronomic evaluation of modified phosphate rock products IPDC's experience. *Fertilizer Research,* **41:** 197-209.

Chapart, J.L. (1983). Elidi du systemi racinaioc du mil (*Pennisetum typhoidis*) dens un soil selslens du Senegal. *Agronomic Tropicale,* **38:** 37-46.

Christianson, C.B., Bationo, A., Henao, J. and Vleck, P.L.G. (1990). Fate and efficiency of N fertilizers in Niger. *Plant and Soil.* **125:** 221-231.

Christianson, C.B. and Vleck, P.L.G. (1991). Alleviating soil fertility constraints to food production in West Africa: Efficiency of nitrogen fertilizers applied to food crops. *In:* Alleviating soil fertility constraints to increased crop productivity. Mokwunye, V. (ed.) Dordrecht, Kluwer Academic Publishers.

Clegg, H.D. (1996). Crop management: Nutrients weeds and rain. *In:* Drought-tolerant crops of southern Africa. K. Leuschner and C.S. Mantha (eds.) International Crops Research Institute for the Semi-Arid Tropics, Patancheru, India, pp. 91-95.

CRIDA (1997). Central Research Institute for Drylands Agriculture—vision 2020—CRIDA perspective plan. Indian Council of Agricultural Research, New Delhi, 80 pp.

Das, D. K., Sathi, A.K., Jena, M.K. and Patra, R.K. (1999). Effect of P sources and dual inoculation of VA mycorrhiza and Rhizobium on dry matter yield and nutrient uptake by Green gram (*Vigna radiata L*). *J. Indian Soc. Soil Sci.,* **47:** 466-470.

Dev Kumar, M. and Gajendra Kumar (1999). Effect of weed control and gypsum application in uptake of N, P, Ca and S by groundnut (*Arachis hypogaea*) and weeds. *Indian J. of Agron.,* **44:** 400-403.

Day, R., Pal, K.I.C., Chauhan, S.M. and Bhatt, D.M. (2000). Field evaluation of plant growth promoting Rhizobacteria of groundnut. *International Arachis Newsletter,* **20:** 77-79.

Dubey, M.P., Sharma, R.S. and Khare, J.P. (1996). Integrated weed management in soybean (*Glycine max*). *Indian J. of Agron.,* **41:** 69-73.

Duraisami, V.P., Raniperumal and Mani, A.K. (2002). Transformations of applied nitrogen in relation to its availability to sorghum in an Inceptisol. *Fertilizer News,* **47:** 13-18.

Fussel, L.K. (1985). Evaluation of millet/cow pea intercropping systems in western Nigeria. Proceedings of the Regime workshop in intercropping in the Sahelian and Sahela-Sudenian zones of West Africa, Niamey, Niger, 26-39 pp.

Fussel, L.K and Serafini, P.G. (1985). Associations de cultures damsles zones tropicales semi airdus d'Afrique ete I'Ovest. *In:* Technologies approprics pour les paysams de zones semi-airedes de lo'Afrique de I'Ovest. Ohm, H.W. and Moges, J.G. (eds.) West Lafayatte, Indiana, pp. 254-278.

Fussel, L.K., Serafini, P.G., Bationo, A. and Klaij, M.C. (1987). Management practices to increase yield and yield stability of Pearl Millet in Africa. *In:* Proceedings of the

International Pearl Millet Workshop. International Crops Research Institute for the Semi-arid Tropics, Patancheru, India, pp. 255-280.

Ganry, F. and Guinard, G. (1979). Mode d'application du fumier et Bilan Azote dans system, Mil-Sol Sableanx du Senegak; Etude all Moyen de L'Azote—15. *In:* Isotopics and Radiation in Research in Soil-Plant Relationships. International Atomic Energy Agency, Vienna, Austria.

Gigoue, J. (1982). Dynamique de l'Azoli Mineral en Sol mi on cultive de Regin Tropiale Sehe du Nord Cameron. These Doc. USTL< Montpellier, France, 171.

Gigoue, J. and Dubernard, J. (1979). Study of the fate of fertilizer nitrogen in a sorghum crop in Northern Cameroon. *In:* Isotopes and radiation in research in soil-plant relationships. International Atomic Energy Agency, Vienna, Austria.

Ghosh, P.K. and Dayal, D. (1997). Variation of yield of succeeding wheat because of various groundnut genotypes. *International Arachis News-letter,* **17:** 64-66.

Hammond, L.L., Chien, S.H. and Mokwunye, U. (1986). Agronomic value of unacidulated and partially acidulated rocks indigenous to the tropics. *Advances in Agronomy,* **40:** 89-140.

Harson, A. and Karsono, S. (1987). Yield and irrigation use efficiency with and without intercropping on Alfisols. *Internationals Arachis Newsletter,* **17:** 56-58.

Hegde, D.M. (2000). Nutrient management in oilseed crops. *Fertilizer News,* **45:** 31-42.

Hundekar, S.T., Badanur, V.P. and Sarangamath, P.A. (1999). Effect of crop residues in combination with fertilizers on soil properties and sorghum yield. *Fertilizer News,* **44:** 59-63.

Hunsigi, G. and Krishna, K.R (1998). Science of Field Crop Production. Oxford and IBH Publishers, New Delhi, 458 pp.

ICRISAT (1983). Annual report—1982. International Crops Research Institute for the Semi-Arid Tropics, Patancheru, Andhra Pradesh, India.

ICRISAT (1985). Annual report—1985. International Crops Research Institute for the Semi-Arid Tropics, Patancheru, Andhra Pradesh India.

ICRISAT (1997). Highlights—1997. International Crops Research Institute for the Semi-Arid Tropics, Patancheru, India, pp. 71-76.

ICRISAT (1998). ICRISAT Highlights—1997. International Crops Research Institute for the Semi-Arid Tropics, Patancheru, Andhra Pradesh, India, pp. 73-75.

IFDC (1988). Annual reports, International Fertilizer Development Centre, Muscle Shoals, Alabama, USA.

IFPRI (2002). Nurturing the soil in Sub-Saharan Africa. International Food Policy Research Institute—News and views. Washington DC, pp. 4-5.

Ikpe, F.N., Powell, J.M., Isiramah, N.O., Wahua, T.A.T. and Ngodigha, E.M. (1999). Effects of primary tillage and soil amendment practices on pearl millet yield and nutrient uptake in the Sahel of West Africa. *Experimental Agriculture,* **35:** 437-448.

Kanwar, J.S and Youngdahl, L.J. (1985). Micronutrient needs of tropical food crops. *In:* Micronutrients in tropical food production. P.L.J. Vleck (ed.) Developments in Plant and Soil Sciences, Martinus Nijhoff, Dordrecht, The Netherlands, pp. 43-67.

Kouyate, Z., Franzleubbers, K., Juo, A.S.R. and Hosner, L.R. (2000). Tillage, crop residue, legume rotation and green manure effects on sorghum and millet yields in the semiarid tropics of Mali. *Plant and Soil,* **225:** 141-151.

Krishna, K.R. (1985). Mycorrhizal symbiosis and phosphorus nutrition of pearl-millet (*Pennisetum americanum* Leeke) in SAT West Africa. International Crops Research Institute for the Semi-Arid Tropics, Patancheru, India, 36 pp.

Krishna, K.R. (2001). Potassium in soils, its transformations. *In.* Soil fertility and crop production, Science Publishers, Enfield, New Hampshire, USA, pp. 141-154.

Krishna, K.R. and Bagyraj, D.J. (1981). Note on the effect of V A mycorrhiza and soluble phosphate fertilizer on sorghum. *Indian J. Agric. Sci.,* **51**: 688-690.

Krishna, K.R and K.K. Lee (1987). Management of vesicular arbuscular mycorrhiza in tropical cereals. *In:* Mycorrhiza in the next-decade. Sylvia, D.M., Hung, L.L. and Graham, J.H. (eds.) University of Florida, Gainesville, Florida, USA, pp. 43-45

Kucey, R.M.N and Paul, E.A. (1981). Carbon flow in plant-microbial associations. *Science,* **213**: 473-474.

Lata, Saxena, A.K. and Tilak, K.V.B.R. (2002). Biofertilizers to augment soil fertility and crop production. *In:* Soil fertility and crop production. K.R. Krishna (ed.) Science Publishers, New Hampshire, USA, pp. 279-312.

Mahadevan, A., Raman, N. and Natarajan, K. (1988). Mycorrhizal for Green Asia. University of Madras, Madras, 351 pp.

Mahakulkar, B.V., Wanjari, S.S., Atale, S.B., Potduka, N.R. and Deshmuka, J.P. (1998). Integrated nutrient management in sorghum (*Sorghum bicolor*) based cropping systems. *Indian J. of Agron.,* **43**: 376-381.

Mandal, C., Mandal, D.K. and Srinivas, C.V. (2000). Agroclimatic classification of sorghum growing areas of India. *J. Indian Soc. Soil Sci.,* **48**: 151-159.

Manjunath A. and Bagyaraj, D.J. (1984). Response of pigeonpea and cowpea to phosphate and dual inoculation with vesicular-arbuscular mycorrhiza and Rhizobium. *Tropical Agriculture,* **61**: 48-32.

Masood Ali and Mishra, J.P. (2000). Nutrient management in pulses and pulse-based cropping systems. *Fertilizer News,* **45**: 57-70.

Mishra, R.K., Choudhary, S.K. and Thripathi, A.K. (1997). Intercropping of cow pea (*Vigna unguiculate*) and horse gram (*Macrotyloma uniflorum*) with sorghum for fodder under rainfed conditions. *Indian J. of Agron.,* **42**: 405-407.

Mishra, M. and Rout, K.K. (1996). Effect of agronomic management practices in soil fertility under jute (*Chorcorus cepularis*)–groundnut (*Arachis hypogeae*)–blackgram (*Phaseolus mungo*) cropping system. *Indian J. of Agron.,* **41**: 546-549.

Mokwunye, U. (1979). Phosphorus needs of soils and crops of the savanna zones of Nigeria. *Phosphorus in Agriculture,* **76**: 87-95.

Mughogho, S.K., Bations, A., Christiansen, B. and Vleek, P.L.G. (1986). Management of nitrogen fertilizers for tropical African soils. Africa. *In:* Management of nitrogen and phosphorus in sub-Saharan Africa. Mokwunye, A.V. and Vleek, P.L.G. (eds.) Developments in Plant and Soil Sciences. Vol. 24 Martinus Nijhoff Publishers, Dordrecht, Netherlands.

Murugappan, V., Santhy, P., Selvi, D., Muthuval, P. and Dhakshina Moorthy, M. (1999). Land degradation due to potassium mining under intensive cropping in semi-arid topics. *Fertilizer News,* **44**: 75-77.

Nagaraju, M.S.S. and Singa Rao, M. (2000). Effect of different plant population of groundnut on soil structure and other physical properties of an Inceptisols. *J. Ind. Soc. Soil Sci.,* **48**: 174-179.

Nicol, R. and Charreau, C. (1985). Travail du sol et economic de l'cau in Africaque de l'Quest. *In:* Technologies approries pour les payans dry zone semi-aircdes de l' Afrique de l'Quest. H.W. Ohm and J.G. Nagraj (eds.) West Lafayette, Indiana, pp. 9-31.

Pal, K.K., Rinku, D., Bhatt, D.M. and Chauhan, S. (1999). Enhancement of groundnut growth and yield by plant growth-promoting rhizobacteria. *International Arachis News Letter,* **19**: 51-53.

Patil, B.M. and Pandey, J. (1996). Chemical weed control in Pigeon pea (*Cajanus cajan*) intercropped with short-duration grain legumes. *Indian J. of Agron.,* **41**: 529-535.

Patil, S.L. and Sheelavantar, M.N. (2000). Yield and yield components of rabi sorghum as influenced by *in situ* moisture conservation practices and integrated nutrient management in Vertisols of semi-Arid tropics of India. *Indian J. of Agron.*, **45**: 132-137.

Patil, R.B. and Varade, P.A. (1998). Microbial population in rhizospheres as influenced by high input rates of fertilizer application to sorghum on a Vertisol. *J. Indian Soc. Soil Sci.,* **46**: 223-227.

Payne, W.A., Drew, M.C., Hossner, L.R. and Lascano, R.J. (1996). Measurement and modeling of photosynthetic response of pearl millet to soil phosphorus addition. *Plant and Soil,* **184**: 67-673.

Pichot, J. and Roche, P.L. (1972). Le Phosphore dans les Sols Tropicaux. *Agronomie Tropical,* **27**: 939-965.

Pieri, C. (1985a). Fertilization des cultures vivirus et fertile des sols en agriculture paysanne sub-sahariene. I'experience de IRAT. *In:* Technologies approprics our les paysons les zones semi-airides del l'Afrique de l'Quest. Ohm, H.W. and Nagy, J.G. (eds.) West Lefayette, Indiana, USA, pp. 254-278.

Pieri, C. (1985b). Conduct de la fertilization les cultures viviries enzones semi-arides. *In:* Secheresse inzones inter tropicale: Prur une lutte integer. Center de co-operation Internationale in Recherché Agronomeque pour le Development. pp 363-381.

Quereshi, A.A. and Narayanswamy, G. (1999a). Quantitative assessment of phosphate solubilizers with phosphates in P deficient Typic Ustochrept. *J. Indian Soc. Soil Sci.,* **47**: 471-474.

Quereshi, A.A. and Narayanswamy, G. (1999b). Direct effect of rock phosphate solubilization with rock phosphates in P deficient Typic Ustochrept. *J. Indian Soc. Soil Sci.,* **47**: 475-478.

Rana, K.S., Mahendra Pal and Rana, D.S. (1999). Nutrient depletion by pigeonpea (*Cajanus cajan*) and weeds as influenced by intercropping systems and weed management under rainfall conditions. *Indian J. of Agron.,* **44**: 267-270.

Raju, P.S., Clark, R.B., Ellis, J.R. and Maranville, J.W. (1990). Efffects of species of VA mycorrhizal fungi in growth and mineral uptake of sorghum at different temperatures. *Plant and Soil,* **121**: 165-170.

Sahrawat, K.L. (1999). Phosphate sorption of bench mark Vertisols and Alfisols profile. *J. Indian Soc. Soil Sci.,* **47**: 144-146.

Sahrawat, K.L. (2000). Criteria for assessment of the residual value of fertilizer phosphorus. *J. Indian. Soc. Soil Sci.,* **48**: 113-118.

Sahrawat, K.L., Rego, T.J., Rahman, M.H., Rao, J.K. and Adam, A. (1995). Response of sorghum to fertilizer phosphorus and its residual value in a Vertisol. *Fertilizer Research,* **41**: 41-47.

Sahrawat, K.L., Rego, T.J., Rehman, M.H. and Rao, J.K. (1998). Phosphorus response effects on macro- and micronutrient removal by sorghum under rainfed cropping on a Vertisol. *J. Indian Soc. Soil Sci.,* **46**: 58-60.

Sanchez, P.A. (1976). Properties and Management of Soils in the Tropics. John Wiley and Sons, New York, USA.

Santhy, P., Velusamy, M.S., Murugappan, V. and Selvi, D. (1999). Effect of inorganic fertilizers and fertilizer manure combinations on soil physico-chemical properties and dynamics of microbial biomass in an Inceptisol. *J. Indian Soc. Soil Sci.,* **47**: 479-482.

Sawadogo. S. and Kobore, M.O. (1984). Le point de la recherché sur les cultures associees en zone Sahelio-Soudanian Burkina-Faso. Proceedings of the regional workshop on intercropping in the Sahelian and Sahelo-Sudanien zones of West Africa. Institut du Sahel, Niamey, Niger, pp. 126-153.

Seetharama, N., Krishna, K.R., Burford, J.R., and Rego, T.S. (1987). Prospects for sorghum improvement for phosphorus efficiency. *In:* Sorghum for acid soils. Proceedings of workshop in evaluating sorghum for tolerance Al toxic tropical soils in Latin America. CIAT, Cali, Colombia, pp. 229-249.

Shapiro, B.I. and Sanders, J.H. (1998). Fertilizer use in semi-arid West Africa: Profitability and supporting policy. *Agricultural Systems,* **56:** 467-482.

Sidhu, M.S., Sikka, R.K. and Kaul, J.N. (1997). Evaluation of summer groundnut based cropping systems for productivity and economic returns in India. *International Arachis Newsletter,* **17:** 54-56.

Singh, M.V. (2001). Evaluation of current micronutrient stocks in different agro-ecological zones for sustainable crop production. *Fertilizer News,* **46:** 25-42.

Singh, K.K. and Mishra, J.P. (1998). Bio-fertilizers in pulses. *Fertilizer News,* **43:** 91-102.

Singh, A.N., Singh, S. and Bhan, V.M. (1996a). Crop-weed competition in summer green gram (*Phaseolus radiata*). *Indian J. of Agron.,* **41:** 616-619.

Singh, D.P., Rajput, A.L. and Singh, S.K. (1996b). Response of French bean (*Phaseolus vulgaris*) to spacing and nitrogen levels. *Indian Journal of Agronomy,* **41:** 608-610.

Singh, H.P., Sharma, K.L., Srinivas, K. and Venkateshwarulu, B. (2000). Nutrient management strategies for dryland farming. *Fertilizer News,* **45:** 43-54.

Singh, K.P., Singh, S.K. and Gautam Kumar (1998). Response of black gram (*Phaseolus mungo*) to sulphur on acid Alfisol of Bihar plains. *J. Indian Soc. Soil Sci.,* **46:** 257-260.

Sharma, R.K. and Rajput, O.P. (1996). Crop geometry and nutrient management in pigeonpea (*Cajanus cajan*) and groundnut (*Arachis hypogaea*) intercropping system. *Indian J. of Agron.,* **41:** 327-331.

Shukla, S.K. and Dixit, S. (1996). Effect of Rhizobium inoculation, plant population and phosphorus on growth and yield of summer green manure (*Phaseolus radiatus*). *Indian J. of Agron.,* **41:** 611-615.

Subba Rao, N. S. and Krishna, K.R. (1988). Interactions between Vesicular-Arbuscular mycorrhiza and nitrogen fixing microorganisms and their influence on plant growth and nutrition. *In:* Biological nitrogen fixation: Recent developments. N.S. Subba Rao. (ed.). Oxford and IBH Publishing Co. pp. 53-70.

Sutaria, G.S. and Patil, N.K. (1999). Run-off and soil losses from loamy sand soils under different cropping systems. *J. Indian Soc. Soil Sci.,* **47:** 129-133.

Suresh, S., Subbramaniam, S. and Chitdeshwar, T. (1999). Effect of long term of fertilizers and manures on yield of Sorghum (*Sorghum bicolor*)—Cumbu (*Pennisetum glaucum*) in rotation on Vertisol and dry farming and soil properties. *J. Indian Soc. Soil Sci.,* **47:** 272-273.

Swaminathan, M.S. (1998). Foreword. *In:* Science of Field Crop Production. G. Hunsigi and K.R. Krishna. (eds.). Oxford and IBH Publishing Co. Pvt. Ltd., New Delhi, pp. 433.

Swinton, S.M., Numa, G. and Samba, L.A. (1984). Les cultures associees en melicu paysan dries deux regions du Nigere. Proceedings of the Regional workshop in intercropping in the Sahelian and Sahelo-Sudanian zone of West Africa. Institut du Sahel, Niamey, Niger, pp. 183-184.

Tandon, H.L.S. (1987). Phosphorus research and agricultural production in India. Fertilizer Development and Consultation Organization, New Delhi.

Tandon, H.L.S. and Kanwar, J.S. (1984). A review of fertilizer use research on sorghum in India. International Crops Research Institute for the Semi-Arid Tropics, Patancheru, Andhra Pradesh, India, pp. 1-60.

Tewatia, R.K., Singh, N., Ghabru, K.K. and Singh, M. (1989). Potassium content and mineralogical composition of some salt-affected soils. *J. Indian Soc. Soil Sci.,* **37:** 687-691.

Thiagarajan, T.M. (1998). Distribution of organic carbon in some Alfisol and Vertisol profiles of Tamil Nadu uplands. *J. Indian Soc. Soil Sci.,* **46:** 446-448.

Tilak, K.V.B.R. (1987). Interactions of vesicular-arbuscular mycorrhiza and nitrogen fixers. *In:* Soil biology. Haryana Agricultural University, Hissar, Haryana, India, pp. 219-226.

Trivedi, S.K., Bansal, K.N. and Singh, V.B. (1998). Important forms of sulphur in profiles of some soil series of northern Madhya Pradesh. *J. Indian. Soc. Soil Sci.,* **46:** 579-583.

Troll, C. (1966). Seasonal climates of the Earth. *In:* World maps of climatology. E. Rodenwelt, and J.H. Juatz (eds.). Springer Verlag, Berlin, New York.

Tripathi, K.P. and Praveen-Kumar (1999). Influence of saline water on growth and nutrient composition of mung bean (*Vigna radiata* L). *J. Indian Soc. Soil Sci.,* **47:** 383-385.

Unger, P.W. and Baumhardt, L.R. (1999). Factors related to dryland sorghum grain yield increases: 1939 through 1997. *Agronomy Journal,* **91:** 870-875.

Venkatesh, M.S. and Satyanarayana, T. (1999). Sulphur fractions and C : N : S relationships in oilseed growing Vertisols of North Karnataka. *J. Indian Soc. Soil Sci.,* **47:** 241-248.

Verma, K.P. and Warsi, A.S. (1997). Production potential of pigeon pea (*Cajanus cajan*) based intercropping systems under rainfed conditions. *Indian J. of Agron.,* **42:** 419-421.

Vleck, P.G. and Mokwunye, A.V. (1990). Soil fertility problems in semi-arid tropics of Africa. *In:* Soil fertility and fertilizer management in the semi-arid tropical India. B. Christianson (ed.). International Fertilizer Development Center, Muscle Shoals, Alabama, pp. 29-46.

Westerman, R.L. and Tucker, T.C. (1979). *In situ* transformations of nitrogen-15 labeled materials in Samoran Desert soils. *Soil Sci. Soc. Am. J.,* **43:** 95-100.

Yadav, J.S.P., Singh, A.K. and Rattan, R.K. (1998). Water and nutrient management in sustainable agriculture. *Fertilizer News,* **43:** 107-117.

Yadav, N.S., Verma, R.S., Trivedi, S.K. and Bansal, K.N. (1999). Vertical distribution of forms of potassium in some soil service of Vertisols of Madhya Pradesh. *J. Indian Soc. Soil Sci.,* **47:** 431-436.

Yadav, R.P., Sharma, R.K. and Shrivastava, V.K. (1997). Fertility management in pigeonpea based intercropping system under rainfed conditions. *Indian J. of Agron.,* **42:** 46-49.

Yahiya, M., Samiullah Khan and Hayat, S. (1996). Influence of potassium on growth and yield of pigeon pea (*Cajanas cajan*). *Indian J. of Agron.,* **41:** 416-419.

CHAPTER 8

The Humid Tropical Agroecosystem
Nutrient Dynamics

Contents

1. Introduction
2. Nutrient Dynamics
    A. Nutrient inputs
    B. Nutrient recovery
    C. Nutrient losses
    D. Long-term studies on soil exhaustion
    E. Fallows in humid tropics
    F. Crop physiological aspects
    G. Mycorrhiza, phosphate rocks (PR) and P dynamics
3. Concluding remarks

1. Introduction

Humid tropics extend on either side of the equatorial belt from 0 to 25° latitude. The vegetation includes a diverse species of perennial forest tree species, shrubs, annuals and farming belts. In this chapter our focus confines to cropping zones. Cereals (e.g. maize, rice, sorghum), root crops (cassava, yam, sweet potato) and legumes (e.g. beans, cowpea, soybean) are common to this agroecosystem. Globally, warm humid and sub-humid tropical agricultural zones extend into nearly 11% of total agricultural area, wherein 30.4% of global population resides (Wood *et al*. 2000). Among the various crops cultivated in humid tropics, the emphasis within this chapter has been confined to cassava, for the sake of brevity of this book. Cassava is an important crop preferred by the farming community because of its low nutrient and energy input requirements. In many parts of humid tropical Africa, Latin America and Asia, roots and tubers (e.g. cassava) constitute important, supplemental sources of carbohydrates, vitamins and amino acids. Frequent shortages of chemical fertilizer inputs, and limited irrigation facilities encountered by farmers in humid tropics makes cassava the appropriate choice. Globally, cassava-based agroecosystem extends into 20 m ha, and is expected to increase to 24 m ha by 2020. The present production at 175 mt is expected to reach 290 mt by 2020 (IFRRI, 1998; FAOSTAT, 1999). According to Scott *et al.* (2000) increase in the cassava growing belt and its consumption has been rapid, at a growth rate of 3.1% per year in tropical Africa.

Agroclimate: The humid tropics may also be defined as those regions, where precipitation and infiltration exceeds crop/soil evaporation for at least 10 months per year. Mean air temperatures in humid tropics may range between 20 and 33°C annually, sometimes reaching as high as 45°C. Seasonal changes in temperatures are generally small. Tropical crops display a broad temperature optimum. According to Norman *et al.* (1995), the approximate cardinal temperatures for tropical crops are 8 to 12°C as basal temperature, below which its physiological functions are affected. Optimum temperatures range between 20 and 38°C, and higher critical temperatures vary between 35 and 45°C depending on the species and its genotypes. Humid tropics enjoy year round positive radiation balance, with relatively small seasonal changes. However, the minor changes in radiation can influence crops in atleast two ways. Mainly, optimum plant population and potential grain/root yield vary, being higher in high radiation cloudless seasons, than in wet cloudy monsoon period (Norman *et al.* 1995). Radiation influences yield components. Diurnal variations are minimal. In the equatorial zones it is 12.1 h during all seasons, and ranges from 10.6 to 13.7 h at 25° latitude. High total rainfall (>2500 mm yr^{-1}), and equally high kinetic energy of rain, lead to high kinetic energy load, which is characteristic of humid tropics, receiving >2000 mm average annual rainfall. Whenever, rainfall intensity exceeds the filtration capacity of soil, and rainfall has high kinetic energy, there is likelihood of erosive runoff. In humid tropics rainfall is erosive at 40%, compared to 5% in temperate zones. This is attributed to heavy rainfall events. For example, in central Nigeria 58% of the annual rainfall is erosive.

Soils: Farmers in humid tropics encounter a wide range of soil types. Generally, it is the Ferralsols (Oxisols) and Acrisols (Alfisols) that predominate in tropical South American and African agricultural zones. Lixisols (Ustalfs) and Nitisols (Alfisols) are frequent in South and Southeast Asia. Vertisols are distributed in parts of peninsular India, northeast Australian tropics and elsewhere in African humid tropics. In terms of expanse, Oxisols and Ultisols are widespread in Latin American countries, mainly in Columbia, Peru and parts of Brazil, that receive >2200 mm annual precipitation. Here, the average annual temperatures fluctuate between 24 and 30°C (Szott and Palm, 1996). According Leon *et al.* (1986), very low P availability and Al toxicity are the soil maladies that influence nutrient dynamics inappropriately in almost 50% of the farming belt in humid South America. In the Asian tropics, >300 m ha, equivalent to 33% of humid agricultural belt are infertile due to rampant acidity. Asadu and Enete (1997) consider soil fertility, mainly impaired nutrient availability as the major limiting factor to crop productivity in tropical Africa. In this zone, land area equivalent to 27% suffers acidity and Al toxicity that disturbs normal levels of nutrient recovery and productivity. Overall, 38% of humid tropical soils need both acidity corrective measure and nutrient inputs to maintain optimum levels of agroecosystem productivity.

Deficiencies of essential nutrients are also encountered in humid tropical soils. Hence, soil nutrient dynamics is influenced inappropriately, causing imbalances in nutrient ratios, chemical fixation etc. Sanchez and Cochran (1980) report that soil nutrient loss due to erosion affects 40 to 50% of the South American humid tropics. Rampant major nutrient deficiency (N and P) may affect optimum nutrient dynamics and productivity in over 80% of Alfisols and Inceptisols. In the Brazilian humid tropics, and in 'Cerrados', nearly 67 to 95% of soils exhibit micronutrient dearth that can retard crop growth (Leon *et al*. 1986). Liming to reduce acidity, Al toxicity, and to enhance P availability in soil is an important soil management technique.

Soils within the humid tropical zones of Africa, Asia and Latin America, wherever annual precipitation is high, the erosional losses of nutrients and organic fraction can be severe. Excessive tillage, nutrient imbalances due to excessive use of major nutrients and poor irrigation management can impair nutrient dynamics and reduce crop productivity (Leon *et al*. 1986). Soil erosion can alter nutrient status detrimentally on as much as 50% of Alfisols and Ultisol areas in America (Sanchez and Cochran 1980). Again, in the humid tropics of South Pacific, Oxisols and Oxic Inceptisols are heavily leached due to high levels of precipitation (2000 to 4000 mm yr^{-1}). Such excessive leaching of top soils decreases exchangeable bases, and leads to deficiencies of Ca and Mg. On the other hand, proportionately higher levels of Al and Mn, along with acidity induces P fixation. Liming alters Ca/Al + Mn ratios favorably enhancing P absorption and restoring productivity. Soils in the humid belts of southern and eastern India, as well as in Sri Lanka are ferruginous, lateritic, acidic in reaction (pH 4.3 to 5.6), and are rich in Al and Fe salts. They are classified as Haplohumults, Humitorepts or Ustic Paleohumults. The heavy rainfall component ranging from 2500 to 4000 mm yearly can cause heavy leaching of essential nutrients (Karche *et al*. 1999). Root growth gets restricted due to Al^+ toxicity. Hence, liming forms an important soil nutrient management practice.

Cassava-based Cropping Systems: Among the various crops cultivated in humid tropics, such as maize, rice, beans, cassava, banana, oil palm, rubber and other tree crops, the major emphasis within this chapter is on cassava-based cropping systems. Nutrient dynamics in humid tropical forests, and that in humid rice culture zones have been dealt separately in chapters 4 and 10. Cassava is cultivated throughout the humid tropics. In general, P is the key element that limits optimum cassava productivity in Latin America (Brazil, Columbia and Peru). Mainly because, its availability is severely diminished due to fixation by Al^{+3} and Mn^{+2} fractions. Whereas, in West Africa and South Asia, N and/or P dearth limits cassava growth and productivity (Howeler and Cadavid, 1990). Actually, in the cassava producing countries, fertile soils are commonly allocated to more profitable crops, leaving those with problems such as high Al^+, low base exchange capacity and high P fixation for cassava cultivation (Lopez *et al.* 1995). Average exchangeable Al may range between 0.5 and 3.1 m eq 1000^{-1} soil and pH <5.2. Percent Al saturation in the cassava belt ranges between 35 and 88%. Cassava is generally well adapted to acid soils, because of its high levels of tolerance to Al in soil solution. However, at very low pH (4.2 to 4.5) conditions, and Al saturation beyond 85% it does suffer from toxicity. It is interesting to note that despite acidity (pH to 3.4 to 4.0) low Al containing peat soils in Malaysian humid belt supports good growth of cassava. Liming at 0.5 to 2.0 t ha^{-1}, often corrects Al toxicity.

Moving water immediately after a heavy rainfall event is the prime soil eroding factor in all of the humid tropics. Primarily, the surface layer rich in mineral nutrients and organic fraction crucial to tropical crops is lost swiftly. Proportionate reductions in nutrient recovery by cassava, beans or other crops should be expected. However in such eroded soil, nutrients are actually deposited down stream and, at that location accumulated nutrients could be recovered by crops. Therefore, restoring the usual nutrient cycling within the agroecosystem. Shifting cultivation too induces soil erosion. Despite it, some farmers practicing shifting cultivation may avoid severe soil erosion and nutrient loss, because of the short duration for which the soils are cultivated.

A wide range of cropping sequences and intercropping procedures are adopted that encompass both, intensive and subsistence farming situations in humid tropics. Broadly, they can be grouped into shifting cultivation, semi-intensive rainfed systems, intensive rainfed systems, irrigated and flooded systems, mixed annual/perennial systems (Norman *et al.* 1995). They involve important cereals (maize, rice) and legumes (beans, cow pea) grown in sequence, or in intercropped situations. Intercropping cereals/legumes with cassava, banana or tree crops is common in humid tropics. Generally, mixed cropping systems are preferred. The timing of crop sequences/mixtures are usually dependent on precipitation patterns. As stated earlier this chapter focuses on cassava, which has reputation as a 'rustic crop' because it grows reasonably well even under adverse soils and humid climatic conditions. Cassava is a perennial crop that stays in the field for 9 to 24 months. For successful cassava cultivation annual temperatures above 18°C and precipitation >100 cm are required. It is grown from 0 up to 2000 m altitude. Once established, this crop can withstand extended periods of drought up to 2 to 3 months (Howeler, 1989). The cassava based cropping ecosystem is more intense in Tropical Africa (12 m ha), Latin America (2.7 m ha) and Southeast Asia (3.5 mm). Globally, this cassava agroecosystem is projected to increase from the present 19 m ha to 23 m ha by 2020 (Scott *et al.* 2000). Proportionate productivity increases from the present 173 mt yr^{-1} may reach 275 mt yr^{-1} in 2020. This necessitates enhanced levels of nutrient inputs, removal by crop, recycling and turnover. The agroecosystem equilibrium in terms of nutrients will get proportionately disturbed. Scott *et al.* (2000) predict that in addition to cassava, area under other carbohydrate fetching crops too will increase during this period. However, for certain other crops such as yam, or sweet potato, both area and productivity levels may dwindle. Overall, inherent soil fertility, nutrient inputs, and competition from other carbohydrates/protein sources will influence cropping patterns in the humid tropics in the future.

2. Nutrient Dynamics

Previous paragraphs contain general information on tropical soils. Acidity, Al and Mn toxicity, erosion and perceivably low soil fertility are major concerns in humid tropics. The physico-chemical transformations of essential nutrients that occur in these soils, are in principle, similar to those explained in chapters 2, 3, 4 and 5. Hence, to avoid repetitions and monotony, discussions here are confined to aspects such as nutrient inputs, recovery by crops, losses, physiological aspects and agronomic procedures that affect nutrient dynamics, and involvement of mycorrhizas in P dynamics. At this point, we may realize that a few other aspects such as carbon turnover rates, symbiotic/asymbiotic N fixation are equally important, although not discussed here. Treatises that deal exclusively with these aspects are available.

A. Nutrient Inputs

Major avenues of nutrient inputs into the tropical cropping ecosystem are from weathering of soil parent material, through atmospheric deposition, chemical and organic manure inputs, fluxes caused by residues and litter.

Nitrogen inputs from atmospheric deposits and/or weathering may not be significant, but sizeable amounts of K, Ca and Mg are derived from such natural processes. Kellman and Tackaberry (1997) estimate that per ha, 0.2 kg P, 7.6 kg K, 7.6 kg Ca and 3.4 kg Mg are added yearly into humid tropical soils through natural weathering process. Similarly, annual inputs from atmospheric depositions are estimated at 0.3 kg N, 0.2 kg P, 3.7 kg K, 2.2 kg Ca and 0.4 kg Mg per ha. Nutrient inputs into the tropical agroecosystem through chemical fertilizers/farm yard manure may vary depending on crop, soil fertility status, yield potential and yield goals. Howeler and Cadavid (1990) summarize that for optimum root yields, N:P:K inputs range at 60:20:90 in West Africa, 40:30:30 in Brazil and 120:22:32 in Southeast Asia. Additional nutrient inputs are also made through FYM, which is applied at 5 to 10 t ha^{-1}. Green manuring, whenever practiced is supposed to supply 200 kg N ha^{-1}, and proportionate quantities of other elements upon incorporation (Tongglum et al. 1992). Nutrients channeled into farms could be based on tissue analysis, in order to avoid imbalances that may occur. If blanket recommendations are adopted for cassava, then sufficiency range is about 5.1 to 5.8% N, 0.4% P, and 1.4 to 1.1% K. To achieve this sufficiency range in tissue, Howeler (1995) suggests N, P, K inputs at 50 to 100 kg N ha^{-1}, 25 to 50 kg P ha^{-1} and 50 to 100 kg K ha^{-1}. Obviously, nutrient inputs vary widely for other tropical crops, and extent of N inputs though symbiotic nitrogen fixation by tropical legumes needs due consideration. A wide range of plant species are capable of N fixation in tropical environments, mainly herbaceous leguminous crops, cover crop (green manure), leguminous shrubs and trees. Cereals fix N asymbiotically through rhizosphere microflora. The ability of plant species to fix atmospheric N varies widely, depending on soil, plant and environmental parameters. Previous summarization indicate that field legumes, such as beans, cowpeas or soybeans may fix 3 to 230 kg N ha^{-1}. Cover crops such as *Centrosema, Peureria,* or *Stylosanthes* may fix up to 150, 23, and 84 kg N ha^{-1} respectively. Shrubs, like *Accacia albida* (20 kg N ha^{-1}), *Allocesurine littoralis* (220 kg N ha^{-1}), *Casurina equistifolia* (60 to 110 kg N ha^{-1}), *Leucana leucociphala* (10 to 500 kg N ha^{-1}) also contribute N to the humid tropical agroecosystems (Peoples and Crasswell, 1992; Lata et al. 2002). Again, N inputs into cereal rhizospheres by asymbiotic organisms are variable. In the Southeast Asian humid tropics, N contributions through rice-cyanobacteria (80 kg N ha^{-1}), rice-bacterial associations (30 kg N ha^{-1}), sugarcane-bacterial associations (100 to 160 kg N ha^{-1}), or rice-azolla (100 kg N ha^{-1}) during cropping often provides a positive N balance. Legumes such as sesbania, clover, or soybean may contribute between 100 and 360 kg N ha^{-1} to Southeast Asian tropical agroecosystem (Kush and Bennet, 1992). Peoples and Crasswell (1992) suggest that N_2 fixation rates are higher in tropical soils with low available N. Therefore, N inputs to legumes are either meager or nil. Giller *et al.* (1994) opine that leguminous crops with comparatively lower productivity could still be very useful, if they leave more residual N in soil for subsequent crops. Cultivating legumes followed by cereals, or legume-cereal intercropping or leguminous trees plus cereal (dry crops) are considered very useful methods to gain N into the agroecosystem.

During such cereal/legume sequences or intercrops, the extent of N_2 fixation, and quantity of N credits derived by the succeeding non-legume crops, such as cereals or cassava vary widely (Peoples and Crasswell, 1992; Hood 2001; Lata *et al.* 2001). In the South Pacific and northeast Australian humid tropics, pastures of guinea grass + Centrosema fixed 67 kg N ha^{-1}. Out of which, 5 kg N ha^{-1} were transferred to grass. Similarly, a tall grass + Calopogonium + Peureria fixed 136 kg N ha^{-1} and transferred 23 kg N ha^{-1} to grass (Jehne, 1980; Reynolds, 1982). Nutrient inputs into tropical soils are also effected via recycling stover/stubble from previous annual or perennial cover or green manure crops. It is believed that nearly half or a third of the nutrients by these can be returned to the field.

B. Nutrient Recovery

The net nutrients removed through the harvested portions of tropical crops vary widely (Table 8.1). Cassava, in particular, and certain other tuber crops extract rather larger amounts of nutrients from the soil. It is attributed to storage roots of cassava, which accumulate relatively larger quantities of mineral elements. Unlike cereals wherein only 20% K is translocated into grains, Howeler (1985) report that 34% N, 60% P and 60% K absorbed by cassava could be traced in roots. Due to high depletion of nutrients, particularly K, most fields may become K deficient if cassava

Table 8.1: Nutrient removal by tropical crops per ton of harvestable products

Crop	Yield (t ha^{-1})	Nutrient removal (kg t^{-1} dry matter)		
		N	P	K
Cassava roots	13.5	4.5	0.8	7
Sweet potato	5.0	12.0	2.6	19
Maize	6.0	17.0	3.2	5
Rice	4.0	17	2.5	4
Phaseolus beans	1.0	40	3.8	23
Soybean	0.9	70	18.0	78
Groundnuts	1.3	81	5.0	27
Sugarcane	20.0	3	1.0	5

Source: Excerpted from a compilation by Howeler, 1991; Keelman and Tackaberry, 1997.

is cultivated continuously without adequate replenishments (Howeler, 1996). Nutrient recovery ratios, not only vary with crop species, but depends much on the stage of the crop, root traits, as well as accumulation and re-translocation aspects. In addition to geographic location and soil type, general soil fertility and the levels of availability of nutrients in the soil, immensely influences the nutrient recovery.

C. Nutrient Losses

Loss of essential nutrients from humid tropical farming zones can severely affect crop productivity. Generally nutrient leaching from soil, volatilization/emissions of gases, large scale removal of nutrients via harvested products are the three major avenues of nutrient loss from the agro-ecosystem. Soil erosion in tropics, including mineral nutrients and organic C mainly occurs through the action of moving water. It can be severe in fields derived from shifting cultivation. Soil erosion rates are also influenced by the intensity of precipitation events and topography (slope), duration of cover crops and mulches if any (see Howeler, 1990; Howeler *et al.* 1999). For example, an increase in slope from 1% to 15% enhances soil loss from 11.2 to 230 t ha^{-1} yr^{-1} in a fallowed field (Lal, 1976a, b). Annual crops and mulches differ in their ability to provide soil cover, and reduce nutrient losses. Tree crops, along with their litter cover, and under story crops can effectively minimize soil and nutrients erosion. A *Leucana* cover in Nigeria reduces erosional loss from 560 to 76 kg ha^{-1} during a single day intense rainfall (Lal, 1991). With regard to cassava in South America and elsewhere in sloped locations, improper terraces in mountainous locations can lead to enormous loss of soil/nutrients. In the Colombian Andes, Howeler (1985) estimated soil loss as high as 100 t ha^{-1} yr^{-1}. It is equivalent to loosing 10 cm top soil per annum. Ruppenthal *et al.* 1997, who studied cassava fields, in humid Andes suggest that between 26 and 1726 kg ha^{-1} organic matter; 0.9 and 65 kg soil N; 0.04 and 2.8 kg ha^{-1} exchangeable K and 0.004 and 0.8 kg ha^{-1} of P (Bray—11) were lost in sediments. Such high nutrient loss via sediments diminishes, if runoff is controlled. Additionally, controlling runoff delays eutrophication in the downstream location where runoff sediments collect heavily.

Gaseous Loss of N: Fertilizer N applied to enhance productivity of tropical crops may be lost due to NH_4 volatilization, or because of denitrification causing nitrous oxide, nitric oxide and N_2 emissions. These factors reduce N recovery and fertilizer N-use efficiency. To quote a few examples, maize fields in Venezuela emitted 3 mg N_2O cm^{-2} h^{-1} and 1 mg NO cm^{-2} h^{-1}, whereas sorghum emitted 3 mg N_2O and 12 mg NO cm^{-2} h^{-1}, sugarcane zones in Hawaii lost <3 mg N_2O and up to 17 mg NO cm^{-2} h^{-1}. Bananas in Costa Rica liberated 31 mg N_2O and 56 mg NO per cm^{-2} ha^{-1}. Native grassland in humid tropical Central America lost 0.7 to 0.9 mg N_2O and 0.1 to 0.3 mg NO $cm^{-2} h^{-1}$ (Erickson and Keller, 1997). On an agroecosystem scale, trace gases N_2O and NO emissions can alter climatic parameters. Mainly because, N_2O reduces stratospheric ozone. Matson *et al.* (1996) opine that N_2O emissions from urea applied to crops in humid tropics (2 to 3%) are higher than those reported from temperate crops (0.5%). Similarly, NO based loss of nitrogenous fertilizers could reach up to 15% in humid tropics (Veldkamp and Keller, 1997). It is common to record higher levels of N_2O and NO emissions immediately after fertilizer N inputs. For example, banana plantations treated with 28 kg N ha^{-1} emitted 45 ng $cm^{-2} h^{-1}$ N_2O. It lessened to 5 to 10 ng N_2O $cm^{-2} h^{-1}$ by 3 to 4 days after N fertilizer inputs, and emissions ceased by 25th day (Veldkamp and Keller, 1997). Soil dependent variation in N_2O and NO emissions are easily discernible. For example, N_2O (31 ng N_2O–N $cm^{-2} h^{-1}$) and NO (56 ng NO–N $cm^{-2} h^{-1}$) emissions were greater in Andisols, when compared with Inceptisols (9 ng N_2O–N $cm^{-2} h^{-1}$; 41 ng NO–N $cm^{-2} h^{-1}$) planted with bananas in the humid tropics of Costa Rica. Cultivation procedures may intensify in the humid tropics. Therefore, fertilizer N usage is expected to increase enormously, which can result in greater N loss through emissions, and may also be deleterious to general climate (Matson *et al.* 1996; Bouman, 1994).

D. Long-term Studies on Soil Exhaustion

Long-term evaluations in Asia (India, Malaysia and Thailand) and Latin America (Columbia) indicate that, if the fields are unfertilized for 30 years from 1955 to 1985, then cassava root yields declined from 28 t ha^{-1} to 15 to 18 t ha^{-1}. Mainly because, nutrient depletions were severe. Soil K declined from 86 to 23 ppm by the 6th crop in sequence. In South India, in order to maintain optimum N, P, K levels in soil and cassava root yield, it needed 100 kg N, 44 kg P and 83 kg ha^{-1} as yearly replenishments (Kaburuthamma *et al.* 1990). Despite yearly K inputs, these soils encountered K depletions from 68 ppm to 25 ppm after 10th crop cycle. Again, in Columbia, soil K depletions were severe. It reached <0.1 m eq 100 g^{-1} soil within 2 to 3 years, and required at least 100 to 125 kg K ha^{-1} annually to maintain soil K levels at 0.2 m eq K 100 g^{-1} soil. If annual inputs were 250 kg K ha^{-1}, then it resulted in K build up at 0.4 m eq K 100 g^{-1} (Howeler, 1991). Tropical soils, particularly those utilized for producing cassava, pine apple, maize etc., get exhausted easily. Restoration of long-term nutrient equilibrium between nutrient removals, and inputs, through soil management procedures are mandatory. Long-term trails in the cassava belts of Latin America and Africa, clearly prove that soils get exhausted rather quickly and nutrient depletion, particularly K can be severe. Generally, restoration of productivity of an exhausted soil required N, P, K replenishments for at least 2 consecutive years. For instance, if exhausted soil yielded 13.3 t ha^{-1}, fertilized plots gave 34.4 t ha^{-1}. Upon restoration of nutrients, particularly K, root yields increased to 29.2 and 41.0 t ha^{-1} respectively. Overall, soil erosion, nutrient depletion and yield decline are commonly the long term problems in the cassava belt. Appropriate soil and nutrient management procedures are crucial to sustain this humid tropical cropping ecosystem. To quote an example, soils prepared with oxen and without fertilizer replenishment suffered 36 t ha^{-1} soil loss, and the resultant root yield was only 7 t ha^{-1}. Soils not ploughed, but fertilized and planted at 80 cm × 80 cm spacing suffered only 10 t ha^{-1} soil loss, and root yields were higher at 18 t ha^{-1}. Obviously, managing soil and its nutrients on a long-term basis needs priority because soil erosion is also rampant in humid tropics.

E. Fallows in Humid Tropics

Fallows are known to bring variability to humid tropical agroecosystem. Vegetated fallows suppress weeds, hence reduce diversion of soil nutrients into unwanted plants. Fallows restore soil fertility, particularly when managed fallows are recycled efficiently (e.g. green manure crop). A legume cover crop during fallow can alter N balance, and provide N credits to the main crop. It is believed that a long duration well managed fallow restores soil fertility and productivity, much better than shorter fallows. Managed fallows, since they recover greater quantity of nutrients than natural vegetation, may actually enhance productivity of shifting cultivation (Szott and Palm, 1996). Evaluation of nutrients recovered by natural and well managed fallows developed on Acrisols of upper Amazonia, indicated clearly that nutrient balances were dependent on crop species planted. Fallows with *Desmodium* or *Peuraria* consistently provided positive N, P, K balances for the cropping system (Szott and Palm, 1996). Nitrogen stocks depleted by a cereal crop, could be replenished in 1.5 to 3.5 years by a well managed, legume-based fallow (Jordan, 1989; Wadsworth, 1990). During fallow period, crops may recover hitherto unavailable pools of P. Plus, organic P could be enhanced by reincorporation of vegetation. Depending on the crop species, net losses of K from the field, and the net increase in K stocks due to absorption/recovery by vegetated fallow may vary. For example, a *Desmodium* based fallow increased K stocks by 30 kg ha^{-1}. Whereas certain other species depreciated it. Overall, nutrient dynamics both during shorter and longer fallows are crucial to farm productivity in humid tropics. Szot and Palm (1996) suggest that, in general, fast growing, and deep-rooted legumes or non-leguminous species are tolerant to acidic, infertile conditions and may prove most beneficial.

F. Crop Physiological Aspects

Physiologically, a sustained growth and productivity despite soil nutrient related constraints seems to be behind farmers' preference for cassava in the humid tropics. The cassava agroecosystem, in fact, flourishes on acid infertile soils, and withstands slightly impaired nutrient dynamics, because of following crop physiology related reasons.
   A. Cassava tolerates low soil pH, and comparatively higher Al and Mn concentrations. These factors otherwise reduce nutrient recovery by the crop.
   B. It tolerates paucity of Ca, N and K in soil. It puts forth good growth despite high P requirements, because it scavenges efficiently.
   C. It is a highly efficient carbohydrate accumulator even on low fertility acid soils. Even under such sub-optimal conditions root yields may reach up to 36 t ha^{-1} yr^{-1}.

On a global basis, average productivity of cassava is 9 t ha^{-1} yr^{-1} but it is < 4 t ha^{-1} yr^{-1} in 56% of humid tropics. Soil fertility status and nutrient availability are indeed crucial, since radiation interception and moisture are not major constraints in humid tropics. By carefully managing nutrient dynamics, it is possible to achieve higher yields. Sometimes reaching as high as 50 to 70 t roots ha^{-1} yr^{-1} (Lopez, 1995).

The pattern of nutrient recovery rates during the different growth phases, re-translocation and accumulation in harvestable parts and stover influence the soil nutrient dynamics of the individual field as well as the cropping ecosystem. In the case of cassava, dry matter accumulation and nutrient removed are slower during the first two months, then becomes rapid in the next four months, but slows down again in the next six months. Leaf fall accounts for lower biomass accumulation at later stages. It is important to note the nutrient concentration in the leaf tissues. Most nutrients accumulate initially in the leaves and stem, but get retranslocated into roots as the crop reaches physiological maturity (Howeler and Cadavid, 1983). A crop yielding 30 t ha^{-1} cassava root removes 200 kg K, and partitions 124 kg K into tops and 76 kg K into tubers (harvestable product). Similarly it partitions 126 kg N into tops and 38 kg N into tubers. Nearly 21 kg P into tops and 10 kg P into tubers (Asher et al. 1980). Information on nutritional aspects of tropical crops can be helpful in devising nutrient schedules that match crops demand with inputs.

G. Mycorrhiza, Phosphate Rocks (PR) and P Dynamics

Both under natural and farming situations, arbuscular mycorrhizas are vital soil biotic factors that influence P dynamics in general, and more appropriately P exploration in soil (Howeler et al., 1982; Saif, 1987). In fact, AM could be partly attributed for better adaptation of cassava to low P, acid infertile soils (Sieverding and Howeler, 1985). Some of the important aspects of cassava-AM symbiosis relevant to P dynamics of the cropping system are:

a. AM inoculated plants absorb higher amounts of P, as well grow and yield better, when compared with non-mycorrhizal plants exposed to similar levels of soil P availability. The P recovery and the magnitude of excess P scavenged and absorbed through AM hyphae, actually varies with fungal isolate, crop species and its genotype, as well as soil characteristics. (Howeler and Sieverding, 1983; Saif, 1987; Howeler *et al.* 1987).

b. Cassava-AM symbiosis imparts ability to absorb P, even though its availability is too low. In other words, the 'Critical-P in soil' required is much lower for plants colonized with symbiotic fungi (AM). For example, Howeler *et al.* (1982) reported that when in association with AM fungi, cassava plants scavenge P available at as low as 3.0 ppm Olsen's P, and almost satisfies the entire P requirements of the crop. Whereas, in non-mycorrhizal state, roots can scavenge P available only at >15 ppm Olsen's P. To reach this 15 ppm Olsen's P level in Colombian soils, nearly 180 kg P ha^{-1} input was needed. Hence, Saif (1987) opines that AM fungi might be crucial to sustain cassava productivity on low-P, infertile acid soils in humid tropics.

c. Mycorrhizal associations also aid in better P recovery from sparingly soluble/available P sources such as phosphate rocks. For example, Howeler and Sieverding (1983) report that in Columbian highlands, depending on AM fungal isolate inoculated, and the chemical nature of phosphate rock source, the cassava root yield enhancements varied between nil and 2 t ha^{-1}.

d. The cumulative effects of different mechanisms related to P recovery by AM fungi can also be perceived as increase in harvestable products. Mainly, as biomass, grains or root yield, which may be either too small, statically insignificant or large enough. The causes for such variation in AM related advantages to P nutrition and crop productivity are many, which have been discussed in great detail by Howeler and Sieverding (1983) and Howeler *et al.* (1987).

Overall, mycorrhiza mediated P transport influences P recovery and harvests in humid tropics, be it beans or cassava. Irrespective of AM inoculation, soils support a diverse and large population of AM propagules, which promptly establish symbiosis with cassava or beans grown in humid tropics. Hence, a sizeable fraction of soil P is mobilized by AM into roots. Such natural advantages need to be quantified and importance of AM symbiosis to the cropping ecosystem need to be highlighted. Inoculant species, if more efficient could be disseminated to improve P recovery by the crops. In fields with leguminous crops, AM-fungi are in constant interaction with nitrogen fixing legumes, and relevance of such tripartite relations on N and P dynamics needs investigation in greater detail.

Both, highly soluble P sources such as single or triple super phosphate or compound fertilizers, and less soluble sources such as basic slag, phosphate rock or thermophosphates are utilized to satisfy P requirements. Phosphates rocks have been efficiently utilized on acids soils in humid tropics. More appropriately, partially acidulated (40% to 50%) version (PAPR) which are treated with H_2SO_4 or H_3PO_4 are known to possess effectiveness approaching soluble fertilizer SSP or TSP (Chien and Menon, 1995). However, field tests in tropical Africa, Latin America and Asia indicate that, if PR source is rich in $Fe_2O_3 + Al_2O_3$, then it might result in unfavorable chemical transformations of P. Water soluble-P released from phosphate rocks through H_2SO_4 treatment might revert to insoluble forms due to reaction with Fe or Al component. A feasible option is to bring a physical compaction between PRs with soluble P fertilizer (at 40 : 60 ratio) to augment P availability in acid tropical soils. Such, compacted PRs have resulted in higher agronomic efficiency, almost equivalent to TSP (Chien and Menon, 1995). Yet another aspect of P dynamics explored for feasibility in humid tropics is to utilize AM fungi, and their ability to exploit sparingly soluble PRs better. Certain combinations of different locally available PRs and AM inoculation tend to provide better P recovery. Detailed evaluations made at different locations in South America using a range of soil types, AM fungal isolates, phosphate rocks and crop genotypes were summarized by Howeler *et al.* (1987). To quote an example, in Colombia, 50 kg P ha^{-1} application as Huila rock phosphate with AM (*Glomus occultum*) inoculation resulted in 30% higher cassava root yield. Inoculation with AM was necessitated because soil possessed very low native AM propagule numbers. Consistently higher P recovery was noticed with AM isolates C-I-I and C-10 of *Glomus occultum*. Clearly, AM fungal isolates differ in their ability to utilize sparingly soluble P sources-rock phosphate. Hence, their influence on P dynamics, particularly P scavenging and recovery will be variable. Such phosphate rock plus AM fungi combinations are known to enhance P recovery by other humid tropical crops such as coffee, beans and pastures.

3. Concluding Remarks

Along with important cereals and legumes, mixed farming is a common feature in humid tropics. Potential productivity levels of crops, particularly root crops such as cassava is very high compared with that attained in farms. Nutrient dynamics seems to be a major limiting factor influencing productivity, because water is not limiting. The following example confirms it. Nearly 56% of tropical cassava belt adopts subsistence farming technique, that yield < 4 t ha^{-1}. The global average is 9 t ha^{-1}, and the potential yield 70 t ha^{-1} at NPK 150:2:150 . This situation could be expected with other crops grown in humid tropics. Soil related constraints impair expected nutrient recovery and turnover rates. Hence, measures that reduce soil erosion, avoid loss of nutrients through chemical fixation or diminish Al and Mn toxicity or acidity need consistent attention. Legumes sequenced with cereals or other crops provide 'N credits', and this needs to be maximized. Regular recycling of stover and other organic residues can stabilize nutrient dynamics and enhance soil quality. Overall, since moisture is not a major constraint, nutrient availability, and general dynamics gains primacy. Management of nutrients efficiently, is crucial to sustenance, as well as to achieve higher productivity goals.

REFERENCES

Asadu, C.L.A. and Enete, A.A. (1997). Food crop yields and soil properties under population pressure in sub-Saharan Africa: The case of Cassava in Southeast Nigeria. *Outlook on Agriculture,* **26**: 29-34.

Asher, C.J., Edwards, D.G. and Howeler, R.H (1980). Nutritional disorders of Cassava. St Lucia, Queensland, University of Queensland, Department of Agriculture, 48 pp.

Bouman, A.F. (1994). Direct emission of nitrous oxide from agriculture soils. Rep. 773004004, Prof. Bilthoran National Institute of Public Health, Environ Netherlands.

Chien, S.H. and Menon, R.G. (1995). Agronomic evaluation of modified phosphate rock products: IFDCs experience. *Fertilizer Research,* **41**: 197.

Erickson, H.E. and Keller, M. (1997). Tropical land use change and soil emissions of nitrogen oxides. *Soil Use and Management,* **13**: 278-237.

FAOSTAT (1999). Food and Agricultural Organization of the United Nations. Statistical databases. Online at http://apps.fao.org.

Giller, K.E., Mc Donagh, J.F. and Cadish, G. (1994). Can biological nitrogen fixation sustain agriculture in tropics. *In:* Soil science and sustainable land management in the tropics. J.K. Syers and D.L. Rimmer (eds.) CAB International, Wallingford, pp. 173-191.

Hood, R. (2002). Use of stable isotopes in soil fertility research. *In:* Soil fertility and crop production. K.R. Krishna (ed.). Science Publishers Inc., New Hampshire, USA, pp. 313-335.

Howeler, R.H. (1985). Potassium nutrition of Cassava. *In:* Potassium in agriculture. ASA-CSSA-SSA, Madison, pp. 819-841.

Howeler, R.H. (1989).Cassava. *In:* Detecting mineral deficiencies in tropical and temperate crops. Plucknett, D.L. and Spragne, H.B. (eds.) West View Press, Boulder, Colorado.

Howeler, R.H. (1990). Phosphorus requirements and management of tropical root and tuber crops. I. *In:* Proceedings of the symposium on phosphorus requirements for sustainable agriculture in Asia and Oceania. International Rice Research Institute, Manila, Philippines, pp. 427-444.

Howeler, R.H. (1991). Long-term effect of cassava cultivation on soil productivity. *Field Crops Research,* **26**: 1-18.

Howeler, R.H. (1995). Diagnosis of nutritional disorders and soil fertility management of cassava. CIAT, Regional Program, Thailand, pp. 191-193.

Howeler, R.H. (1996). Mineral nutrition of Cassava. *In:* Mineral disorders of root crops in the South Pacific. *ACIAR Proceedings,* **65:** 110-116.

Howeler, R.H. and Cadavid, L.F. (1983). Accumulation and distribution of dry matter and nutrients during a 12-month growth cycle of Cassava. *Field Crops Research,* **7:** 123-127.

Howeler, R.H. and Cadavid, L.F. (1990). Short and long-term fertility trials in Columbia to determine requirement of Cassava. *Fertilizer Research,* **26:** 61-80.

Howeler, R.H., Cadavid, L.F. and Burckhardt, E. (1982). Cassava response to VA mycorrhiza inoculation and phosphorus application in greenhouse and field experiments. *Plant and Soil,* **69:** 327-340.

Howeler, R.H., Oates, C.G. and Costa Allen (1999). Strategic environmental assessment: An assessment of the impact of small holder cassava production and processing on the environment and biodiversity. International Fund for Agricultural Development (cited in IFPRI paper No. 31).

Howeler, R.H. and Sieverding, E. (1983). Potentials and limitations of mycorrhizal inoculation illustration by experiments with field-grown cassava. *Plant and Soil,* **75:** 245-261.

Howeler, R.H., Sieverding, E. and Saif, S. (1987). Practical aspects of mycorrhizal technology in some tropical crops and pastures. *Plant and Soil,* **100:** 249-283.

IFPRI (1998). International Food Policy Research Institute, IFPRI-Impact. June-1998. Washington, D.C.

Jehne, W. (1980). Endo-mycorrhizae and the productivity of tropical pastures: The potential for improvement and its practical realization. *Tropical Grasslands,* **3:** 202-209.

Jordan, C.F. (1985). *Nutrient Cycling in Tropical Moist Forest Ecosystems.* John Wiley and Sons, New York, USA. pp 190.

Kaburuthamma, S., Mohan Kumar, B., Mohan Kumar, C.R., Nair, G.M., Prabhakar, M., Nair, P.G. and Piller, N.G. (1990). Long range effects of continuous cropping and manuring on cassava production and soil fertility of soil. *In:* 8th International Symposium on Tropical Root crops. R.H. Howeler (ed.) Bangkok, Thailand, pp. 259-269.

Karche, V.K., Raja, P. and Sahgal, J. (1999). Sand mineralogy of soils representing three agroecological sub-regions of peninsular India. *J. Indian Soc. Soil. Sci.,* **47:** 775-780.

Kellman, M. and Tackaberry, R (1997). Tropical environments. Routledge, New York and London, 375 pp.

Kush, J.S. and Bennet, J. (1992). Nodulation and nitrogen fixation in rice: potential and prospects. International Rice Research Institute, Manila, Philippines, pp. 3-5.

Lal, R. (1976a). Soil erosion on Alfisols of western Nigeria nutrient element corses and eroded sediments. *Geoderma,* **16:** 403-417.

Lal, R. (1976b). Erosion on Alfisols of Western Nigeria 1. Effects of slope, crop rotation and residue management. *Georderma,* **16:** 363-375.

Lal, R. (1991). Soil erosion in the tropics: Principles and management. McGraw Hill, New York.

Lata, Saxena, A.K. and Tilak, K.V.B.R. (2002). Biofertilizers to augment soil fertility and crop production. *In:* Soil fertility and Crop Production. K.R. Krishna (ed.) Science Publishers, Enfield, New Hampshire, USA, pp. 279-312.

Leon, L.A., Fenster, V.E. and Hammond, L.L. (1986). Agronomic potential of eleven phosphate rocks from Brazil, Columbia, Peru and Venezuela. *Soil Sci. Soc. Am. J.,* **50:** 798-802.

Lopez, J., Mabrack, M. and El Sharkawy, A. (1995). Increasing crop productivity in cassava by fertilizing production of plant material. *Field Crops Research,* **44:** 151-157.

Matson, P.A., Billow, S.H. and Zacheriassen, J. (1996). Fertilization practices and soil variations control nitrogen oxide emissions from tropical sugarcane. *J. Geophys. Res.,* **101:** 18533-18515.

Norman, M.J.T., Pearson, C.J. and Searle, P.G.E. (1995). The ecology of tropical food crops. Cambridge University Press, New York, pp. 4-8.

Peoples, M.B. and Crasswell, E.J. (1992). Biological nitrogen fixation: investments expectations and actual contributions to agriculture. *Plant and Soil,* **141:** 13-39.

Reynolds, S.G. (1982). Contributions to yield, nitrogen fixation and N transfer by local and exotic legumes in tropical grass-legume mixtures in Western Samoa. *Tropical Grasslands,* **2:** 76-80.

Ruppenthal, M., Leihener, D.E., Steinmiller, N. and El-Sharkawy, M.A. (1997). Losses of organic matter and nutrients by water erosion in cassava-based cropping systems. *Experimental Agriculture,* **33:** 487-498.

Saif, S.R. (1987). Growth responses of tropical forage plant species to vesicular-arbuscular mycorrhizae 1. Growth, mineral uptake and Mycorrhizal dependency. *Plant and Soil,* **97:** 25-35.

Sanchez, P.A. and Cochran, T.T. (1980). Soil constraints in relation to major farming systems of tropical America. *In:* Soil-related constraints to food production in the tropics. International Rice Research Institute, Manila, Philippines, pp. 107-140.

Scott, G.J., Rosegrant, M.W. and Ringler, C. (2000). Roots and tubers for 21st century. International Food Policy Research Institute/Centro de la Papa, Discussion paper No. 31, pp 1-65.

Sieverding, E. and Howeler, R.H. (1985). Influences of species of VA mycorrhizal fungi on cassava yield response to phosphorus fertilization. *Plant and Soil,* **88:** 213-221.

Szott, L.T. and Palm, C.A. (1996). Nutrient stocks in managed and natural humid tropical fallows. *Plant and Soil,* **186:** 293-309.

Tongglum, A., Vichukit, V., Jantawat, S., Sitterbusaya, C., Tiraparan, C. Sinthuprama, S. and Howeler, R.H. (1992). Cassava cultural practices research in Thailand. *In:* Cassava breeding, agronomy and utilization research in Asia. Howeler, R.H. (Ed) Proceedings of 3rd Regional workshop, Malang, Indonesia, pp: 199-223.

Veld Kamp, E. and Keller, M. (1997). Nitrogen oxide emissions from a banana plantation in the humid tropics. *Journal of Geophys. Res.,* **102:** 15889-15898.

Wadsworth, G., Keisenances, M.M., Gordan, D.R. and Singer, M.J. (1990). Effects of length of forest fallow on fertility dynamics in a Mexican Ultisol. *Plant and Soil,* **122:** 151-156.

Wood, S., Sebastian, K. and Schess, S.J. (2000). Pilot analysis of global ecosystems-agroecosystems. International Food Policy Research Institute. Washington, D.C. 110 pp.

CHAPTER 9

The Citrus Agroecosystem of Florida
Nutrient Dynamics and Productivity

> *The Citrus agroecosystem in Florida, as we perceive it today, maintains, utilizes and perpetuates what is actually derived through a historically recent, well intended and focused human endeavor (by Floridians), spanning over past four centuries. This agroecosystem developed predominantly through human ingenuity, presently contributes a major portion of global citrus produce.*

CONTENTS

1. The Citrus Agroecosystem of Florida
   A. Development and Expanse
2. Soils and Agroclimatology
   A. Soils
   B. Agroclimatology: Moisture and Nutrients
3. Nutrient Dynamics in the Citrus Belt
   A. Nitrogen in the citrus grooves
   B. Phosphorus in citrus plantation
   C. Tree physiology and nutrient dynamics
4. Concluding Remarks

1. The Citrus Agroecosystem of Florida

A. Development and Expanse

Citrus was introduced into Florida through the Spanish conquests around the 16th century. It first appeared in Florida sometime between the arrival of Ponce de Leon in 1513 A.D. and the establishment of St. Augustine in 1565 A.D. (Mack, 1985). Grierson (1995) opines that citrus culture in a good scale, got initiated at St. Augustine in North Central Florida. Then, it spread southwards into the peninsula, to establish itself into the present day intensively cultivated citrus mono-culture zone. Initially, citrus stocks were mainly the sour or sevile orange (*Citrus aurentium*), unlike the present day sweet orange (*C. sinensis*). The early citrus culture effort in Florida was also confined to hammock soils that are rich in organic fraction, and wherever farmers encountered well-drained scrub Oakland along the Indian River basin. However, by mid 1800's citrus culture had spread well into sand hills, flat land and southern muck lands. Early settlers moving into Florida in 1820's, that is after the United States of America obtained this territory from Spain, started selecting sweet oranges among the wild orange trees. Then, by saving these seeds established grooves with sweet orange stocks. They also used budding and grafting techniques to selectively propagate and develop sweet oranges based orchards. The commercial citriculture in Florida, and first shipments of grape fruits outside of this eco-zone began around 1880 to 85 (Grierson, 1995). The spread of such commercially viable citrus farms was then distracted by two factors, one of them relates to soil and nutrients. Soils with organic matter had to be preferred, meaning that hammocks were needed. The second factor meant that citrus grooves had to be close, situated within cartable distance from loading docks for better marketability and profitability. Therefore, plantations were confined along the St. John's river, or Indian river (East-coast lagoons) and between St. Augustine and Jacksonville. Severe freezes in 1890's brought down ambient temperatures to as low as –2°F. It resulted in movement of citrus belt towards central and southern Florida. Citrus growers had to abandon certain portions in North Florida around Marion country. This drift of citrus agroecosystem southwards also meant that citrus grooves will then encroach into sandy soils not so fertile. In addition to soil fertility constraints, the vast monoculture of citrus had to face adverse effects of pestilence, disease and the mysterious 'citrus decline'. This citrus decline was partly related to micronutrient imbalance.

For a considerable length of time, that is a few decades, the vastly developed citrus monoculture in Florida faced three main problems. Firstly, it concerned with farmers' ability to find new plots with hammock soils, then avoiding pestilence and diseases, thirdly transportation and marketing. Basically soils, their fertility status and the resultant nutrient dynamics dictated the rapidity of establishment of citrus grooves, and its productivity. The citrus tree nutrition then became a major concern, as well as important research item for scientists from the University of Florida's, Citrus experimental stations at Lake Alfred, and at Fort Pierce on Indian river. 'Ammoniation' was the first tree nutrition problem solved. It was a problem related to soil nutrient dynamics, namely copper imbalance in tree tissues and its deficiency in soils. Grierson (1995) states that in 1940's, scientists then guided citrus planters to mix as many essential micronutrients to their fertilizer schedules. Zinc was to be supplied in chelated form. Later it was the N:K imbalance that affected the citrus grooves and their productivity. 'Orange splitting' caused by an impaired K availability was corrected using K fertilizers. Here, the uncongenial major nutrient dynamics, directly affected the market value of orange, but upon correction with K supplements fruits had softer peel.

The historical development of citrus in Florida, as discussed above focuses on soil fertility, mineral nutrition of trees and their influence on productivity. However, as this agroecosystem developed to attain the present state of ecosystematic and nutrient equilibria, it did pass through interference by several other factors. For example, major diseases and pests, cold temperatures, and drought had all solely adverse effects. Political and legislative changes, fluctuations in market value, migrations etc., had both adverse or positive influences depending on situation. Agronomic improvements, development of high yielding genetic stocks, quality control and post-harvest techniques had positive influences on the groove productivity or profitability. These important aspects are however, not within the purview of this chapter on citrus agroecosystem.

2. Soils and Agroclimatology

A. Soils

Depending on the geographical location, a wide range of soil types have been utilized to culture citrus. Whatever be the soil type, for a thriving citrus mono-culture optimal combination of soil physical, chemical and biological properties are a necessity. Adequate nutrient reserve and/or nutrient inputs through fertilizers are almost mandatory. To a great extent, soil type preferences are dictated by the growth habit, particularly the root distribution (Alva and Tucker, 1997). Majority area of the citrus belt in Florida is occupied by soils classified as Spodosols (Aquods, Aquolls, Aquults and Humods). The other soil types encountered are Histosols (Fibrists, Hemists, Saprists with Psammaquents) and Psamments with Aquults or Udults (Brady, 1995).

In Florida, citrus cultivation was initially localized around the central ridge (Fig. 9.1). Here the soils are predominantly Entisols with deep and well drained vadose zone, and no confining horizon. These soils support extensive, and deep root growth due to good aeration. Consequently, nutrient recoveries would be better, because soil nutrients are explored efficiently. However these Entisols are sandy (90 to 96% sand), inherently poor in nutrients and organic matter. Mineralization is also rapid. Even the mineral nutrient applied through fertilizers are not retained properly (Alva and Tucker, 1997). Excessive drainage may result in moisture deficits. Lack of hard pan results in low water table. Under such deep

Fig. 9.1. A) Citrus bloom; B). Fruit; C). Citrus groove in the Central Florida (Photo courtesy: The Citrus Research and Education Center, CREC, University of Florida at Lake Alfred, Florida, USA).

sandy soils, citrus trees tend to develop deep roots, so that, nutrients and water are scavenged efficiently. Such roots penetrate to a depth of 5.8 to 7.6 m. However, in recent times, increased use of micro-irrigation has immensely affected root distribution, nutrient recovery and physico-chemical transformations of nutrients in soil. In contrast to traditional methods, micro-irrigation causes localization of citrus roots within top 15 to 30 cm. It helps citrus roots to scavenge nutrients made available at shallow depths via fertigation.

Forced by successive freezes, up in north and north central Florida, this citrus monoculture zone drifted southwards. Such citrus grooves in south had then to flourish on low land soils, which are sandy, fine textured and calcareous due to marine deposits. Most of the early citrus grooves planted on the coastal zone often occupied such Alfisols. However, in due course, Spodosols with spodic horizon (organic hardpan) were also utilized.

Sizeable fraction of recently planted citrus grooves occupy such strongly acidic Spodosols requiring liming. Both Spodosols, and to a certain extent Alfisols are comparatively higher in nutrient status, cation exchange capacity and water holding capacity. High CEC helps better utilization of soil nutrients by citrus trees. Certain disadvantages infrequently linked with these soils are alkalinity in the coastal belt, clay contents, poor drainage and deposition of subsurface material during bedding process. Soil pH affects nutrient dynamics in the citrus grooves. Primarily, it influences availability of macro and micronutrients (Alva and Chen, 1995; Alva et al. 1995). He et al. (1999b) reported that for each unit decrease in pH, NO_3–N, P and K concentrations in soil increased by 0.67, 0.55 and 2.53 mg L^{-1} respectively. Sometimes routine nitrogen input itself can cause soil acidification. The major mechanism for this acidification relates to H^+ ion release through nitrification of NH_4^+ and subsequent leaching of NO_3. Significant correlations were also reported for soil pH versus Zn and Mn concentrations. A decrease in soil pH by one unit decreased the extractable Zn and Mn by 2.6 and 1.87 mg kg^{-1} respectively. In contrast, decrease in soil pH by an unit increased Fe availability by 18.9 mg kg^{-1} soil. Depending on the extent of soil acidity encountered, the pH and its interrelationships to Ca and Mg availability are important. According to Alva and Tucker (1997), liming the soils is a relatively recent program that began around 1960's. It was intended to correct Ca deficiency, which primarily affected tree physiology, reduced nutrient recovery, fruit quality and productivity (Spencer, 1960; Spencer and Koo, 1962). On Lakeland fine sands, Ca application had raised and stabilized soil pH at 6.4 over a 7-year period. Compared with it, trees in uncorrected plots recovered less nutrients and produced low yields. Normally, nutrients are delivered as sprays, and the quantity of lime to be used is calculated to raise the soil pH from 6.5 to 7.0 (Reitz et al. 1964). However, during practical citrus farming, liming not only raises pH, but has beneficial effects on nutrient recovery from soil, on the physiology (leaf and canopy growth) and most importantly the fruit yield. Anderson (1987) reported gradual, steady increase in canopy volume, nutrient uptake and fruit yield of citrus grooves aged between 1 to 12 years due to liming. Such productivity advantages due to liming were proportionate to the age of citrus grooves.

Production practices such as liming acid soils, and in some cases use of sewage sludge have modified several other soil characters, in addition to pH. According to Zhu and Alva (1993) such practices predominantly influence nutrient distribution in the soil profile and availability patterns. Where sewage is used, distribution of metal ligands in the profile, and around the root zone needs careful attention. It can unfavorably alter tree physiology and transformation of essential nutrients in the ecosystem. Distribution of micronutrients and metal ligands such as Al in soil profiles is mainly an outcome of pedogenic processes, and to a certain extent on fertilizer programs. Quite often, micronutrient distribution pattern in surface horizon (Ap) differs from sub-surface horizon. This could be attributable to liming, fertilization, irrigation etc. Recently Zhang *et al.* (1997) have grouped soils in the Florida citrus belt based on the relative distribution of various forms of micro-elements in soil surface. The two distinct categories are: a) soils with pH > 6.5, and b) those with pH < 6.5. For example, analysis of surface horizon of soils with pH > 6.5 in the citrus belt indicates that Fe occurs predominantly in Fe oxide and amorphous forms. Whereas, organically bound Fe content was insignificant. Similarly, organically bound Al was too small at only 3 to 10% of the total Al in soils. On the other hand, in soils with pH $\leq$ 6.5 organically bound Mn constituted substantial fraction of total Mn pool. Now, along the depth of soils, both amorphous and crystalline Fe oxide occurred throughout the profile. Organically bound Fe was mostly confined to surface horizons (Ap), and in spodic horizon in Spodosols.

Subsoil Fe was greater in Alfisol compared with Spodosols. In general, all forms of Fe were low in the E horizon. Zhang *et al.* (1997) observed a general similarity between Fe and Al distribution. Aluminium accumulated in Bh horizon in Spodosols, and Bt in Alfisols. Enhancement in soil pH correlates negatively with exchangeable forms of both Fe and Al. Alva and Obriza (1993) state that high rainfall, high temperature and frequent irrigations induce leachings of cations, thus increasing acidity of sandy soils. Such soil acidification, although slow, increases exchangeable and organically bound forms of micronutrient Fe, Mn, also Al. However, in certain groups of Floridian soils, acidification was comparatively rapid due to carbonates in parent materials. Salinity might be encountered during citrus culture in Florida. Since fertilizer programs have direct influence on total dissolved solids, fertilizers with low concentration of solids are preferred. The primary effect of salinity is reduction in citrus tree nutrient recovery and growth. However, citrus root stocks vary in their ability to tolerate salinity. In general, the decreasing order of salinity tolerance among citrus varieties is Cleopatra Mandarin > Sour Orange > Sweet Orange > Swingle Citramelo > Rough Lemon > Carrizo Citrange (Tucker *et al.* 1995a).

B. Agroclimatology: Moisture and Nutrients

The average annual rainfall received by the citrus belt in Florida ranges form 1270 to 1530 mm, which is greater than annual evapo-transpiration (1000 to 1200 mm). Although, peak annual rainfall at 1400 mm is sufficient, it is poorly distributed. Nearly 60% of annual precipitation occurs between June to September. Such uncongenial rainfall distribution pattern, coupled with seasonal non-uniformity necessitates supplemental irrigation to minimize risk of water stress (Ali Fares *et al.* 1997). Irrigation and water management strategies are crucial to citrus production in Florida. Firstly, it avoids water stress effects on nutrient acquisition and yield formation. Secondly, optimized water management avoids undue leaching losses of nutrients, particularly the highly mobile NO_3-N, and agrochemicals into soil layers below the root zone (Ali Fares *et al.* 1997). The rates of water drainage below the root zone are higher immediately after an irrigation or rainfall event, and with it proportional nutrient losses through percolation can be expected. Rainfall pattern, which is not well distributed but concentrated during 3 months from June, aids NO_3 leaching. Tucker *et al.* (1995b), after estimating leaching potentials of Entisols from this citrus belt have classified them as most vulnerable to nutrient leaching. Most often, leached NO_3 could be traced in drinking water wells, indicating that NO_3-N leaching is significant. As a precautionary long-term measure, minimizing NO_3 leaching to ground water was suggested. Additionally, N application methods that minimize NO_3-N leaching beyond the citrus root zone were also advocated. Ritter *et al.* (1991) argue that water transport through soil profile is a crucial parameter that controls NO_3-N leaching in citrus grooves. Hence, optimized irrigation schedules may provide better management of nutrients. Alva and Paramasivam (1998a) have pointed out that we are yet to understand fully, the fate of NO_3-N that leaches beyond the citrus root zones, particularly in sandy soils. According to Alva and Ali Fares (1998), integrated soil moisture management (based on depth of rooting) at a given time during the year compares well with:

a) Full point—equivalent to maximum water content that can be stored, with minimum leaching and;
b) Refill point—maximum allowable depletion of soil water such as 1/3 or 2/3 depletion. In the field, often a sensor placed below the rooting zone records changes in soil moisture content and indicates extent of leaching losses, so that irrigation schedules can be fine tuned. Therefore, any change in established irrigation schedules may directly affect nutrient leaching, availability and accumulation in soils. The recent trend with citrus growers is to switch to low volume irrigation system. These micro-irrigation systems are used to deliver fertilizers (fertigation). Since, fertigation distributes nutrients efficiently near the citrus roots, nutrient recovery and fertilizer-use efficiency is enhanced. In a nutshell, shift to fertigation brings about changes in soil nutrient dynamics and productivity of citrus grooves. We may also achieve better control over water versus nutrient interactions by utilizing real time monitoring of soil moisture levels through tensiometers, capacitance probes and neutron scattering.

These methods allow accurate distribution of moisture and nutrients at times crucial to citrus trees, namely at flowering, fruit set and fruit development stages. Equally so, it is helpful during February to May period of the year, when moisture and nutrients are critical to new flush expansion, bloom, fruit set and enlargement (Ali Fares *et al.* 1997)

3. Nutrient Dynamics in the Citrus Belt

A. Nitrogen in the Citrus Grooves

Soil Depth, Nitrogen Availability to Roots and Leaching Losses: Mobility of NO_3–N through soil profile is a crucial aspect of N dynamics in citrus grooves. As such, the citrus root system is relatively shallow, well branched with woody lateral and fine fibrous roots (Castle 1980a; 1980b). Fibrous roots are concentrated around soil surface, and decreases substantially below 30 cm soil depth, with very few functional roots below 120 cm depth (Zhang *et al.* 1996b). Hence, Alva and Paramasivam (1998b) inferred that NO_3–N in soil solution at depths up to 120 cm only, represents N available to citrus roots. Typically, NO_3–N concentration in the soils of this citrus belt peaks around 60 cm depth. Since, fibrous roots are practically absent between 120 and 240 cm depth, NO_3–N at this depth cannot be absorbed (Fig. 9.2). It may not be immobilized either in soil micro-flora, or nitrified, because microbial population is limited, as well the soil carbon is scanty in deeper layers of these sandy soils (Carlisle *et al.*1989; Yeomans *et al.* 1992). Since NO_3–N at depths of 120 to 240 cm or more is unavailable to citrus roots for absorption, this pool actually is vulnerable to leaching, and a potential source for ground water contamination. Clearly, the downward movement of water regulates, to a certain extent the NO_3–N distribution and availability to citrus roots. Hence, irrigation schedules are to be optimized to minimize NO_3–N leaching below the root zone.

Through careful fertilization schedules excessive mobility of NO_3–N could be regulated, and allowed to localize within 90 cm soil depth. If controlled release fertilizers (CRP) are utilized, potentially leachable NO_3–N becomes minimal because of the slow-release nature of fertilizer (Alva, 1992; Alva and Paramasivam, 1998b). In practice, ammonium nitrate and urea are the commonly used fertilizer N sources in the citrus grooves. In this regard, citrus planters in Florida are advised to adopt a package termed, best management procedures (BMP) for two main reasons. One of them relates to achieving higher N-use efficiency. Whereas, the other is to reduce leaching and ground water contamination with NO_3–N. Both of these above aspects are determined by the rates of nitrate or urea transformation in sandy, poorly buffered soils with low organic matter

The Citrus Agroecosystem of Florida

Nitrogen dynamics in Citrus grooves

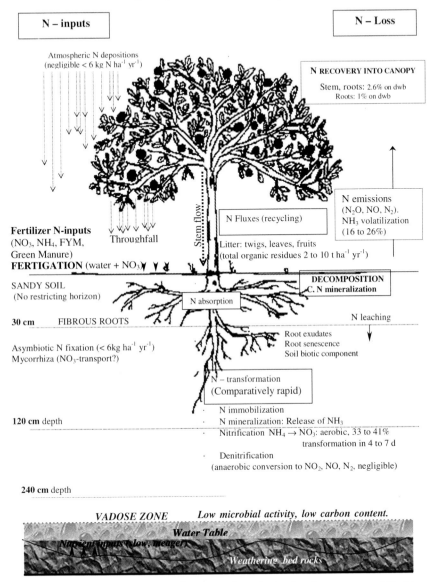

Fig. 9.2. A summarized depiction of nitrogen dynamics in the citrus grooves of Florida
Note: dwb = dry weight basis; soil depth not to scale

(Khakural and Alva, 1995; 1996). Firstly, such N transformations regulate availability of NH_4^+ and NO_3^- forms to plant roots, therefore, partly decide net N recovery by citrus roots. Nearly, 33 to 41% of NH_4–N is transformed into NO_3–N in 4 to 7 days depending on soil type. Secondly, higher levels of NH_4^+ or NO_3^- in freely available inorganic state induces rapid leaching losses, volatilization (NH_3) or emissions of N_2, NO_2 or NO evolved through denitrification process.

Nitrogen Volatilization: Thorough knowledge regarding extent of NH_3 volatilization is needed for optimal management of nitrogen in citrus grooves (He et al. 1999a). They summarize that nearly 40 to 60% of N applied goes unaccounted in any cropping system, including citrus plantations. Out of it, nearly 10 to 40% loss could be due to NH_3 volatilization. In the citrus grooves of Florida, potential maximum volatilization rates are greatest with $NH_4 HCO_3$ (23.2% of applied NH_4–N) > $NH_4 SO_4$ (21.7%) > $CO(NH_2)_2$ (21.4%) > NH_4NO_3 (17.6%). Normally, loss through NH_3 volatilization increases as N inputs are enhanced. Volatilization of N in the citrus grooves are immensely influenced by temperature. For example, N loss through volatilization doubled or tripled as temperature increased from 5°C to 25°C. Generally, NH_4^- retention on these sandy soils is insignificant. Soil and environmental conditions that favor nitrification may tend to decrease NH_3 volatilization. However, in citrus grooves, higher temperature, say 25°C to 45°C inhibited nitrification. Hence, it enhances NH_4–N availability for volatilization to occur (see Fig. 9.2).

Nitrogen Mineralization: Primarily, the organic matter generated under citrus grooves varies depending on tree age, species, and soil type (Table 9.1). Total organic residues produced may be as high as >10,000 kg ha^{-1} annually. Incidentally, the Florida climate, which is characterized by high rainfall and temperature favors rapid decomposition of organic residues such as litter. Inorganic nitrogen derived through mineralization of such an accumulation of organic residue is sizeable. Such NH_4–N released via mineralization under aerobic conditions is often rapidly converted to NO_3–N via nitrification process (Duo et al. 1997). In a citrus groove,

Table 9.1: An example of organic residues generated annually (kg ha^{-1}) under the canopy and beneath in soils by citrus groove in Florida

Citrus type (soil)	Groove age (yr)	Dry leaf	Dry root	Total organic residue
Cleopatra Mandarin	4	401	4249	4650
	20	1269	9018	10287
Temple orange	7	377	1891	2268
	40	640	4536	5176

Source: Excerpted from Duo et al. 1997
Roots extracted from 0 to 30 cm soil depth

organic residues such as dry leaves, other vegetative parts, fibrous roots and soil organics contribute N, and serve as substrate for mineralization. Therefore, a measure of total N in the residues provides an idea regarding potentially mineralizable nitrogen (PMN). In nature, all the N in the PMN fraction is not mineralized within a year. Still then, net mineralization rates were generally proportionate to PMN (Duo *et al.* 1997). Some of these recent studies demonstrate that climatic parameters influence N mineralization (N min) in citrus grooves. Nitrogen mineralization rate was optimum at 25°C. Thus, N mineralization is generally rapid during summer owing to higher soil temperatures. The average N mineralization rates are also dependent on soil depth. Duo *et al.* (1997) have reported that, if in surface horizon (0 to 15 cm) N-min ranged between 50 and 286 kg N ha^{-1}, at 15 to 30 cm depth it ranged from 7 to 144 kg N ha^{-1}.

Generally, N replenishment levels aim at providing for N removed annually in harvested fruits, plus to satisfy N needed for annual tree growth. In this regard, citrus experts in Florida have reported that a well managed groove may produce around 94 Mg ha^{-1} grape fruit and 79 Mg ha^{-1} oranges on a fresh weight basis. It corresponds to 74 and 110 kg N ha^{-1} removed into harvested fruits respectively. However, current recommendations of N inputs range between 200 and 270 kg N ha^{-1} (Tucker *et al.* 1995a), which will be above the net N removed by citrus grooves. We may also realize that, annually, N-min contributes partly towards N requirements of the citrus groove. Typically, contributions through N-min in citrus grooves can vary between 40 and 153 kg N ha^{-1} y^{-1} depending on tree age, soil type and groove management practices. The N-min from residues under citrus trees varied between 58 to 84 kg N ha^{-1} for a 4-year-old groove, and between 126 and 153 kg N ha^{-1} for a 20-year-old groove. At these above rates of N-min, it accounts for one-third to one half of the annual N recommended for mature, bearing grooves. Therefore, while developing citrus N budgets and appropriate N fertilizer schedules, contribution by N-min needs due consideration. It may be helpful in reducing N inputs, attain higher N-use efficiency, and reduce N leaching losses thus leading to a favorable N dynamics. Seen from a different angle, Duo *et al.* (1997) remark that precise N management in citrus grooves is one that duly considers N-min rates. It is almost mandatory, because it can minimize NO_3–N contamination of ground water. Under U.S. environmental protection laws, maximum NO_3–N contamination allowed is 10 mg NO_3–N L^{-1} in drinking water. Alva and Paramasivam (1998a) state that this limit is being exceeded in several locations in the citrus belt because of rampant N inputs without consideration to N-min. In addition to inorganic N sources, Florida citrus grooves receive a good share of nitrogen and other nutrients through bio-solids, yard wash, compost etc. Their application influences citrus leaf nutrient status, fruit yield and quality (Stofella *et al.* 1996a; Stofella, 1996b; He *et al.* 2000b).

Field evaluations of nutrient release from such composts is important to estimate nutrient contribution to citrus trees, potential leaching loss of nutrients, and finally to determine fertilizer rates, placement as well as application schedules. He *et al.* (2000a) have reported that organic N mineralization rates vary between 23 and 48% of total organic N in composts. Mineralized N recovered from composts/bio-solids were at best equivalent to 36 to 43% of total mineralized N (He *et al.* 2000a, b).

Citrus Agronomy and N-use Efficiency: Our evaluations of citrus grooves for N inputs, N recovery and resultant productivity are greatly influenced by plant densities. For example, Alva and Paramasivam (1998b) pointed out that in the past, experimental evaluations on nutrient dynamics were all made on unirrigated low-density planting (< 247 trees ha^{-1}) of robust growing root stocks with a fruit yield potential 30 to 40 Mg ha^{-1}. In contrast, current commercial grooves are high density plantings (>345 trees ha^{-1}) maintained under regular irrigation, fertilization or controlled release fertilizers leading to high potential yields between 60 and 80 Mg ha^{-1}. It is imperative that N dynamics will be enormously influenced by planting densities and groove management techniques. Primarily, shift to fertigation or controlled fertilizer release leads to enhanced N recovery into fruits thus supporting higher productivity. Fertilizer-N efficiency is also significantly influenced by the methods of application. Broadcasting dry granular fertilizers (DGF) was less efficient (0.38 Mg fruits kg N^{-1}) in stimulating fruit yield compared with fertigation (FRT) (0.42 Mg fruits per kg N) or controlled release fertilizers (CRF) (0.49 Mg fruits per kg N^{-1}). At such prevailing fertilizer N efficiencies in the grooves, Alva and Paramasivam (1998a, b) reported that N requirement to produce 1.0 Mg fruit varied between 2.75 kg N for DGF, 2.58 kg N for FRT and 1.79 kg N for CRF. Increased fertilizer N-use efficiency, leading to reduced N applications can also minimize leaching losses and ground water contamination. Above all, enhanced N efficiency reduces cost of cultivation and energy usage. Fertilizer N uptake efficiency, and in general N dynamics are maintained favorably by following certain improved practices such as:

a) Placement of granular fertilizer followed by a light irrigation that allows dissolution and movement towards roots. It also minimizes NH_4^+ volatilization and N loss through gaseous emissions;

b) Fertilization within wetting area under the canopy, followed by short washing cycles allows fertilizer to stay slightly below ground, thus lessening losses;

c) Avoiding fertilizer N application during heavy rainfall months—June through August, reduces risks of NO_3 leaching; and

d) Careful scheduling of irrigation by monitoring soil moisture using tensiometers minimizes N loss, and improves N uptake efficiency in sandy soils. Finally, having followed various efficiency measures, a tree crop such as citrus, attains 20 to 40% N-use efficiency (Dasberg, et al. 1984a, b; Dasberg, 1987). Whereas, annual field crops may attain between 40 to 50% N-use efficiency (see chapters 2, 3, 4 and 5).

Nitrogen Budgets in Citrus Grooves: The nitrogen budgets may vary with each unit of citrus groove, or even within demarkable portions of a groove (see Fig. 9.1). Nitrogen budgets will then depend on yield expectations, soil N reserve, N fluxes and transformations through mineralization steps, environmental parameters, as well as groove management procedures adopted. Similar factors may affect N budget calculations even at broader scale, say the citrus agroecosystem in Florida as a whole unit. Whatever be the scale of evaluation, N budgets will be guided by N inputs, internal fluxes and recycling (see Fig. 9.2).

According to Alva and Paramasivam (1998a) the major nitrogen inputs into citrus grooves that need consideration, while developing N budgets and appropriate fertilizer schedules are:
  a. Fertilizer N as inorganic source is a major input under most conditions of citrus groove management procedures. Whenever organic N inputs occur, the extent mineralized and available to citrus roots during first and subsequent years need to be ascertained. If incessant inorganic N inputs occur, then residual or carryover effects need accounting.
  b. Nitrogen fluxes are also created by decomposition of leaf litter and turnover of fibrous roots. The magnitude of such N fluxes normally increase with age of the groove.
  c. Atmospheric N depositions occurs whenever gaseous forms of N compounds are dissolved in precipitation and returned to soil. It is believed that such contributions are marginal at < 6 kg N ha^{-1} y^{-1}.
  d. Nitrogen fixation, mediated mainly by asymbiotic soil microbes is marginal, and is often < 6 kg N ha^{-1} y^{-1}.

The net nitrogen removals from the citrus agroecosystem are also variable, but generally determined by the following basic aspects:
  a. Nitrogen in fruits: For a bearing perennial tree, this is a major net N removal component. More often, this a net loss of N to citrus agroecosystem itself, because the fruits are exported out of the groove. Table 9.2 provides a summarized view on net removals of N and other nutrients through fruits.
  b. Nitrogen required for annual re-growth of leaf and root tissue. Normally, two distinct flushes in spring and summer occur which utilizes N that is applied as fertilizer.
  c. Flowering and fruit set utilizes a certain portion of stored N, which needs replenishment.

Table 9.2: Fruit weight and concentrations of various mineral nutrients in four different citrus varieties grown in Central Florida

Citrus varieties	Fruit weight (g)	Nutrient removal into citrus fruits									
		N	P	K	Ca	Mg	Fe	Mn	Zn	Cu	Al	Na
		← %dw →			← %dw →		← µg g dw →					
Hamlin	249	1.1	0.14	1.26	0.36	0.10	19	3	12	5	19	0.03
Parson Brown	221	1.0	0.14	1.19	0.44	0.11	26	5	13	5	21	0.03
Valencia	262	1.1	0.16	1.19	0.35	0.10	40	4	11	5	16	0.04
Sunburst	141	1.2	0.15	1.20	0.29	0.09	33	4	18	6	15	0.05
SE$_\pm$	±8	0.03	± 0.0	±0.05	±0.02	±0.02	±5	±0	±3	±1	±1	±0.01

Source: Excerpted from Hanlon et al. 1995; Alva et al. 1998; Alva and Paramasivam, 1998a, b.
Note: Values are means from 3 different trials during a three-year evaluation (1994 to 1996)

d. Gaseous N losses occur. However, it is meager because soils in this citrus belt are well drained and aerobic. Anaerobic microsites in soil are sparse, hence, denitrification reactions are minimal and negligible, particularly in the central ridge and coastal zones.

In addition to above net removals stated, N losses occur due to application efficiency, fertilizer formulation etc. Using above basic suggestions, a tangible approximation of N budget might be arrived at— i.e. net N inputs needed and removals expected for a citrus groove.

B. Phosphorus in Citrus Plantation

Much of Florida citrus is grown on extremely sandy soils with inherently low fertility (Hoogweg and Hornsby, 1997). Hence, a combination of improved fertilization and irrigation programs are mandatory, if satisfactory production levels are to be achieved. Normally, such fertilization schedules are based on N requirements, with a fixed ratio adopted for other two major nutrients, for e.g. N:P:K is maintained at 1:0.2:1. Clearly, N seems to be the key component that drives nutrient dynamics, namely inputs/recovery in the citrus grooves. However, knowledge regarding P dynamics in soil is important, because it influences productivity of citrus grooves (He et al. 1999c; He et al. 2000c). Phosphorus dynamics has to be regulated, if not, an excess can result in contamination of subsurface water. Run off and/or excessive leaching can cause eutrophication of water bodies. Basically, the sandy soils in citrus grooves possess very low P retention capacity. In addition, the annual removal of P through fruit harvest may account for only 30 to 50% of applied P (He et al. 1999c). Also, there are no accurate estimates of the extent of P storage in different tissues of the tree on an yearly basis. Hence, it is imperative that P leaching, as well as surface runoff will be prominent factors that affect, firstly the P availability to citrus, then water quality.

Phosphorus leaching can be significant (Calvert et al. 1981), but it is dependent on the presence of argillic horizon with higher clay content, which retards downward movement of P. Zhang et al. (1996a, b) suggest that formation of raised beds allows nutrients to be drained laterally as surface flow, therefore, it reduces chances of downward movement of P and accumulation just above argillic layers (hard pan).

He et al. (1999c) believe that a thorough knowledge regarding P sorption-desorption characteristics of all the different horizons in the soil profile, particularly at the surface layers, just above the argillic layer (hard pan), and below it can be useful in deciphering P movement and accumulation in soils. They conclude that Riverine fine sands (Alfisols) have low P sorption capacity, hence P mobility in the profile is relatively high. The southern Florida wood soils are calcareous. They possess one or more calcareous horizons, and the accumulated $CaCO_3$ usually lies below in the profile. The pH values reach > 7.0 due to $CaCO_3$ deposition which dominates the chemical dynamics of these soils. In terms of P and their dynamics, different forms of P exist in combination with Ca in soils. In these soils, P exists as mono and dicalcium phosphates, which are readily available to plants, and do not easily leach out of root zone. Added P fertilizers undergo a series of chemical reactions with Ca. Some of these reactions decrease P solubility with time, which is termed as P fixation. Hence, Obriza et al. (1993) point out that P availability to citrus roots will be determined by rates of soluble P applied, and rates of dissolution of fixed P if any. Fertilizer P will then be available, only until it is not fixed chemically into less soluble forms. Generally, to maintain high P availability in calcareous soils, water soluble P fertilizers are replenished periodically (Tucker et al. 1995b). Obviously, incessant P inputs result in its accumulation and fixation, but with time a portion of it becomes partially available. Incidentally, phosphorus deficiency is not a common feature in Florida citrus grooves. Newly planted grooves receive P annually, and as trees approach maturity P inputs subside to once in few years. Phosphorus inputs into citrus grooves are usually based on tissue and soil analysis (see Tucker et al. 1995a). Soil P test values < 10 ppm (Mehlich's extract) are considered very low, 15 to 30 ppm as medium and > 60 ppm as very high. Ordinarily, a box of orange (41 kg) removes 64 g P, which is equivalent to 1.5 g P in 1 kg of fruits harvested and transported out of the agroecosystem. However, P removal ratios vary depending on the citrus cultivar. For example, Alva and Paramasivam (1998b) found that Valencia citrus removed 33 kg P in 500 boxes of orange, whereas Hamlin removed only 30 kg P (see Table 9.2).

Arbuscular Mycorrhizas and Citrus P nutrition: A wide range of AM fungal species colonize citrus roots naturally in the outfields in Florida, namely, *Glomus fasciculatum, G. mossea, G. desserticola, G. etunicatum, G. intraradicis,*

G. calospora etc. These AM fungi are directly linked with P nutrition of citrus trees, and indirectly they influence P dynamics to a certain extent, particularly in soils with low P availability (Nemec *et al.* 1981; Nemec, 1987; Graham, 1987). In nature, they act as extended conduits, other than roots, which mediate P flow from soil to citrus tree roots. It is useful to quantify such P nutritional benefits accrued to citrus grooves. We may then be able to understand the role of mycorrhiza in P dynamics better. Accurate estimates of P transport effects are not available, owing mainly to large and varied native soil AM population, variability in fungal efficiency with regard to P transport, and environmental parameters. Inconsistencies in P equivalence of AM are encountered. Albeit, citrus as a crop species responds to external inoculation with AM fungi and derives P nutritional benefits. The critical P limits in soil are much lower for mycorrhizal citrus plants compared with those in non- mycorrhizal state (Menge *et al.* 1978a, b; Johnson 1984; Graham, 1984; Graham *et al.* 1987). Under greenhouse culture, AM inoculation of Troyer Citrange seedlings substituted for up to 56 ppm P, equivalent to adding 105 kg P ha^{-1} and with Brazilian Sour orange, P benefits equaled 278 ppm (i.e., adding 550 kg ha^{-1}) (Menge *et al.* 1978a, b). Extrapolation to outfield situations in Florida are difficult because of wide soil and environmental variabilities that are encountered, and reduction in AM effect due to interacting factors in field soil. Graham (1984) suggests that dependency of citrus on AM fungi for P nutrition and growth is greater compared with other crop species. This is partly attributable to scarce root patterns and scanty root hairs.

Since, AM fungi can enhance P recovery by citrus seedlings, inoculating AM fungi may provide favorable P dynamics, both in nursery and in outfield after transplantation. Unfortunately, many commercial citrus nurseries practice fumigation which eliminates useful AM fungi. However, methods to multiply, inoculate and evaluate these fungi on citrus have been devised (Nemec, 1984a, 1984b, 1985), such as fluid drills, slurry inoculation, placement etc.

Mycorrhizasts have generally expressed that soil nutrient status, particularly P, can immensely influence AM activity and P nutritional benefits. Hence, Nemec (1987) cautions that while composing potting mixes, their nutrient status or external P inputs should be compatible with AM species inoculated. Peat mixes with least P inputs, are known to provide healthy AM-colonized citrus seedlings, so that P nutritional benefits could be appropriately high. The primary effect of AM symbiosis is on P recovery by citrus, but as a consequence other beneficial effects may also be perceived. For example, Johnson (1984, 1985) demonstrated that AM, firstly enhances P recovery from soil and raises leaf tissue P concentration. Higher P in leaf tissues, then stimulates higher rates of photosynthesis—leading to better biomass accumulation. Douds *et al.* (1987) reported that AM roots require nearly 4 to 17% more photosynthetically fixed carbon than non-mycorrhizal root systems. Also, there seems to exist a limit in the extent of AM colonization in citrus, beyond which AM mediated P nutritional benefits may cease; this despite continued C drain into AM fungi. Obviously, P and C dynamics in citrus tree are interlinked through AM symbiosis. Physiologically, the crux is in the extent of C utilized per unit P transported by AM into citrus plants.

There is no doubt that AM fungi have co-evolved with this citrus agroecosystem in Florida. During these 400 years of citrus cultivation in Florida, this agrobelt must have drawn substantial quantities of P from soil, whose transportation, definitely was partly mediated by AM fungi. We may agree that in natural soils, their contributions to P dynamics could be either feeble or significant. However, it is necessary that we understand the P nutritional benefits of this symbiotic phenomenon better, and utilize it profitably in future. In particular, there is great need to evaluate AM in outfield situations in Florida. Recent studies, especially in 1990's, have been conspicuously very few. Even at a marginal 1 kg P ha^{-1} y^{-1} benefit, AM symbiosis must have already mediated transport of tons of P into citrus trees during the past decade. Perhaps, enough to gauge their importance to P dynamics of the citrus agroecosystem in Florida.

Other Nutrients: Potassium deficiency is not a common feature in Florida citrus grooves, but may appear if N inputs are made without appropriate K replenishments. To avoid K imbalance, N:K ratios in fertilizer schedules are maintained at 1:1, although 1:1.25 is required in calcareous soils (Jackson *et al.* 1995). On an average, 7.7 to 8.2 kg K are removed in 100 boxes (41 kg each) of fresh orange fruits. It is equivalent to 1.2% K on dry weight basis (Table 9.2). In general, the quantity of micronutrients removed by harvested fruits is extremely small even in high producing grooves. However, micronutrient availability is influenced immensely by soil chemical nature and transformations. Soil reaction (pH) is a key factor influencing micronutrient dynamics. Overall, in a high pH soil, availability of micronutrients such as Zn, Mn, Fe, Cu and B decreases, excepting Mo (Jackson *et al.* 1995).

Severe Zn deficiency can influence tree physiology and recovery of several other nutrients. Soil pH is a crucial factor limiting Zn supply to citrus. Under alkaline conditions, Zn precipitates and becomes unavailable, hence, soil pH is regulated to less than pH 7.0 (Jackson *et al.* 1995). Zinc inputs to soils are not made exclusively, instead it is channeled along with pesticide sprays, totaling 3.7 to 5.0 kg Zn ha^{-1} as ZnO or $ZnSO_4$. It is sprayed as foliar application. Organically chelated forms combined with lignin sulfonate N gluco-heptonate are also utilized (Jackson *et al.* 1995).

As a precaution, soil application of Zn is not practiced in calcareous soils, because higher pH renders it unavailable immediately. In central Florida, soils under citrus grooves may contain 340 to 450 kg Cu ha^{-1} in the top 25 cm. A moderate disease control spray in citrus grooves adds Cu. Since, moderate to highly acidic soil reactions can solubilize sizeable quantities of Cu, managing its dynamics by regulating soil pH becomes a good option. Copper levels as low as 1.0 ppm in soil solution can be toxic, affecting citrus root growth and physiological activity. Therefore, it affects nutrient recovery in general. While soils with moderate levels of Cu are frequent, there are virgin sandy soils in Florida which are deficient in Cu. Citrus cultivation in these locations needs Cu replenishment. Copper inputs, if made as foliar sprays, then it is not included in fertilizers applied to soil. Manganese deficiency in acid sandy soils of Florida can retard tree growth, nutrient recovery and fruit yield. Manganese inputs at 1.5 to 2.5 kg per acre corrects deficiency, if any. However, on calcareous soils Mn precipitates due to alkalinity, and becomes unavailable. Hence, Mn foliar sprays are preferred. Citrus grooves on acid soils receive 8 to 11.5 kg of Mn as MnSO$_4$ per ha, once in couple of years (Jackson *et al.* 1995). Boron (although needed in small quantities by citrus grooves) deficiency can cause severe fruit drop and can reduce productivity. Once again, like other micronutrients, soil pH dictates B availability. Its availability decreases as alkalinity sets in and accentuates. Boron deficiency, although not frequent, can appear if high analysis fertilizers are used continuously without micronutrients being included in the schedules. Imbalances, however, are easily corrected via B inputs, which are generally preferred as soluble borate sprays or soil application. It persists in soil for 2 to 3 years. Soil chemical dynamics and availability of molybdenum is different from other micronutrients discussed above. It is less available in soils acidic in reaction. Hence, Mn deficiency may be frequent in soils that are unamended and allowed to become excessively acidic. Thus, in acid soils it may not be feasible to enhance Mo availability through soil application or spray. Perhaps, liming the soil to raise pH could be apt and effective. In general, foliar sprays are quick and effective in correcting Mo deficiencies, but soil application seems unsatisfactory (Jackson *et al.* 1995).

C. Tree Physiology and Nutrient Dynamics

The citrus belt in Florida flourishes on two distinct types of soil. The deep well drained Entisols, support robust canopy, deep roots and high nutrient recovery. Whereas, in poorly drained Spodosols characterized by high water table, the tree growth is smaller, nutrient acquisition is comparatively low and productivity is commensurate. In either case nutrient inputs and agronomic procedures that regulate nutrient dynamics are aimed at rapid tree growth. In well established grooves tree fertilization programs are aimed at achieving maximum fruit production of acceptable quality. Overall, soil nutrients, their chemical transformations and availability dictates citrus growth physiology and productivity. Conversely, the physiological aspects of citrus plantation also influences the fertilizer schedules, extent of nutrient inputs, removal, recycling and storage—i.e. overall nutrient dynamics in the groove. Firstly, tree age, size and fruit bearing nature affect nutrient inputs, their frequency and timing. For example, a one-year-old non-bearing sapling receives 70 to 140 g N tree^{-1} y^{-1} in six equal splits, and at 3 years it receives 70 to 420 g in 4 splits. The ratios of nutrients applied too vary depending on physiological stage of the plantation. Young non-bearing trees are replenished with N:P:K at 1:1:1, whereas fruit bearing plantations receive N:P:K at 1:0.2:1 (Ferguson *et al.* 1995). Although, nutrient inputs increase with age and size of the tree, N inputs are reduced at containment stage, because it then regulates vegetative growth.

Nutritional management of bearing trees may be similar to non-bearing trees on several points. Although, tree growth continues to be an important factor determining nutrient schedules even after fruit bearing has begun, fruit production receives primacy during fertilizer disbursement and deciding on the nutrient management practices (Tucker *et al.* 1995a; Duo *et al.* 1997). Nitrogen continues to be the main factor, though other elements too regulate groove productivity. Although, nutrient removal via fruit harvest can be sizeable, still it accounts for only part of fertilizer input program.

'Crop load' is a factor that determines inputs of N and other nutrients into the grooves. A fruit bearing groove with average productivity is not replenished heavily with nutrients (Fig. 9.1). Those producing > 2000 boxes ha^{-1} may receive between 225 and 275 kg N ha^{-1}, but depends on soil test and leaf analysis (see Hanlon *et al.* 1995). A grape fruit groove producing >1600 boxes ha^{-1} receives 205 kg N ha^{-1}. Nutrients, particularly N inputs, are clearly influenced by the citrus variety. For some special cases, such as 'Orlando tangelos' or 'Honey tangerines' 285 to 340 kg N ha^{-1} is added yearly. On an average, N rates at 200 g N per box of orange or 135 g per box of grape fruit harvested is useful (Ferguson *et al.* 1995). Whichever be the citrus variety, physiologically the internal nutrient dynamics sought in a tree aim at minimum diversion of nutrients to vegetative growth (Duo *et al.* 1997). Nutrient inputs into citrus agroecosystem should also be proportionate to physiological variations in fruit bearing nature. Perhaps it is not yet practiced stringently. Pattern of N removal, storage, retranslocation into fruits, N-use efficiency and productivity of citrus grooves are actually affected by the alternate bearing habit. The 'off years' or low fruit yields alternate with 'on years' or high fruit yield years (Alva and Paramasivam, 1998a). Obviously, long-term assessment of the frequency of 'on' and 'off' years and related nutrient dynamics will be a useful exercise.

Nutrient inputs in citrus grooves is generally modified to suit canopy loss following pruning, freezes or extensive loss of root system due to flooding (Ferguson *et al.* 1995). Fertilizer rates, particularly N, P and K are reduced proportionately with canopy size. Root volume decreases in response to canopy size, hence nutrient recovery would also decrease. Proportionately, fruit productivity too decreases. Citrus grooves exposed to freezes are also not fertilized or irrigated, until regrowth is evident. Generally bearing trees store sizeable quantities of N and other nutrients that can be mobilized to newly regenerating shoots, twigs and leaves. Bearing trees that undergo freeze are usually not given fertilizer-based nutrients during first post-freeze season, particularly if it has received sufficient nutrients earlier (Tucker *et al.* 1995b).

4. Concluding Remarks

In four centuries, citrus culture in Florida has evolved into an intensively managed plantation ecosystem. Nutrient inputs, mainly N, and to a certain extent P, K, as well as micronutrients have been steadily increasing. Proportionate removals of nutrients through fruit harvest have also occurred. Inappropriately high N input has resulted in accumulation and ground water contamination in certain pockets. Nutrient imbalances have also appeared. However, timely soil chemical evaluation, judicious recommendations regarding nutrient schedules by citrus experts, and adoption of efficient methods of nutrient applications (e.g., fertigation) have all thwarted undue disturbance in soil nutrient balance and productivity equations. Overall, careful management of N and P dynamics in soil, and better orchard management options should lead to steadily higher harvests in future, without drastic effects on nutrient equilibria and ecosystematic functions.

REFERENCES

Ali Fares, Alva, A.K., Paramasivam. S. and Nkeddi-Kizza, P. (1997). Soil moisture monitoring techniques for optimizing citrus irrigation. *Trends in Soil Science,* **2:** 153-180.

Alva, A.K. (1992). Differential leaching of nutrients from soluble versus controlled release fertilizers. *Environmental Management,* **16:** 769-776.

Alva, A.K. and Ali Fares (1998). A new technique for continuous monitoring of soil moisture content to improve citrus irrigation. *Proc. Fla. State Hort. Soc.,* **111:** 113-117.

Alva, A.K. and Chen, E.Q. (1995). Hydrogen ion inhibition of copper uptake by citrus seedlings. *In:* Plant interactions at low pH. R.A. Date (ed.), Kluwer Academic Publishers, The Netherlands, pp. 631-634.

Alva, A.K., Graham, J.H. and Anderson, C.A. (1995). Soil pH and copper effects on young 'Hamlin' orange trees. *Soil Sci. Soc. Am. J.,* **59**: 481-487.

Alva, A.K. and Obriza, T.A. (1993). Variation in soil pH and calcium status influenced by micro-sprinkler wetting pattern for young citrus trees. *Hort. Science,* **28**: 1166-1167.

Alva, A.K. and Paramasivam, S. (1998a). Nitrogen management for high yield and quality of citrus in sandy soils. *Soil Sci. Soc. Am. J.,* **62**: 1335-1342.

Alva, A.K. and Paramasivam, S. (1998b). An evaluation of nutrient removal by citrus fruits. *Florida State Hort. Sci.,* **111**: 126-128.

Alva, A.K., Paramasivam, S. and Graham, W.D. (1998). Impact of nitrogen management practices nutritional status and yield of Valencia orange trees and ground water. *J. Environ. Quality,* **27**: 904-910.

Alva, A.K. and Tucker, D.P.H. (1997). Impact of soil pH and cations availability on citrus production in low buffered sandy soils. *Trends in Soil Science,* **2**: 37-57.

Anderson, C.A. (1987). Fruit yields, tree size, and mineral nutrition relationships in Valencia orange trees as affected by liming. *Journal of Plant Nutrition,* **10**: 1907-1916.

Brady, N.C. (1995). *The Nature and Properties of Soils.* Prentice Hall of India Ltd., New Delhi, 637 pp.

Calvert, D.V., Stewart, E.H., Mansell, R.S., Fiskell, J.G.A., Rogers, L.H., Allen Jr, and Grutz, D.A. (1981). Leaching losses of nitrate and phosphate from a Spodosol as influenced by tillage and irrigation level. *Soil and Crop Science Society of Florida Proceedings,* **40**: 62-71.

Carlisle, V.V., Sodek, F., Collins, M.E., Hamond, L.L., and Harris, W.G. (1989). Characterization data for selected Florida soils. *Soil Sci. Res. Rep.* – 89. University of Florida, Gainesville, Florida, USA.

Castle, W.S. (1980a). Citrus root systems; their structure, function, growth, and relationship to tree performance. *In:* Fourth Proc. Int. Soc. Citrus. Sydney, Australia, pp. 62-67.

Castle, W.S. (1980b). Fibrous root distribution of orange trees on rough lemon root stock at three tree spacings. *J. Am. Soc. Hortic. Sci.,* **105**: 478-480.

Dasberg, S. (1987). Nitrogen fertilizers in citrus orchards. *Plant and Soil,* **100**: 1-9.

Dasberg, S., Biclorai, H. and Erner, J. (1984a). Nitrogen fertilization of Shamati Oranges. *Plant and Soil,* **75**: 41-51.

Dasberg, S., Erner, Y and Bielorai, H (1984b). Nitrogen balance in a citrus orchard. *J. Environmental Quality,* **13**: 353-356.

Douds, D.D., Koch, K.E. and Johnson, C.R. (1987). Carbon pertitioning and P uptake in split root mycorrhizal and non-mycorrhizal citrus. *In:* Mycorrhizal in the next decade. D.H. Sylvia, L.L. Hung, and J.H. Graham (eds.). University of Florida, Gainesvalle, Florida USA, 247 pp.

Duo, H., Alva, A.K. and Khakural, B.R. (1997). Nitrogen mineralization from citrus tree residues under different production conditions. *Soil Sci. Soc. Am. J.,* **61**: 1226-1232.

Ferguson, J.J., Davics, F.S., Tucker, D.P.H., Alva, A.K, and Wheaton, T.A. (1995). Fertilizer guidelines. *In:* Nutrition of Florida citrus trees. D.P.H. Tucker, A.K. Alva, L.K. Jackson, and T.A. Wheaton (eds.). University of Florida, Lake Fl, USA, pp. 27-32.

Graham, J.H. (1984). Application of vesicular-arbuscular mycorrhizal fungi in greenhouse crops: fruits and vegetables. *In:* Application of mycorrhizal fungi in crop production. J.J. Ferguson (ed.). University of Florida, Gainesville, USA, pp. 61-68.

Graham, J.H. (1987). Non-nutritional benefits of VAM fungi—do they exist? *In:* Mycorrhiza in the next decade. D.M. Sylvia, L.L. Hung, and J.H. Graham (eds.). University of Florida, Gainesville, Florida, pp. 237-239.

Graham, J.H., Syverstein, J.P. and Smith, M.L. (1987). Water relations of mycorrhiza and phorphorus fertilized non-mycorrhizal citrus under drought stress. *New Phytologist.,* **105**: 411-419.

Grierson, W. (1995). 400 years of Florida Citrus. *Citrus Industry,* **76:** 28-36.
Hanlon, E.A., Obriza, T.A. and Alva, A.K. (1995). Tissue and soil analysis. *In:* Nutrition of Florida citrus trees. D.P.H. Tucker, A.K. Alva, L.K. Jackson and T.A. Wheaton, (eds.). University of Florida, Lake Alfred, USA, pp. 13-16.
He, Z.L., Alva, A.K., Calvert, D.V. and Banks, D.J. (1999a). Ammonia volatilization from different fertilizer sources and effects of temperature and soil pH. *Soil Science,* **164:** 750-758.
He, Z.L., Alva, A.K., Calvert, D.V., Li, Y.C. and Banks, D.J. (1999b). Effects of nitrogen fertilization of grape fruit on soil acidification and nutrient availability is a Riviera fine sand. *Plant and Soil,* **206:** 11-17.
He, Z.L., Alva, A.K., Li, Y.C., Calvert, D.V. and Banks, D.J. (1999c). Sorption-desorption and solution concentration of phosphorus is a fertilized sandy soil. *J. of Environ. Quality,* **26:** 1804-1810.
He, Z.L., Alva, A.K., Li, Y.C., Calvert, D.V, Stoffilla, P.J. and Banks, D.J. (2000a). Nutrient availability and changes in microbial biomass of organic amendments during field incubation. *Compost Science and Utilization,* **8:** 293-301.
He, Z.L., Alva, A.K., Li, Y.C., Calvert, D.V., Stoffilla, P.J. and Banks, D.J. (2000b). Nitrogen mineralization and transformation from composts and biosolids during field incubation in a sandy soil. *Soil Science,* **165:** 161-167.
He, Z.L., Calvert, D.V, Alva, A.K., Banks, D.J. and Li, Y.C. (2000c). Nutrient leaching potential of mature grape fruit trees in a sandy soil. *Soil Science,* **165:** 748-758.
Hoogweg, C.G. and Hornsby, A.G. (1997). Simulated effects of irrigation practices on leaching of citrus herbicides in Flat woods and Ridge type soils. *Soil and Crop Science Society of Florida Proceedings,* **56:** 98-108.
Jackson, L.K., Alva, A.K., Tucker, D.P.H. and Calvert, D.V. (1995). Factors to consider in a nutrition program. *In:* Nutrition of Florida citrus trees. D.P.H. Tuicker, A.K. Alva, L. K. Jackson and T.A. Wheaton (eds.). University of Florida, Lake Alfred, USA, pp. 1-12.
Johnson, C.R. (1984). Phosphorus nutrition in mycorrhizal colonization, photosynthesis, growth and nutrient composition of *Citrus aurentium. Plant and Soil,* **80:** 35-42.
Johnson, C.R. (1985). Phosphorus nutrition of mycorrhizal colonization and photosynthesis of *Citrus aurentium* L. *In:* Proceedings of the 5th North American conference on mycorrhizal. R. Molina (ed.). Oregon State University, Corvallis Oregon, USA, p. 383.
Khakural, B.R. and Alva, A.K. (1995). Hydrolysis of urea in two sandy soils under citrus production as influenced by rate and depth of placement. *Commun. Soil. Sci. Plant Analysis,* **26:** 2143-2156.
Khakural, B.R. and Alva, A.K. (1996). Transformation of urea and ammonium citrate in an Entisol and a Spodosol under citrus production. *Commun. Soil. Sci. Plant Anal.,* **273:** 3045-3057.
Mack, B.T. (1985). Citrifacts, I and II, A portion of Florida Citrus history. Associated Publications Corporation, Orlando and New York, USA, pp. 13-17.
Menge, J.A., Johnson, A.L.V. and Platt, R.G. (1978a). Mycorrhizal dependency of several citrus cultivars under three nutrient regimes, *New Phytologist,* **81:** 553-559.
Menge, J. A., Labanuskas, C.K., Johnson, A. L. V and Platt, R. G. (1978b). Partial substitution of mycorrhizal fungi for phosphorus fertilization in the greenhouse culture of citrus. *Soil Sci. Soc. Am. J.,* **42:** 926-930.
Nemec, S., Menge, J.A., Platt, R.G. and Johnson, E.L.V. (1981). Vesicular-arbuscular mycorrhizal fungi associated with citrus in Florida and California and note on their distribution and ecology. *Mycorrhizal,* **73:** 112-127.
Nemec, S. (1984a). Application of VA mycorrhizal fungi in fruit crops—field evaluations. *In:* Applications of mycorrhizal fungi in crop production. J.J. Ferguson (ed.). University of Florida, Gainesville, USA, pp. 35-45.

Nemec, S. (1985). Growth of mycorrhizal citrus out-doors in containers. *In:* Proceedings of 5th North America conference on Mycorrhiza. Molina R. (ed.), Oregon State University, Corvallis, Oregon, USA, pp. 243.

Nemec, S. (1987). Gromus intro radium infection and citrus growth in soil free potting media. *In:* Mycorrhiza in the next decade. D.M. Sylvia, L.L. Hung and J.H. Graham (eds.). University of Florida, Gainesville, Florida, USA, 282 pp.

Nemec, S. (1984b). Soil environment, and some biotic factors affecting citrus root health. *In:* Proc. of Indian Society of Citrus Culture, New Delhi, pp. 24-33.

Obriza, T.A., Alva, A.K. and Calvert, D. . (1993). Citrus fertilizer management on calcareous soils. *University of Florida Co-op. Extension Service Circular,* **1127:** 1-9.

Reitz, H.J., Leonard, C.D., Stewart, I., Koo, R.C.J., Calvert, D.V., Anderson, C.A., Smith, P.F., and Resmusen, G.F. (1964). Florida Agriculture Experimental Station Bull. 536B.

Ritter, W.F., Scarborough, R.W., and Chiruside, A.E. (1991). Nitrate leaching under irrigation in coastal plain soil. *J. Irrig. Drain. Eng.,* **117:** 490-502.

Spencer, W.F. (1960). Effects of heavy applications of phosphate and lime on nutrient uptake, growth, freeze injury, and distribution of grape fruit trees. *Soil Science,* **89:** 311-318.

Spencer, W.F. and Koo R.C.J. (1962). Calcium deficiency in field citrus trees. *Amer. Soc., Horti. Sci. Proc.,* **81:** 202-208.

Stoffella, P.J., Calvert, D.V., Li, Y.C., and Hubbel, D.H., (1996b). Municipal solid waste compost (MSW) influence in citrus leaf nutrition, yield and fruit quality. *Proc. of Inter American Soc. for Tropical Horticulture,* **40:** 157-160.

Stofella, P.J., Yuncong Li, Calvert, D.V. and Graetz, D.A. (1996a). Soilless growing media amended with sugarcane filter cake compost for citrus root stock production. *Compost Science and Utilization,* **4:** 21-25.

Tucker, D.P.H., Alva, A.K., Jackson, L.K. and Weaton, T.A (1995a). Nutrition of Florida citrus trees. *Univ. of Florida Co-op. Extn. Serv. Sp.,* **169:** 1-61.

Tucker, D.P.H., Wutscher, H.K., Alva, A.K., Obriza, T.A., Calvert, D.V., Syverstien, J.P., Boman, B.J., and Ferguson, J.J. (1995b). Recommendations for special situations. *In:* Nutrition of Florida citrus trees. D.P.H. Tucker, A.K. Alva, L.K. Jackson, T.A. Wheaton (eds.). University of Florida, Lake Alfred, Florida, USA, pp. 27-33.

Yeomans, J.C., Bremmer, J.M., and Mc Carty, G.V. (1992). Denitrification capacity and denitrification potential of sub surface soils. *Commun. Soil. Plant Anal.,* **23:** 919-927.

Zhang, M., Alva, A.K., Li, Y.C. and Calvert, D.V. (1996a). Broadcast versus fertigation method. *Plant and Soil,* **183:** 79-84.

Zhang, M., Alva, A.K., Li, Y.C and Calvert, D.V. (1996b). Root distribution of grape fruit trees under granular broadcast versus fertigation method. *Plant and Soil,* **183:** 79-84.

Zhang, M., Alva, A.K., Li, Y.C. and Calvert, D.V. (1997). Fractionation of Iron, Manganese, Aluminium and Phosphorus in selected sandy, soil under citrus production. *Soil Sci. Soc. Am. J.,* **61:** 794-801.

Zhu, B. and Alva, A.K (1993). Distribution of trace metals in some sandy soil under citrus production. *Soil Sci. Soc. Am. J.,* **57:** 350-355.

CHAPTER 10

Tropical Forest Ecosystems
Nutrient Dynamics

Contents

1. Tropical Forests: An Introduction
   A. Expanse
   B. The forest climate
   C. Soils
2. Nutrient Dynamics in General
   A. Nutrient inputs and fluxes
   B. Emissions and deposits
3. Land Use Change and Nutrient Dynamics
   A. Gaseous emission and land use change
   B. Clearing and burning versus nutrient dynamics
   C. Nutrient dynamics during land use change in reverse, from farming to forestry
4. Concluding Remarks

1. Tropical Forests: An Introduction

A. Expanse

Tropical forests, in general, play an important role in global energy, biomass and nutrient turnovers. Such tropical evergreen lush forests and woodland ecosystems are extensive. They occupy nearly 25×10^6 km², which is equivalent to 17% of Earth's land surface, and contribute as much as 40% terrestrial net primary production (Keller and Matson, 1994). The humid forests account for approximately 50% total tropical forest belt, with a net primary productivity that ranges between 13 and 28 Mg ha^{-1} yr^{-1}. Whereas, dry tropical forests and woodlands possess a relatively lower net primary productivity, ranging from 8 to 21 Mg ha^{-1} yr^{-1}. Clearly, dynamics of tropical forests, that is their spread, perpetuation or shrinking, as well as variations in productivity levels are partly attributable to underlying nutrient dynamics in both soil and aboveground portions. Some of these aspects are highlighted within this chapter. Globally, three distinct types of humid tropical forests can be identified, namely the

American, African and Indo-Malaysian. The American neo-tropical forest formation covers predominantly the Amazonia, foothills of Andes and Guyanian plateau extending into nearly 40,000 km². Destruction of such neo-tropical forest formation have been greatest in parts of Amazonia, in Costa Rica, Belize and the Caribbean. Peripheral areas include the southeast of Brazil, Central American locations in Equador, Panama and southern Mexico, and small tracts in the Caribbean Islands. The African tropical forests are confined to humid zone in the Congo basin, comprising locations in countries such as Zaire, Cameroon, Gabon, Equatorial Guinea etc. The West African tropical zones are located in Liberia, Eastern Sierra Leone, Coté de Ivore, South West Ghana etc. The African humid forests are also located in Mozambique, Mauritius and Madagascar. The Indo-Malaysian forest formations spread into small areas in Peninsular India (Western Ghats), Northeastern India, and Sub-Himalayas. In Southeast Asia, tropical forests are profuse in the Indonesian archipelago, Malaysian highlands, Sri Lanka, Philippines, Thailand and Vietnam. They are also frequent in Australian locations in Fiji, Papua New Guinea, Queensland, New South Wales, Pacific Islands etc. (Reading *et al.* 1995). In Australasia, humid tropical forests cover over 25,000 km^{-2}.

B. The Forest Climate

Defining the Tropical Forest Environment: There are several different ways of defining tropical zones and their characteristic vegetations. Firstly, an astronomist defines 'tropics' as that part of the globe which occurs between the tropic of Cancer (23.5°N) and tropic of Capricorn (23.5°S). Since, these are latitudinal limits, wherein the sun can lie on the zenith, tropics receive large amounts of solar radiation throughout the year. Distinct seasonal fluctuations are minimal, with perhaps no perceivable winter season, if topography does not confound local temperatures. Holdridge (1967) proposed a classification system based on quantitative aspects of climatic parameters. In this scheme, a climax vegetation with distinct ecology including nutrient dynamics occurs in each forest type. For a given bio-temperature (BT), the vegetational variations will then be a function of precipitation and nutrient status. Similarly, for a given latitude, the climax tree vegetation, and nutrient dynamics are a function of, say, elevation, soil type etc. Longman and Janik (1987) believed that a useful classification for tropical zones should consider a combination of physiognomy of tropical vegetation along with environmental parameters. One such system of classification proposed for tropical vegetation, by Holdridge (1967) is shown in Table 10.1.

Tropical forest zones can also be defined and identified based on mean annual precipitation. The mean annual precipitation needed to define a humid tropical vegetation depends on temperature-altitude zone. It is

Table 10.1: Humid tropical forest zones grouped based on physiognomy and climato-geomorphology

Bioclimatic zone	Rainfall (mm)
1. Lowland (L) zones (BT > 24°C)	
L Moist forest	1500-4000, less than 4 months with less than 200 mm
L Wet forest	4000-8000, less than 2 months with less than 200 mm
L Rain forest	>8000, no months with less than 200 mm
2. Premontane (P) zones (BT 18-24°C)	
P Moist forest	1000-2000, 2-4 months with less than 100 mm
P Wet forest	2000-4000, less than 2 months with less than 100 mm
P Rain forest	>4000, no months with less than 100 mm
3. Lower montane (LM) zones (BT 12-18°C)	
LM Moist forest	1000-2000, 2-4 months with less than 100 mm
LM Wet forest	2000-4000, less than 2 months with less than 100 mm
LM Rain forest	>4000, no months with less than 100 m
4. Montane (M) zones (BT 6-12°C)	
M Moist forest	500-1000, 2-4 months with less than 50 mm
M Wet forest	1000-2000, less than 2 months with less than 50 mm
M Rain forest	> 2000, no months with less than 50 mm

Source: Holdridge (1967) as quoted by Reading *et al.* 1995

Note: Longman and Jenik' (1987) system also makes more detailed demarcation of vegetation, even within the subclassifications of forests shown in Holdridge's 1987 system—Examples, lowland moist forests could be split further into broad-leaved, needle-leaved, riparian-flooded, swampy, seasonal evergreen, deciduous etc.

highest (>2000 mm) for low lands that occur between 0 and 500 meters above sea level, and lowest (125 mm) for montane zone (>4500 meters above sea level). Using this classification, we can relate vegetation communities directly to climatological parameters. However, variations in vegetation due to topographic, hydrological and edaphic factors are not provided primacy.

Yet another approach, as outlined by Walter (1979) utilizes combination of altitudinal and rainfall gradients to classify tropical forests. It was effectively employed to demarcate and study the Venezuelan humid vegetation. Main constraint identified in this system is that, temperature zones which alter through altitudinal gradient or other factors, may confound perceptions on tropicality.

Considering the theme of this book, it should be realized that general fertility status, nutrients, their transformations, recovery and recycling ratios, i.e., nutrient dynamics and equilibrium levels reached within tropical forest ecosystems are immensely important. It can confound the sub-classification arrived at, using any of the above definitions. Hence, soil nutrients and their dynamics in forest, that is soils and aboveground portions should be provided due weightage, while defining a tropical vegetation. On many occasions the dynamics of tropical forests, their spread or extinction could be directly related to nutrient dynamics, and perhaps comparatively less to other factors. In fact, tropical forest development and spread is intricately woven with the variations in soil nutrient dynamics.

The Forest Microclimate: A well developed tropical forest canopy will definitely modify the general climatic parameters to create its own unique microclimate within the forest canopy that differs, at times markedly compared with the outside environment. Modifications perceived in microclimate usually relate to climatological parameters such as solar radiation, temperature, soil moisture, humidity etc. Alterations in climate will have enormous effects on vegetation, its growth physiology, nutrient recovery, and productivity etc. Simultaneously, a range of forest soil related parameters too are altered, such as soil temperature, moisture, aeration etc. which, in turn, regulates volatilization and emissions of N and C, rates of physico-chemical transformations etc. In greater detail, the impinging solar radiation (insulation), its interception and pathway through the forest canopy is of great importance to nutrient dynamics. In this regard, both the intensity and spectral properties of the impinging radiation are of value. The modification of insolation pathway allows stratification into "euphotic zones" at the top of the canopy which receives 25 to 100% insolation. Under it, the oligotrophic zone receives 3% irradiance or even less. The ratio of visible red ($\approx$ 600 to 700 mm) and infra-red (700 to 1000 mm) radiation that impinges and infiltrates is known to influence vegetation, from seed germination to growth until senescence. This ratio is high in the canopy but less in under story. Such variation in the flux of photosynthetic irradiance will directly influence net primary productivity, carbon accumulate and recovery of other nutrients into forest plantation and related transformations in soil. Temperatures vary markedly within a forest stand. Under story temperatures are generally lower than above the canopy or in clearing. Soil temperatures too are altered by a forest stand. Such variation in air and soil temperatures can be discerned even between forest types such as evergreen, riparian, and deciduous forests and within them. For example, differences in diurnal fluctuations are common. The peak temperatures recorded at 14.00 hr in a Ghanian closed forest was 26°C, but that in adjacent clearings was 33°C (Longman and Janik, 1987).

Precipitation is a crucial input that influences several aspects of tropical forest, among which nutrient dynamics is important. Precipitation is highly variable, both in spatial and intensity terms. Generally, partitioning of rainfall within a tropical forest is one of the prime determinants of vegetation pattern, its productivity as well as the nutrient dynamics that ensue. Interception of rainfall is high in tropical forests, mainly due to thickly foliated crown, epiphytes and climbers. Stem flow, which is low often gets harvested by epiphytes. Throughfall is enhanced due to dripping. However, on rainless days, dew is an important source of moisture. The dew component contributes between 100 and 300 ml m^{-2}. The extent of partitioning effect on precipitation received by a tropical forest varies (Vitousek and Sanford, 1986). For example, Puerto Rican rain-forests intercepted 27 to 38% of total precipitation, stem flow was 0 to 1% and throughfall 62 to 73%. Whereas, Malaysian low land forests intercepted 22% precipitation, stem flow varied between 0 to 0.5% and throughfall accounted for 77%. In a natural teak forest of Thailand, interception was 63% and throughfall 37% (Kellman and Tackaberry, 1997). Variation in precipitation and its partitioning actually influences a wide set of functions in tropical forests. Soil moisture levels are equally important, if not more on certain occasions. Again, local variations in soil moisture profoundly affects nutrient availability, and their dynamics in general. In turn, forest productivity is also affected. It is said that within a space of twelve months, a forest location may go through several periods of wilting points that alternates with saturation of soils with moisture. It is also common to encounter variations, both in relative humidity and saturation vapor pressure. Usually, such differences are conspicuous in locations above or below the canopy and between the mature forest and clearings, on a diurnal basis or depending on seasons (Reading *et al.* 1995).

C. Soils

Basically, pedogenesis in humid tropics is immensely influenced by intense rainfall periods, topography, altitude and chemical nature of parent material. High rainfall aids soil processes such as base leaching, clay and nutrient translocations. Obviously, tropical forest soils show wide variations in physico-chemical properties, fertility in general, and specific nutrient contents. In Southeast Asia and Central America, fertile volcanic Andosols support large expanses of tropical forests. Whereas, forests soils of Amazonia and Central Africa (Congo) are mainly Oxisols and Spodosols deficient in several essential nutrients. According to Sanchez (1981), Oxisols and Ultisols predominate in the humid tropics occupying nearly 63% of the area. These soil types are more prevalent in tropical American forest belt accounting for 82% of the forest area. Whereas, in Asia they constitute 38% of forested area. Both, Oxisols and Ultisols are acidic, infertile, and derived from prolonged weathering of silicate parent material. Kaolinites form the predominant clay fraction which determines the CEC and water holding capacity of soils.

Dense foliage and canopy in tropical forests interrupt light efficiently, allowing a mere 2% radiation to infiltrate to the forest floor. Also, the micro-climate in general, temperature and humidity more specifically are held well buffered under such thick forest canopies. Temperature variations at the forest floor (soil surface) are limited to 2°C daily, while humidity stays saturated. Soil temperatures generally raise in places where forests are cleared for shifting cultivation. Clear cutting and/or burning definitely enhances day time temperature. Lal (1986) reported that even at 10 cm soil depth, temperatures under agricultural crops were 5°C greater than forest floor. Clear cutting followed by agricultural cultivation disturbs soil structure enormously. It creates soil compaction, increases bulk density and reduces porosity. For example, in Nigerian tropical forests, mechanical clearing increased bulk density of top 10 cm layer by 22% over a forested control plot. Whereas, traditional clearing methods increased bulk density by 14% over forested control (Ghuman *et al.* 1991). In Brazilian forests, plots cleared for agricultural purposes showed an ever increasing trend in bulk density as field crop cultivation proceeded. After few years, bulk densities nearly doubled over forested controls. Soil compaction and reduced porosity effected via clear cutting or farming, immensely affects water infiltration and soil aerobicity. For example, water infiltration rates were only 36% of the forested control plots in Nigeria. Similar results were obtained in Costa Rica (Keller, 1994). Incidentally, infiltration of water and soil air percentage together control severe, if not all nutrient transformations that occur in the tropical soils. It also includes tree root activity and nutrient absorption.

2. Nutrient Dynamics in General

The trend in the previous chapters has been to discuss physico-chemical aspects of soil nutrients, basic transformation reactions involving major and micronutrients. In other words, much of the relevant literature related to tropical soils, their chemistry and nutrient transformations are already presented in previous chapters (1, 2, 3, 4, 7). Both, to avoid monotony, and to maintain brevity of this book, discussions in this chapter are confined to recent progress in specific aspects of nutrient dynamics in tropical forests. Mainly, studies on C and N dynamics have been highlighted. At this point, it is suggested that background informations and greater insights into other nutrients (P, K and micronutrients) are available in textbooks on tropical soils or forest soils. Again, I have preferred not to dwell excessively on the influence of tropical forest composition, plant species, their physiology etc. on nutrient dynamics. This does not reflect on their importance to nutrient dynamics in tropical forest, they are crucial too. Let us now consider some of the recent, and pertinent aspects published on nutrient (C and N) dynamics on tropical forests.

A. Nutrient Inputs and Fluxes

Major avenues of nutrient input in a humid tropical forest ecosystem are precipitation, aerosol deposition, nitrogen fixation and natural weathering. Such nutrient inputs will vary, both qualitatively and quantitatively, depending on extent of dust or aerosol deposition, regional dust storms, trends in precipitation, forest fires, volcanic eruptions, soil type, its parent material, and management practices such as shifting agriculture. Reading et al. (1995) suggest that sometimes nutrient inputs through precipitation varies even between two rainfall events that occur in short intervals. Actually, in many tropical forest stands, major fraction of the yearly total rainfall occurs within a few storms. Therefore, much of nutrient inputs are received within a short span of few rainfall events. Table 10.2 depicts the extent of yearly nutrient inputs through precipitation.

Table 10.2: Annual rate of nutrient precipitation in humid tropical forests

Location	Nutrient inputs via precipitation (kg ha^{-1} yr^{-1})		
	N	P	K
Ghana	14	0.4	17.5
Ivory Coast	21	2.3	5.5
Venezuela: location 1	21	25.0	23.4
Venezuela: location 2	9.9	1.1	2.6
Terra firme	21.7	24.9	24.6
Papua New Guinea	6.5	1.1	2.6

Source: Excerpted from Reading et al. 1995.

In a given time, soil nutrients recovered by tropical trees may not necessarily tally with inputs derived from natural chemical weathering and atmospheric depositions. As such, nutrients derived through rock weathering is a slow process, and depends on factors that influence weathering processes and rates. Quite often, nutrients released into soil via natural weathering stays well below in the lower horizons of soil, and out of reach for tree roots. In fact, a large fraction of nutrients absorbed and utilized by forest vegetation, actually belongs to recycled portions, and not from natural weathering. Often, the nutrient fluxes attained through recycling processes, such as litter fall, root production, senescence etc. are much higher than that achieved by natural chemical weathering (Table 10.3).

Litter and Nutrient Flux: A major component of forest biomass that drives the nutrient cycling process is the decomposition of organic matter (litter), that contains valuable quantities of nutrients. Since, measurements of fine litter and its decomposition rates provide an indirect measure of nutrient turnover, this parameter has been used to compare nutrient

Table 10.3: Weathering inputs and minerals recycled annually in litter flux in a tropical forest situated in Venezuela, South America

Nutrient	Weathering input (kg ha^{-1} yr^{-1})	Forest nutrient flux (kg yr^{-1})		
		Litter fall	Root production	Total
Calcium	7.6	13.0	4.2	17.2
Magnesium	3.4	5.2	5.2	10.4
Potassium	7.6	15.4	13.5	25.9
Phosphorus	0.22	2.2	2.3	4.4

Source: Kellman and Tackaberry, 1997.

cycling rates in different forest stands. In a given forest section at steady state, the litter decomposition rate (k) is often assessed using the ratio of annual fine litter fall divided by standing stock of litter on forest floor. For example, a ratio 'k' of value 1.0 indicates total litter turnover in one year, while >1 indicates more rapid decomposition than litter deposition rates; while < 1.0 means slower turnover i.e. litter will be left on soil surface at the end of year. We can expect that litter turnover coefficient will influence the nutrient turnover rates proportionately. Kellman and Tackaberry (1997) suggest that litter decomposition could be considered as a cascade of complex degradative processes mediated by the biotic component. Litter turnover ratios and quantities of nutrients recycled are dependent on litter quality. Generally, litter rich in lignin and secondary metabolites, or woody material low in nutrients decomposes slowly, and contributes meagerly to nutrient cycling. These are commonly termed low quality litter. Leaves, twigs and succulent portions, rich in nutrients decompose rapidly, therefore are called 'high quality' or fine litter. Climatic factors and soil microflora also influence nutrient turnover rates.

Rapid and Slow Nutrient Fluxes: Nutrient fluxes in tropical forests may be either rapid or relatively slow. However, Procter (1987) suggests that accurate distinction between rapid and slow nutrient fluxes might be difficult, mainly because, flux rates within forests are variable and differences are hazy. Typically, rapid fluxes are contributed by small litter fall composed of comparatively succulent debris of leaves, twigs, flowers and fruits. Nutrient drips caused by throughfall from the forest canopy and stem flow along the tree stand are other contributors to rapid flux of nutrients. The small litter fall estimates in low land tropical forests range from 34 t ha yr^{-1} (Reading *et al.* 1995). Specific nutrient fluxes caused by such small litter vary spatially and temporally, as well possess strong seasonality. Seasonal effects are less marked in perennial moist forests, compared with those prone to drought or seasonal fluctuation in moisture, temperature, humidity etc. Leaf fall is also influenced enormously by the seasonal changes in tree physiology versus environmental interactions. Leaf shedding is dependent on soil moisture, deciduous nature of forest stand, photosynthetic functions etc. The timing of leaf senescence varies widely depending on forest stand, tree species composition and environment. It means that, concomitant nutrient flux derived through litter fall caused by small branches, twig fall, flower and fruit drop shows less seasonality than leaf fall. Nutrient deposits retained in forest canopy in general, more specifically on leaf surfaces, vegetation mats created by climbers and twiners are efficiently leached via throughfall and stem flow. Nutrients leached into soil from canopy depends on temperature, precipitation, leaf area index, and extent of contact or retention of water on leaf surface. Essentially, water that drips through the canopy carries with it dissolved nutrients and sediments/dust particles. The nutrient element and its quantity fluxed via throughfall or stem flow, obviously depends on extent of atmospheric deposition and selective removal by epiphytes, phyllosphere microflora etc. Preferential movement of certain nutrients via a certain pathway is possible. For example, more of potassium is known to traverse via throughfall. Indeed, nutrients fluxed via small litter decomposition, throughfall and stem flow are very important to tree nutrition and forest growth, mainly because nutrients are made available rapidly and in solution. Most humid, hot low land forests support rapid nutrient fluxes, because of high rates of microbial activity and organic matter decomposition. According to Reading *et al.* (1996), slow fluxes caused by branch falls are much less investigated. Relatively small pools of mineral nutrients may be stored in large branches, trunks and wood material, and they may stay unreleased for longer durations.

B. Emissions and Deposits

Nitrogen depositions into tropical forest ecosystem from atmosphere has increased enormously. In addition, chronic inputs of fertilizer N into forest plantations, which were previously N deficient, definitely alters both nutrient dynamics and productivity. It has been pointed out that such chronic additions, which have spanned for over decades, also allow retention of large quantities of N in soil. Such retentions usually subside, and stabilize only as 'N saturation' levels are reached and N losses become conspicuous. In tropical Hawaiian forests, such anthropogenic N retention may last for over decades, before the onset of N losses as N_2O emission or NO_3 leaching. There are several estimates and extrapolations regarding anthropogenic N depositions in tropical forests, extent of N sink and emissions possible. Even considering tropical forests as weak N sinks, and that only 2% of it is lost to atmospheric as N_2O or NO emissions, such anthropogenic N fluxes will substantially increase by the year 2020. At that time, it would be about 0.4 Tg N yr^{-1} for N_2O and 0.2 Tg N yr^{-1} for NO. This represents a 13% and 18% increase over current levels of

N_2O (3 Tg N yr^{-1}) and NO (1.1 Tg N yr^{-1}) emissions in tropical evergreen forests (Davidson and Kingerlee, 1997). It could be detrimental to global atmospheric composition and regional tropospheric photochemistry.

NO Emissions: Methane and nitrous oxide are long-lived gases, whereas, nitric oxide (NO) is a short-lived, reactive gas, and global atmospheric budgets are not well deciphered. Also, proper estimation of NO flux from soil to atmosphere is difficult, because, NO can react and be deposited within plant canopies. According to Bakwin *et al.* (1990) internal deposition of NO within forest canopy may account for 80% of NO generated. Overall, NO emissions from undistributed tropical forests are greater in quantity than temperate vegetation (Keller and Matson, 1994; Hall and Matson, 1999). In the tropical forests, N dynamics is significantly influenced by soil N status, its anthropogenic inputs and other major nutrients. In P limited forest plots, short term N inputs may itself cause substantial N_2O and NO emissions. Such soil N_2O emissions reached 1 kg N ha^{-1} during the first 2 weeks after N inputs, which are almost equivalent to that reported for agricultural sites. Whereas, ^{15}N investigations suggests that unfertilized forests soils, which are N deficient are extremely good sinks for nitrogen. Nearly, 68% of $^{15}NH_4^+$–N is removed from the inorganic fraction within first 15 min. after addition, compared with only 13% removed form soils fertilized with N for long durations, or those which had reached N saturated conditions. Further, it was observed that repeated N inputs significantly alters N immobilization by microbes, and reduces net N demand by soil microbes. At this stage, microbial growth and turnover gets limited by factors other than N, if any. For example, under P limited conditions, chronic N additions did not alter N availability, gross N mineralization or even net NH_4^+ immobilization.

Methane Emission/Utilization: Well drained, aerobic soils are typically good sinks for methane. Methane consumption is regulated by its diffusions from atmosphere into soil pore space. Based on diffusivity estimates, soil methane consumption is calculated to hover around 29 Tg CH_4 yr^{-1}. Tropical forests or savannas may, on an average, absorb 1 mg CH_4 m^{-2} d^{-1} (Keller *et al.* 1986; Dalmas *et al.* 1992). Actually, methanogens which are obligate anaerobic bacteria mediate methane production in tropical forest soils. Commonly, acetate or hydrogen plus CO_2 are utilized during this process. However, a range of other substrates such as formals, methanol, methyl amines, carbon monoxide and dimethyl sulfide may also be utilized. Within the tropical ecosystem, methane production proceeds under severely anaerobic conditions in soil, in fresh and salt water sediments, flooded soils (wetlands) etc. Accumulation of NO_3 and SO_4 may inhibit methanogenesis (Cicerone and Oremland, 1989). On the other hand, methylotrophs are obligate aerobic microbes known to utilize

Cl compounds as substrate for energy. In terms of soil chemical transformations they are similar to any other chemolithotrophs such as nitrifiers.

Regulation of Soil-Atmosphere Fluxes of N_2O and CH_4: Measurements of fluxes of NO_2 and CH_4 between soil-atmospheric in the tropical forests has helped in gaining insights into biological, and physico-chemical controls on emissions of two major essential elements. Keller (1994) has summarized inferences drawn from over 70 different experiments in tropical South America. He actually segregates data on N_2O and CH_4 emission into 4 quadrants in a graph, wherein positive values indicate net emissions away from soil, and negative fluxes indicate net consumption of gas by soil (Fig. 10.1). Nitrous oxide fluxes in these experiments ranged between 0.8 and 11.7 Ng N cm^{-2} L^{-1}. Methane fluxes fluctuates between— 1.4 and 64 J mb m^{-2} d^{-1}. Since most measurements were located within first quadrant, it is clear that, predominantly NO_2 was emitted, but CH_4 was consumed by soils. Simultaneous emissions of both CH_4 and NO_2 do occur, however, simultaneous absorption of CH_4 and NO_2 were not recorded (quadrant III). Soil aerobicity and microbial activity form the crux of the mechanisms involved in NO_2 and CH_4 flux. Clearly, aerobic soils support moderate or high NO_2 emission and consumption of CH_4 (quadrant I) production of N_2O by nitrifiers or denitrifiers is possible under low oxygen tension. Occurrence of anaerobe microsites in soil

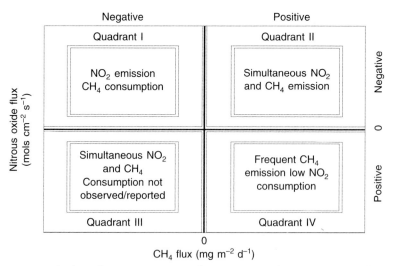

Fig. 10.1: Method to depict gross NO_2 and CH_4 emissions/consumption in relation to aerobicity/anaerobiosis in tropical forest soils.
Note: Fluxes depicted are based on data reported by Keller (1994). Positive indicates net emissionsof CH_4, NO_2 away or from the soil into atmosphere and negative indicates net consumption of these gases by soil.

explains NO_2 emissions, even though bulk soil may be aerobic. Keller (1994) explains that large CH_4 emissions in quadrants 3 and 4 correspond to anaerobic conditions.

Interlinked to soil aerobicity is the soil moisture. In fact, level of available soil moisture filled air pore space directly regulates aerobicity and gas exchange, hence NO_2 and CH_4 emission. Three appropriate soil moisture regimes were identified, namely (a) dry to moist, (b) moist to wet, and (c) saturation. In all three regimes, the extent of soil moisture dictates gas exchange. In the first case, that is under dry conditions soils are highly aerobic. Therefore, occurrence of anaerobic microsites in soil are limited, so the productivity of N_2O and CH_4 too are limited. Under very dry conditions, N_2O emission may become negligible because the microbial activity needed for it is highly restricted. In moist soils, pore space is filled up with water, thus restricting gas exchange. Aerobic packets are relatively less, and anaerobic microsites increase. Methane consumption by soil gets limited because of restricted gas diffusion. On the other hand, methanogenesis in anaerobic soils, actually balance out any emissions and leads towards zero flux for methane. In moisture saturated soils, anaerobiosis predominates therefore soils emit CH_4. Microbe-aided denitrification continues, but NO_2 emission might be impeded, because of restrictions to gas exchange at saturated moisture conditions. Keller (1994) further explains that in addition to soil moisture versus soil air interactions and their influence on NO_2 and CH_4 dynamics, soil properties such as texture, nutrient conditions especially nitrogen and carbon status, soil mineraliation potential etc. are equally important. For example, in Brazilian tropical forests, N mineralization rates directly correlated with nitrous oxide fluxes (Matson and Vitousek, 1987). Nitrogen fertilization may cause higher NO_2 fluxes. In fact spatial and temporal variations of NO_2 and CH_4 fluxes are common, particularly after fertilization of tropical forest soil. Soil water contents together with soil structure (aggregates) also play an important role in NO_2 and CH_4 dynamics in soil, due to their influence on soil redox conditions (Weitz et al. 1998; Weitz et al. 1999). Light textured soils support less anaerobic sites, hence NO_2 and CH_4 are appropriately influenced based on extent of aerobicity. Clayey soils are known to cause greater levels of NO_2 emissions (Matson et al. 1990).

Emissions of Volatile Organic Compounds: Iseprene and Mono-Terpene:
Tropical forests emit different "volatile organic compounds (VOC)", which play an important role in the regulation of oxidants in atmosphere, balancing global carbon cycle, producing organic acids and inducing acidic depositions (Fehsenfeld et al. 1992). Among the organic compounds, emissions of isoprene and mono-terpene have been frequently investigated. Rasmussen and Khalil (1998) state that perhaps, tropical forests contribute to more than half of the global isoprene emissions. Within tropical forests,

around Manaus in Brazil, isoprene emissions were estimated at 25 to 38 mg m^{-2} d^{-1}, and monoterpenes at 6 mg m^{-2} d^{-1} (Zimmerman *et al.* 1988). Tropical forest vegetation also emits carbon monoxide. Kirchoff and Marinto (1990) reported strong sources of CO emission from tropical forest vegetation. It ranges from 4 to 10^{11} mols cm^{-2} s^{-1}. However, effects of land use change on CO emission, and the mechanism(s) of CO emission have not been understood clearly. Isoprene emission by forest vegetation accounts for 40% of global hydrocarbon emissions to atmosphere (Guenther *et al.* 1995). Certain other studies indicate that 70% of global isoprene emission (350 Tg out of 500 Tg) is contributed by tropical forest plants. In the atmosphere isoprene is then oxidized to carbon monoxide and organic acids. Isoprene emitting plant species have been traced and reported from several natural tropical vegetation zones. The extent of isoprene emissions may however vary. Table 10.4 provides an idea regarding isoprene emissions at standard temperature ranges by a few tree and vine species identified in Panamenian tropical forests. Obviously, the isoprene emission rates are influenced by plant species richness and diversity, growth stage, habit, foliage etc. Basal isoprene emission rates may also vary depending on season, temperature and photosynthetically active radiation (PAR). Higher temperature in tropical vegetation and greater intensity of PAR induces isoprene emissions. Sometimes, reaching as high as 105 mm m^{-2} s^{-1} in tropics. Mathematical models to predict isoprene emissions, which consider a range of natural factors such as temperature, plant species, radiation, season and several others have been

Table 10.4: Isoprene emission rates under standard condition of temperature, and photosynthetically active radiation in a tropical zone in Republic of Panama

Species and family	Isoprene flux (n mols m^{-2} s^{-1})
Acrocomia vinifera (tree), Palmae	20.2
Annona hayasii (tree), Annonaceae	8.9
Astronomium graneolus (tree), Anacardiaceae	26.0
Bonamia maripoidus (tree), Convolvulaceae	18.0
Cissampelos pariera (vine), Menispermaceae	27.8
Cnestidium rufescens (vine), Connaraceae	26.2
Dioclea guianensis (vine), Papilionoideae	43.3
Dolicocarpus major (vine), Dillinacea	32.4
Ficus insipida (tree), Moraceae	36.5
Ficus spp (tree), Moraceae	15.9
Leuhea sumanii (tree), Tiliaceae	23.6
Spondias mombin (tree), Anacardiaceae	32.5
Stigmaphyllon hypargyreum (tree), Malpighiaceae	36.3
Xylopia frutiscens (vine), Annonaceae	14.6

Source: Keller and Lerdau, 1999.
Note: Growth habit is shown in the parentheses

proposed (see Guenther *et al.* 1993; 1995; Keller and Lerdau, 1999). Keller and Lerdeu (1999) opine that the global model proposed by Guenther *et al.* (1993) which considers temperature and irradiance as major factors may underestimate actual emissions. Whereas, their recent model may too underestimate the actual isoprene emission rates, but only slightly because it considers several other parameters.

3. Land Use Change and Nutrient Dynamics

Deforestation in the tropics is one of the most important changes in the Earth's agricultural ecology and soil processes. Satellite and field surveys indicate that, globally approximately 57 to 75000 km^{-2} of tropical broad-leaved forests were converted to shifting agriculture or prominent farming. Deforestation related land use change seems most rampant in Brazil (80,000 km^2 yr^{-1}), whereas in South Asian countries such as India and Indonesia, deforestation is estimated at 10,000 km^2 yr^{-1} (Repatto, 1990). Houghton (1994) has remarked that deforestation is widespread in tropical forest belts, and approximately 20 to 30% has already been converted. According to FAOSTAT (1999) statistics, deforestation rates in tropics may range between 0.8 and 1.8% per year. By definition, deforestation involves a wide range of processes, such as selective removal of trees for timber, clear cutting and burning, creation of pastures, agricultural fields, meadows etc. Such deforested zones, quite often remain unused, or at times reforestation is achieved via replanting trees. In the tropics, deforested zones may be converted pastures, fallows, shifting or permanent agriculture. Burning adopted during deforestation may increase soil pH in humid tropical soils. Higher weathering ratios that ensue after burning, may increase soil acidity (Reiners *et al.* 1994). Most importantly, nutrient loss as consequence of land use change decreases soil fertility and productivity in the long run.

A. Gaseous Emission and Land Use Change

Land use trends in humid tropics have been dynamic, and sometimes drastic with high rates of forest clearing to establish pastures, or agricultural fields. Such cleared land may be abandoned as unused fallow. Depending on the nature of change in land use, its intensity and expanse, the gaseous emissions of N_2O, NO or those such as CO, CH_4, or isoprene may vary Nepstead *et al.* (1999). It is important to quantify land use dependent changes in N_2O, and/or NO fluxes, because these gases play an important role with *in situ* N cycles and atmospheric chemistry (Verchot *et al.* 1999). N_2O is a green house gas with radiative forcing strength per molecule being 200 times greater than CO_2 (Shine, 1995). N_2O causes stratospheric ozone destruction (Cicerone, 1987). Globally, forests in humid tropics may account for 20 to 50% of total N_2O emission into atmosphere (Riley and Vitousek, 1995; Potter *et al.* 1996). According to Verchot *et al.* (1999) deforestation is a plausible cause for increasing trends in N_2O and NO emission from tropics. There are explicit instances to support this hypothesis. Pastures developed after forest clearing emitted larger doses of N_2O and NO in Manaus, Brazil (Matson *et al.* 1990). Similarly, Keller *et al.*(1993) observed 5 to 8-fold increase in N_2O emission (6.7 kg N ha^{-1} yr^{-1}) in young pastures in Costa Rica, after forest clearing. Such changes in N_2O emission were drastic during the initial stages of change to pasture, than after longer durations. However, on an average, Brazilian Amazonian forest may emit about 2 (± 0.5) kg N_2O–N ha^{-1} yr^{-1} (Verchot *et al.* 1999). Additionally, Verchot *et al.* (1999) have compared N_2O fluxes in different land use configurations in eastern Amazonia, and reported that N_2O emission followed the order: primary forest (2.4 kg N ha^{-1} yr^{-1}) secondary forest (0.9 kg N ha^{-1} yr^{-1}) > active pasture (0/3 kg N ha^{-1} yr^{-1}); > degraded pasture (0.1 kg N ha^{-1} yr^{-1}). Whereas the NO fluxes followed the order: primary forest (1.5 kg N ha^{-1} yr^{-1}) > degraded posture (0.7 kg N ha^{-1} yr^{-1}) $\geq$ active pasture (0.5 kg N ha^{-1} yr^{-1}) > secondary forests (0.3 kg N ha^{-1} yr^{-1}). Further, they inferred that N_2O : NO emission rates could be related to soil water contents. Overall, it is believed that small patches of forest clearing in Amozonia may not significantly contribute to increases in soil emission of N_2O or NO into atmosphere. Based on several studies in Amazonia, it was concluded that tropical deforestation as a factor may contribute 0.4 Tg N yr^{-1} to global N_2O budget, which is equivalent to 7% of annual anthropogenic source.

Land use change from tropical forestry to pastures/cropping can result in decreased rates of N cycling and induce changes in soil inorganic—N pools. If NO_3–N dominated forest soils, it was NH_4^+—N in pastures. Such shifts could be related to changes in nitrification rates. Overall, conversion of primary forests to man made pasture comparatively reduced net mineralization, and net nitrification rates (Reiners *et al.* 1994; Matson *et al.* 1990; Piccolo *et al.* 1994; Null *et al.* 1995; Verchot *et al.* 1999). In this regard, models may be helpful in obtaining better predictions on N_2O : NO emissions. Recently, Verchot *et al.* (1999) evaluated one such conceptual model depicting nitrification, denitrification sequence, and the extent of N_2O and NO emissions. It is called the "Hole-in-the-pipe conceptual model", that simulates soil emission of N_2O and NO (Firestone and Davidson, 1989). They report that observed N_2O and NO fluxes agreed with predicted values that were derived from the model equations for several locations.

Land Use Change and Methane Emissions: Conversion of natural tropical forest soil to agriculture purposes, generally diminishes the sink capacity of soil for methane. For example, conversion of tropical forest to pastures in Panama decreased soil methane consumption by three times (Keller *et al.* 1990; Fearnside and Barbosa, 1998). The reason suggested is that pastures or agricultural soils are compacted, and impede drainage. Therefore, wetter soils allow less methane diffusion and absorption. Some observations suggest that conversion of forest to pasture may directly alter methane fluxes. Forest soils which are generally sinks to CH_4, may actually become small CH_4 emitters due to land use change (Table 10.5). Recently, Houghton *et al.* (2000) reported that annual flux of carbon from Brazilian Amazonia was about 0.2 Pg C yr^{-1} over the period 1989-1998 (1 Pg is 10^{15} g).

Table 10.5: Methane and nitrous oxide fluxes from soil to atmosphere or vice versa from different land use categories in Costa Rica (Central America)

Land use	CH_4 flux (mg m^{-2} d^{-1})	N_2O flux (ng-N m^{-2} d^{-1})
Forest	− 1.22	6.75
Young pasture	+0.65	48.95
Old pasture	+0.65	2.20
Crops	−1.22	29.00

Source: Keller and Matson, 1994

Note: Negative fluxes indicate net transfer of a gas from atmosphere to soil (i.e., consumption by soil). Positive fluxes indicate net transfer of a gas from soil to atmosphere (emission from soil).

This estimate considered deforestation rates, regrowing forests and biomass. Further, the annual emission of carbon from land use change and fire approximately offsets the sink calculated for natural forest ecosystem. Therefore it is inferred that fluxes in such a large belt as Brazilian Amazonian forests are nearly balanced with respect to carbon, with an annual variability of ±0.02 P, C yr^{-1}.

B. Clearing and Burning versus Nutrient Dynamics

In tropical forests, clearing and burning is a common practice due to which substantial amounts of nitrogen and other gases are emitted from the ecosystem (soil plus vegetation). Volatilization during fire, and emissions from soil are the main avenues of loss. Cuertzen and Andrae (1990) state that each year upto 60 m ha of tropical forest and savannas may be cleared and burned, hence may have series of implications to nutrient turnover, atmospheric chemistry, climate and ecology. Approximately 4 to 14% of nitrogen emissions after clearing and burning occurs as N_2O + NO. Globally, it translates to 2.5 to 1.3 Tg yr^{-1}, equivalent to 10 to 54% of annual atmospheric input of NO_x from fossil combustion (Cuertzen and Andrae, 1990; Watson *et al.* 1992). NO_2 deposited on tropical forest soils can be significant. Based on approximations on ash depositions (670 gm^{-2}), by Ewel *et al.* (1981; 1986), and recorded concentrations of 22 mg NO_2 g ash^{-1} immediately after burning a secondary forest site in Costa Rica, Neff *et al.* (1999) derived that roughly 1474 mg NO_2 m^{-2} were deposited as ash. On the soil it may roughly create 2 ppm of NO_2 in upper 1.0 cm soil layer. On a regional scale, tropical forest burning also leads to elevated levels of ozone during the dry seasons over South America and Equatorial Africa (Andreay *et al.* 1988; 1994).

Gaseous emissions from vegetation and soil may be dependent on the clearing and burning method. Such procedures are in turn dependent on type of forest vegetation, location, local practices etc. In general, burning process initially removes small vegetation and litter fall. Then, remaining larger woody material are ashed. For convenience during experimental evaluations and considering relationships between burning and nutrient dynamics, Neff *et al.* (1999) identified three stages namely:

i. Pre-clearing stage before the removal of forest canopy;
ii. Post-clearing stage, but before initial burning episodes and
iii. Post-burning stage that follows immediately after complete burning.

Such characterizations may be helpful in deciphering the effects of clearing and burning, more accurately and efficiently. Now, let us fit an example to the pattern of clearing and burning stages. In tropical Atlantic coasts of Costa Rica, soil-atmosphere NO fluxes average only 0.5 mg N cm^{-2} hr^{-1} at stage-1-prior to clearing and burning. It increased to 4.1 mg N cm^{-2} hr^{-1} following clearing, but before burning. At stage-3 following burning, emission tripled to 12.0 mg N cm^{-2} hr^{-1}. Neff *et al.* (1999) report that such elevated NO emissions may proceed for up to 3 to 4 days after a burning event. Table 10.6 provides an example about fluxes of N_2O and NO from biomass burning and that from soil, as influenced by burning and conversion to pasture in a Costa Rican tropical forest. These magnitudes of NO fluxes reported for La Selva in Costa Rican forests, tally well with those found at California (Anderson *et al.* 1988), Mexican tropical forests (Davidson *et al.* 1991), or those in Venezuela (Johansson *et al.* 1988). Since, such phenomenon of enhanced N_2O and NO emissions are known from large number of tropical forest locations, it would be useful to understand the underlying mechanisms, and any possibilities for regulating the process. As such information on effect of clearing and burning on biotic components (microbes etc.) that mediate N_2O or NO production is meager (Neff *et al.* 1999).

We may realize that nitrogen loss or emissions from tropical forest soils occur due to biological nitrification, denitrification and chemodenitrification. As we know already, nitrification is an aerobic process mediated by microbes, wherein NH_4^+ is converted to NO_2 and

Table 10.6: Nitrogen oxide emissions after one-time burning event, and that from soil after ten-year period since conversion of forest to pasture in Costa Rica

	Flux from biomass burning (kg N ha^{-1})	Flux from soil (kg N ha^{-1})	Flux ratio (Soil: Burning)
N_2O	3.7	429	115
NO_x	65	60	0.7

Source: Erickson and Keller, 1997.
Note: Biomass burning flux was calculated as F = M × Y × (N:C) × K, where M is the above ground carbon in biomass (191 t C ha^{-1}), at La Selva, Costa Rica (from Jordan, 1985), assuming a ratio of carbon to biomass of 0.5. E is the burning efficiency (0.28). N:C is the ratio nitrogen to carbon. K is the emission factor. Factors E, N:C and K are derived from Cuertzen and Andrea 1990. Flux from soil was measured 10 years after first conversion to pasture.

then to NO_3. Whereas, denitrification is an anaerobic process, again mediated by microbes leading to conversion of NO_3 into N_2O, NO or N_2. Incidentally, NO emissions may be caused by both nitrification and denitrification. Additionally, NO is produced during an abiotic chemical decomposition process, which is collectively termed chemo-denitrification (Galbally, 1989). Laboratory studies suggest that such chemo-denitrification proceeds more rapidly under acidic conditions (Blackmer and Cerrato, 1986). After evaluating several suitable substrates and inhibitory factors, Neff *et al.* (1999) concluded that microbial denitrification may not constitute a major pathway for NO release after burning. Instead, microbial nitrification (aerobic process) seems critical to NO release. A nitrogenous soil medium, naturally supports all three processes—biological nitrification, denitrification as well as chemo-denitrification. Perhaps, in different proportions and combinations depending on aerobicity of soils. However, Neff *et al.* (1999) cautioned that with the techniques utilized, we do not discriminate between NO production through biotic nitrification, from that caused via chemo-denitrification. Laboratory evaluations have shown that nitrogen emissions from tropical forest is enhanced with increasing soil moisture levels. Soil wetting produced larger pulses of NO emissions, which disappears as soil dries up after the wetting cycle. According to Cardenas *et al.* (1993) NO emissions follows a curvilinear pattern in response to soil moisture between 0 and 100%, peaking at 30 to 70% water holding capacity.

C. Nutrient Dynamics versus Land Use Change in Reverse; Farming to Forestry

In the preceding discussions, aspects of land use change and resultant modifications in soil nutrients, gaseous exchange parameters and nutrient dynamics, actually considered a change from forest through clearing into pastures/farming. Let us now consider a diametrically opposite situation. In several parts of the tropical world, intensive and/or exhaustive farming has lead to soil deterioration. Several of such degrative practices reduces productivity. In order to control deterioration and ameliorate soil nutrient values, land used for farming is converted to agro-forests or into forest plantations. It would be interesting to understand the ameliorative mechanism involved in such land use change. Tree plantations improve tropical soils (Fisher, 1990; Brown and Lugo, 1992; Lugo, 1992). 'Nutrient pumping' from subsoil by trees may be key to restoration of soil fertility via forest plantation in nutrient depleted soils of humid tropics. Particularly, where nutrient recovery is hampered due to sodicity (Gupta and Abrol, 1990; Bhojavid et al. 1996). On a generalized basis, Fisher (1990; 1995) proposed five different mechanisms which contribute to soil amelioration through land use change from farming to forest plantations. They are:

a. Increase in soil organic matter, as a result of C fixation via photosynthesis, which results in litter fall and greater root turnover (Cairns et al. 1997);

b. Nitrogen fixation by leguminous trees that replenish soil nitrogen to a certain extent and alters N dynamics favorably (see Imo and Timmer, 2000);

c. Tree rhizosphere that extends to deeper horizon induces mineralization and increases microbial biomass and activity in subsoil;

d. Tree canopies modify micro-climate, especially temperature and gas exchange;

e. Most importantly, nutrient pumping that reflects lifting of nutrient from lower horizons (hitherto unavailable to crop roots) by deeper tree roots, later their accumulation in upper horizon and in canopy. Finally leading to re-incorporation into upper soil layers.

In a nutshell, effective nutrient pumping will lead to nutrient replenishment in upper layers/soil surface with a concomitant, verifiable depletion of nutrient in sub-soil (Bhojavid and Timmer, 1998). Essentially, long tree roots act as conduits for lifting nutrients from lower horizon to surface. Effectivity of such tree-induced nutrient pumping may depend on a variety of soil and tree related factors. For example, tree species, its root traits, nutrient recovery levels, extent of nutrient depletion already seen in surface soil, nutrient availability in lower horizons of soil, and actual rates of nutrients transferred from lower horizon to soil surface. It would also be interesting to investigate and understand the changes in gas exchange and fluxes of other nutrients due to a change from cropping to forestry in humid tropics.

4. Concluding Remarks

The tropical forest ecosystem is well distributed in America, Africa and Asia. Both, natural vegetation and plantations are important sinks for C, N, and other nutrients. Among the various types of vegetation, tropical forests are known to possess the highest rates of biomass conversion/productivity. In this regard, both inherent soil fertility and nutrient dynamics are crucial for the sustenance of plantations. Forest clearing for shifting agriculture/pasture is rampant. It affects nutrient dynamics appreciably. Nutrient loss through soil erosion, if excessive needs attention, particularly in areas cleared for shifting agriculture. Excessive emissions from tropical forest plantations, mainly N_2O, CH_4, CO, isoprene etc., need to be regulated. Incessant high N fertilizer application should be minimized to avoid N saturation. Nutrient imbalances may appear if other nutrients are not replenished in appropriate proportions. Hence, this aspect needs attention. Investigations on methods to curb emission of radiative gases NO_2, CH_4, CO will be highly useful. It can delay atmospheric pollution. At the bottom line, we should strive to enhance productivity of tropical forest plantations, without accentuating polluting effects and disturbing nutrient equilibria.

REFERENCES

Anderson, I.C., Levine, J.S., Popth, M.A. and Riggan, I.J. (1988). Enhanced biogenic emissions of nitric oxide and nitrous oxide following surface biomass burning. *J. Geophys. Res.*, **93**: 3893-3898.

Andreay, M.O., Anderson, B.E., Blake, D.R., Bradshau, J.D., Collins, J.E., Gregory, G.L., Sachse, G.W and Shipman, M.C. (1994). Influence of plumes from biomass burning on atmospheric chemistry over equatorial and tropical South Atlantic during CITE3, *J. Geophys. Res.*, **99**: 12793-12808.

Andrea, M.O., Browell, E.V., Garstang, M., Gregory, G.L., Sachse, G.W., Setzer, A.W., Silva–Dias, P.L., Talbot, R.W., Torres, A.L. and Wofsy, S.C. (1988). Biomass burning, emissions and associated haze layers over Amazonia. *J. Geophys. Res.*, **93**: 1509-1527.

Bakwin, P.S., Wofey, S.C. and Fan, S.M. (1990). Measurement of reactive nitrogen oxides (NO_2) within and above a tropical forest canopy in the wet season. *J. Geophys Res.*, **95**: 16765-16772.

Bhojavid, P.P and Timmer, V.R. (1998). Soil dynamics in an age sequence of *Prosopis juliflora* planted for sodic soil restoration in India. *Forest Ecology and Management*, **106**: 181-193.

Bhojavid, P.P., Timmer, V.R. and Singh, G. (1996). Reclaiming sodic soils for wheat production by *Prosopis juliflora* (Swartz) afforestation in India. *Agroforestry Systems*, **34**: 139-150.

Blackmer, A.M. and Cerrato, M.C. (1986). Soil properties affecting the formation of nitric oxide by chemical reactions of nitrate. *Soil Sci. Soc. Am. J.*, **50**: 1215-1220.

Brown, S. and Lugo, A.E. (1992). Above ground biomass estimates for tropical moist forests of the Brazilian Amazon. *Inter Science*, **17**: 8-18.

Cairns, M.A., Brown, S., Helmer, E.H. and Baumgasdner, G.A. (1997). Root biomass allocation in the World's upland forests. *Oecologia*, **111**: 1-11.

Cardenas, L., Rondon, A., Johansson, C. and Sanheuza, E. (1993). Effects of soil moisture, temperature, and inorganic nitrogen on nitric oxide emissions from acidic tropical savannah soils. *J. Geophys. Res.*, **98**: 14783-14790.

Cicerone, R.J. (1987). Changes in stratospheric ozone. *Science*, **233**: 867-869.

Cicerone, R.J. and Oremland, R.L (1989). Biogeochemical aspects of atmospheric methane. *Global Biogeochemical. Cycles*, **2**: 299-327.

Cuertzen, P.J and Andrea, M.O. (1990). Biomass burning in the tropics: Impact on atmospheric chemistry and biogeochemical cycles. *Science*, **250**: 1669-1678.

Davidson, E.A. and Kingerlee, W.A. (1997). A global inventory of nitric oxide emissions from soils. *Nutrient Cycling in Agro-ecosystem*, **48**: 37-50.

Davidson, E.A., Vitousek, P.M., Matson, P.A. Riley, R., Garcia-Mandiz, G. and Meass, J.M. (1991). Soil emissions of nitric oxides in a seasonally dry tropical forest of Mexico. *J. Geophys. Res.*, **96**: 15439-15445.

Dalmas, R.A., Servant, J.B., Tathy, J.P., Cross, B. and Labat, M. (1992). Sources and sinks of methane and carbon-di-oxide exchange in forest and Equatorial Africa. *J. Geophys. Res.*, **96**: 145439-15445.

Erickson, H.E. and Keller, M. (1997). Tropical land use change and soil emission of nitrogen oxides. *Soil Use and Management*, **13**: 278-287.

Ewell, J.J (1986). Designing agricultural ecosystems for humid tropics. *Ann. Rev. Ecol. Systems*, **17**: 245-271.

Ewel, J.C., Berish, C., Brown, B., Priu, N. and Raiel, J. (1981). Slash and burn impacts on a Costa Rica wet forest site. *Ecology*, **62**: 816-829.

Fearnside, P.M. and Barbosa, R.I. (1998). Soil carbon changes from conversion of forest to pasture in Brazilian Amazonia. *Forest Ecology and Management*, **108**: 147-166.

Fehsenfeld, F. Calvert, J., Fall, J., Goldan, P., Geunther, A.B., Hewitt, C.N., Lamb, B., Liu, S., Trainer, M., Westberg, H. and Zimmerman, P. (1992). Emissions of volatile organic compounds from vegetation and the implications of atmospheric chemistry. *Global Biogeochemical Cycles*, **6**: 351-388.

Firestone, M.K. and Davidson, E.A. (1989). Microbiological basis of NO and N_2O production and consumption in soil. *In:* Exchange of trace gases between terrestrial ecosystems and the atmosphere. M.O. Andreae and D. Schimel (eds.). John Wiley and Sons, New York, pp. 7-21.

Fisher, R.F. (1990). Amelioration of soil by trees. *In:* Sustained productivity of forest soils. S.P. Gesel, D.S. Lacale, G.P., Wheelman and R.F. Powers (eds.). University of British Columbia, Vancouver, Canada, pp. 290-300.

Fisher, R.F. (1995). Amelioration of degraded rainforest soils by plantations of native trees. *Soil Sci. Soc. Am. J.*, **59**: 544-549.

FAOSTAT (1999). Food and Agricultural Organisation of the United Nations. Rome, Online: http://fao.org.

Galbally, I.E. (1989). Factors controlling NO_x emissions from soils. *In:* Exchange of trace gases between terrestrial ecosystems and the atmosphere M.O. Andreau and D.S. Schimel, (eds.). John Wiley, New York, pp. 23-37.

Ghuman, B.S., Lal, R. and Shearer, W. (1991). Land clearing and use in the humid Nigerian tropics. 1. Soil physical properties. *Soil Science Society of America Journal*, **55**: 184-188.

Guenther, A., Zimmerman, P., Hashy, R., Manson, R. and Fall, R. (1993). Isoprene and monoterpene emission rate variability. Model evaluation and sensitivity analysis. *J. Geophys. Res.*, **98**: 12609-12617.

Guenther, A., Zimmerman, P., Harley, R., Monson, R. and Fall, R. (1995). A global model of natural volatile organic compound emission. *J. Geophys. Res.*, **100**: 8873-8892.

Gupta, R.K and Arbol, I.P. (1990). Salt affected soils: their reclamation management for crop production. *Advances in Soil Science,* **11**: 223-288.

Hall, S.J. and Matson, P.A. 1999 Nitrogen Oxide emissions after nitrogen additions in tropical forests. *Nature,* **400**: 152-155.

Holdridge (1967). Cited in Humid Tropical Environments. 1995. A.J. Reading, R.D. Thomson, and A.C. Hillington (eds.). Blackwell Publishers, Oxford, England.

Houghton, R.A. (1994). The worldwide extent of land use change. *BioScience,* **44**: 305-313.

Houghton, R.A., Skole, D.L., Nobre, C.A., Hackler, J.L., Lawrence, K.T. and Chomeutowski, W.H. (2000). Annual fluxes of carbon from deforestation and regrowth in the Brazilian Amazon. *Nature,* **403**: 301-304.

Imo, M. and Timmer, V.R. (2000). Vector competition analysis of a Leucana-maize alley cropping system in Western Kenya. *Forest Ecology and Management,* **126**: 255-268.

Johansson, C., Rodhe, H. and Sanheuza, E. (1988). Emission of NO in a tropical savanna and cloud forest during the dry season. *J. Geophys. Res.,* **93**: 7180-7192.

Jordan, C.F. (1985). Nutrient cycling in tropical forest ecosystem: John Wiley and Sons, New York.

Keller, M. 1994 Controls on soil atmospheric fluxes of nitrous oxide and methane: effects of tropical deforestation. *In:* Climate biosphere Interaction. R.G. Zepp (ed.). John Wiley and Sons. Inc. New York, pp. 121-138.

Keller, M., Kaplan, W.A. and Wofry, S.C. (1986). Emissions of N_2O, CH_4 and CO_2 from tropical forest soils. *J. Geophys. Res.,* **91**: 11791-11802.

Keller, M. and Lerdau, M. (1999). Isoprene emissions from tropical forest canopy leaves. *Global Biogeochemical Cycles,* **13**: 19-29.

Keller, M. and Matson, P.A (1994). Biosphere-Atmosphere exchange of trace gases in the tropics: Evaluating the effects of land use changes. *In:* Global atmospheric-biosphoric chemistry R.G. Prinn (ed.). Plenum Press, New York, pp. 103-117.

Keller, M., Matric, M. and Stallard, R.F. (1990). Consumption of atmospheric methane in soils of Central Panama: Effects of agriculture development. *Global Biogeochemical Cycles,* **4**: 21-27.

Keller, M., Veldkemp, E., Weitz, A.M. and Reiners, W.A. (1993). Effect of pasture age on soil trace-gas emissions from a deforested area in Costa Rica. *Nature,* **365**: 244-246.

Kellman, M. and Tackaberry, R. (1997). Tropical environments: The functioning and management of tropical ecosystems. Routeledge, London/New York, pp. 61-177.

Kirchoff, V.W.J.H and Marinho, E.V.A. (1990). Surface carbon monoxide measurements in Amazonia. *J. Geophys. Res.,* **95**: 16933-16943.

Lal, R. (1986). Conversion of tropical rain forest. Agronomic potential and consequences. *Advances in Agronomy,* **39**: 173-264.

Longman K.A. and Janik, J. (1987). *Tropical Forest and Its Environment.* Longman Publishers, London, U.K.

Lugo, A.E. (1992). Comparison of tropical true plantations with secondary forests of similar age. *Ecological Monographs,* **62**: 1-41.

Matson, P.A. and Vitousek, P.M. (1987). Cross system comparison of soil nitrogen transformations and nitrous oxide flux in tropical ecosystem. *Global Biogeochem. Cycles,* **1**: 163-170.

Matson, P.A., Vitousek, P.M., Livingston, G.P. and Swanberg, N.A. (1990). Sources of variation of nitrous oxide flux in Amazonian ecosystems. *J. Geophysical Research,* **95**: 16789-16798.

Neff, J.C., Keller, M., Weitz, A.W. and Veldkamp, E. (1999). Fluxes of nitric oxide from soils following clearing and burning of a secondary tropical rain forest. *J. Geophys. Res.,* **100**: 25913-25922.

Nepstad, D.C., de Carvalho, C.R., Davidson, E.A., Jipp, P.H., Lefevbre, P.A., Negrieros, G.H., de Silva, E., Stone, T.A., Trumbore, S.E. and Viera, S. (1999). Large scale impoverishment of Amazonian Forests by logging and fire. *Nature,* **398**: 505-508.

Null, C., Piccols, M.C., Stendler, P.A., Melills, J.M., Feigh, B.J and Cerri, C.C. (1995). Nitrogen dynamics in soils of forests and active pastures in the western Brazilian Amazon soils. *Soil. Biol. Biochem.,* **27**: 1167-1175.

Piccolo, M.C., Neill, C. and Cerri, C. (1994). Nitrogen mineralization and net nitrification along a tropical forest to pasture chromosequence. *Plant and Soil.,* **162**: 61-70.

Potter, C.S., Matson, P.A., Vitousek, P.M. and Davidson, E.A. (1996). Process modeling of controls on nitrogen trace gas emissions from soils worldwide. *J. Geophys. Res.,* **101**: 1361-1377.

Procter, J. (1987). Nutrient cycling in primary and secondary rainforests. *Applied Geography,* **7**: 135-152.

Rasmusen, R.A. and Khalil, M.A.K. (1998). Isoprene over the Amazon Basin. *J. Geophys. Res.,* **93**: 1417-1421.

Reading, A.J., Thompson, R.D. and Millington, A.C. (1995). *Humid Tropical Environments.* Black Well, Oxford and Cambridge, U.S.A., pp. 165- 171.

Reiners, W.A., Bouwman, A.F., Persons, W.F.J. and Keller, M. (1994). Tropical rain forest conversion to pasture: Changes in vegetation and soil properties. *Ecological Applications,* **4**: 363-377.

Repatto, R. (1990). Deforestation in the tropics. *Scientific American,* 262: 18-24.

Riley, R.H. and Vitousek, P.M. (1995). Nutrient dynamics and nitrogen trace gas flux during ecosystem development in montane rain forest. *Ecology,* **76**: 292-304.

Sanchez, P.A. (1981). Soils of the humid tropics. *In:* Blowing in the wind: Deforestation and long range implications. V.H. Sutline, N. Altshuler and M.D. Zamoza (eds.). College of William and Mary, Williamsburg, V.A. pp. 347-410.

Shine, K.P., Fonquart, Y., Ramaswamy, V., Solomon, S. and Srinivasan, J. (1995). Radiative forcing in climate change-1994. J.T. Houghton *et al.* (eds.) Cambridge University Press, New York, pp. 163-203.

Verchot, L.V., Davidson, E.A., Cattanio, J.H., Ackerman, I.L., Erickson, H.E. and Keller, M. (1999). Land use change and biogeochemical controls of nitrogen oxide emissions from soils in eastern Amazonia. *Global Biogeochemical Cycles,* **13**: 31-46.

Vitousek, P.M and Sanford, R.L. (1986). Nutrient cycling in moist tropical forest. *Annual Review of Ecology and Systematics,* **17**: 137-167.

Walter, H. (1979). Vegetation of earth and ecological systems of the Geobiosphere. Springer Verlag, New York.

Watson, R.T., Meira Filho, L.G., Sanheuza, E. and Jawtos, A. (1992). Greenhouse gases: Sources and sinks. *In:* Climate change-1992—The supplementary report to the IPCC scientific assessment. J.T. Houghton, G.L. Callande and S.K. Varney (eds.). CambridgeUniversity Press.

Weitz, A. M., Keller, M., Linder, E. and Crill, P.M. (1999). Spatial and temporal variability of nitrogen oxide and methane fluxes form a fertilized tree plantation in Costa Rica. *J. Geophys. Res.,* **104**: 30097-30107.

Weitz, A.M., Vildkump, E., Keller, M., Neff, J. and Crill, P.M. (1998). Nitrous oxide, nitric oxide and methane fluxes from soils following clearing and burning of tropical secondary forests. *J. Geophys. Res.,* **103**: 28047–28058

Zinmerman, P.R., Grunberg, J.P., and Westberg, C.E. (1988). Measuremens of atmospheric hydrocarbons and biogenic emission fluxes in the Amazon boundary layer. *J. Geophys. Res.,* **93**: 1407-1416.

CHAPTER 11

The Temperate and Boreal Forests
Nutrients and Productivity

CONTENTS

1. Introduction: Expanse, Climate and Productivity
2. Nutrient Dynamics
    A. Rhizosphere bio-geochemistry in temperate forests: Swedish studies
    B. Nutrient loading of tree seedlings in nursery: Canadian examples
    C. Nitrogen Saturation in temperate Forests
3. Nutrient Dynamics in Boreal and Peat Land Forests
    A. Drainage and nutrient dynamics in peat land
    B. Carbon, other nutrients and biomass
4. Concluding Remarks

1. Introduction: Expanse, Climate and Productivity

The temperate landscape on earth is believed to have developed millions of years ago during Pleistocene ice ages. At present, about 1.0 billion ha of temperate forests and 1.4 billion ha of boreal zones contribute to the overall global forest cover estimated at 4.2 billion ha. The temperate forests occur in eastern North America, western and central Europe and north eastern Asia. The boreal (or 'Taiga') forests, which represent one of the largest terrestrial biomes occur between 50° and 60° N latitudes. These boreal zones are found in North America, Eurasia, Siberia, Scandinavia, and are predominant in Alaska and Canada.

Climate and Soils: The temperate forests experience well defined seasons with a distinct winter. The growing season with moderate climate extends for only 140 to 200 days depending on the length of frost-free period, which is generally 4 to 6 months. Temperatures vary between –30°C and +30°C, and precipitation that ranges between 75 and 150 cm annually is distributed evenly throughout the year. For precipitation levels ranging between 50 to 150 cm, the interception of rainfall could be between 10 and 35% with braod-leaved species, and 25 to 45% with coniferous stands. Transpiration rates may vary between 270 and 430 mm yr^{-1} for conifers aged 50 to 80 years, while that for broad-leaved forests of similar age varies between 150 and 350 mm yr (Roberts, 1999). The evapotranspiration from understory depends on the composition of vegetation. For example, pine stands with understory comprising heather and other shrubs contributed 10 to 50% of total evapotranspiration. Whereas an understory of predominantly *Acacia* sp. found below Pseudotsuga contributed only 3 to 8% of evapotranspiration (Fig. 11.1a and b). Within boreal zones, seasons can be identified as short, moist or moderately long summers. The winters are long, very cold and dry. The growing season is generally less than 130 days. Precipitation is mostly in the form of snow, ranging between 40 and 100 cm annually.

Fig. 11.1 A) Natural Pine Forest in the Northwest United States of America (Corvallis, Oregon). B) A typical Pine plantation and under story vegetation in the Southeastern United States of America (Gainesville, Florida)

Continent-wise, the major soil types encountered in termperate zones, that is mid-latitudes (48°-60° N or S) which support natural forests, forest plantations and cropping systems are as follows:

North America — Ultisols (Aquults, Humults), Mollisols (Udolls, Xerolls), Undifferentiated Spodosols, Cryic Spodosols, Alfisols, Inceptisols (Aquepts), Entisols;

South America — Mollisols (Borolls), Andisols and Ustolls

Europe and Asia — Undifferentiated Spodosols, Alfisols (Boralfs), Mollisols (Boralls), Inceptisols (Andepts), Histosols, Xeralfs;

South Africa — Ardisols, Xeralfs, Entisols (Psamments);

Australia — Alfisols (Udalfs, Xeralfs), Inceptisols (Ochrepts).

In terms of vegetation, soil orders encountered among the temperate broad-leaved and mixed forests are mainly Ultisols and some Alfisols. Whereas, pine stands in North America, Europe and elsewhere are supported mostly by Spodosols, Histosols, Inceptisols and Alfisols. In other words, soil type may influence tree stand. For example, on sandy soils pines replace broad-leaved species, hence we encounter pine woods in Georgia, and deep south in USA.

Generally, the structure of a temperate forest can be discerned into five prominent layers, namely:
I. a tree stratum, which is 60 to 100 feet high, and composed of broad-leaved and/or coniferous species;
II. a small tree or sapling layer with younger trees of the tall species and others;
III. a shrub layer, often with members of the heath family such as rhododendron, azaleas, huckleberries etc.;
IV. a herb layer of perennials that bloom in spring; and
V. a ground layer of lichens, mosses etc.

The temperate forests are quite variable. If deciduous species predominate in some places, it could be conifers elsewhere. Mixed forest of conifers and broad-leaved trees are common in transition zones, for example, in eastern and central Massachusetts in USA (Aber, 2001). The tree flora of temperate forests, generally comprises 3 to 5 species km^{-2}. They may include broad-leaved species such as oak (*Quercus*), maple (*Acer*), elm (*Ulmus*), Walnuts (*Juglans*) etc. Whereas, the tree flora within coniferous stands may comprise pine, fir and spruce. The moist conifer and evergreen broad-leaved forests localize in zones with wet winters. The dry conifers dominate in higher elevation zones with low precipitation. Whereas, temperate conifers confine to zones with mild winters, but high precipitation (>200 cm).

Productivity: The average primary productivity of temperate forests is comparatively high at 600 to 2500 $gm^{-2}\ yr^{-1}$, but less than other biomes such as tropical forests, nutrient rich zones in oceans, or swamps and marshes. The temperate forests are key biomass accumulators on the globe. They play a vital role in carbon cycle by sequestering large quantities of carbon, therefore minimizing glass house effect and global warming. The published average C content is 150 t C ha^{-1} in above and below ground portions of temperate forests, compared with 400 t C ha^{-1} for tropical forests (Mathews *et al.* 2000). The proportion of total ecosystem carbon in below ground biomass is greatest in boreal forests (84%), followed by temperate forests (63%), and least with tropical forests (<50%). In general, the above-ground biomass of all types of temperate and boreal forests may vary between 40 and 250 t C ha^{-1}. On an average, 12 to 17.5 kg C M^{-3} and 620 to 930 g N m^{-3} are held within temperate woods at a C:N ratio 21 to 15 (Crawley, 1986). Obviously, the biomass accumulation rates within temperate forests vary depending on geographic location, vegetation and species composition, soil fertility and environmental parameters. To quote an example, for different conifer species such as Scots pine, Norway spruce, Corsican pine, Sitka spruce and others grown in the United Kingdom, the biomass accumulation rates ranged from 2.4 to 3.9 t $ha^{-1}\ yr^{-1}$. Whereas, these same species accumulated 7.6 to 24.6 $ha^{-1}\ yr^{-1}$ in northeast of USA.

The emphasis within this chapter confines to nutrient chemistry in the rhizosphere, nutrients during seedling development, biomass (C) accumulation and N dynamics in the outfield. Certain aspects, such as symbiotic relations, both N fixation and ectomycorrhizas are important, however, they have not been dealt here. Excellent recent treatises are available on these special topics concerned with microbial ecology.

2. Nutrient Dynamics

In simple terms, whenever a forest is developed around a new site or regenerated within temperate latitudes, firstly nutrients are fed into the vegetation/soil system from the atmosphere, weathering of rocks or through fertilizers. In turn, nutrients are lost to the atmosphere due to biogeochemical processes, or leached from soil into streams or ground water. During this process, a certain quantity of nutrients are held within the temperate forest ecosystem, both in soil and above ground tree stand. Nutrient inputs via throughfall from canopy, and stem flow along the branches stems enrich the soil. Simultaneously, the below-ground nutrient inputs occur through root leachate and exudates, but these are often not quantified, because it is difficult and tedious. For similar reasons, nutrient inputs derived through senescing fine roots and mycorrhizas are not worked out routinely. However, there are reports which account for such nutrient inputs by fine roots, nitrogen fixation, and mycorrhizal senescence. Many researchers recognize that such inputs can be important. On the whole, most reports deal with aerial litter deposition and decomposition, especially leaves and branches. Elaborate descriptions on various aspects of rhizosphere biogeochemistry, nutrient transformation, or nutrient management in nursery and outfield are available. In any case, during the early phase of forest growth, nutrients are channeled towards canopy development, foliage and roots. Once the canopy closure occurs, it leads to a decline of nutrient translocation into foliage. Consequently, demand for nutrients from soil too declines. In addition, fluxes, internal re-absorption and recycling involving mainly litter decomposition, ground flora etc. leads to establishment of consistent nutrient cycling and equilibrium (Table 11.1). To quote an example, such internal fluxes and

Table 11.1: Nutrient cycling and balance sheets for major nutrients within a Corsican pine stand (40 years old) in northern Scotland

	Rate of process (kg ha^{-1} yr^{-1})		
	N	P	K
OVERVIEW			
Total uptake	69	6.0	28
Accumulation in tree	18	2.1	11
Litter fall	51	3.9	11
Crown leaching	Trace	Trace	11
Accumulation in humus	12	1.5	1
SINKS			
In new needles	92	9.4	45
To replace lost structures	20	1.7	4
Net structural increase	26	3.0	13
TOTAL REQUIREMENT	138	14.1	62
SOURCES			
Uptake from soil	69	6.0	28
Re-absorption from needles	61	7.2	32
Re-absorption from other tissues	8	0.9	2
TOTAL	138	14.1	62

Source: Excerpted from Packham et al. 1992; Cole, 1995

accumulations of nutrients involving needles/leaves, branches, and twigs, it seems can maintain maximum growth rates of 20 m^3 ha^{-1}y^{-1} in Corsican pine stands (Packham et al. 1992). This is equivalent to one half of nutrient requirement for fresh growth within temperate forest stands. Packham et al. (1992) also provide another comprehensive example that depicts various nutrient fluxes and stores in a temperate broad-leaved forest. It is as follows (all values in kg ha^{-1} yr^{-1}):

a. Atmospheric input   N = 6, K = 3, Ca = 7, Mg = 6, P = 0.2
b. Aerosols   N = 12, K = 3, Ca = 2, Mg = 8, P = 0.1
c. Nutrients in ground N = 22, K = 16, Ca = 8, Mg = 3, P = 0.2
   flora
d. Nutrients cycled by N = 24, K = 23, Ca = 11, Mg = 3, P = 3
   ground flora
e. Nutrients held in   N = 716, K = 416, Ca = 828, Mg = 145, P = 44
   canopy
f. Nutrients absorbed   N = 85, K = 60, Ca = 111, Mg = 16, P = 5
   by trees/vegetation
g. Nutrients recycled   N = 180, K = 49, Ca = 237, Mg = 24, P = 8
   through litter
h. Soil nutrients held in N = 5550, K = 249, Ca = 840, Mg = 64, P = 9
   reserve

Obviously, firstly soil, then tree canopy are the largest reservoirs of nutrients in the temperate broad-leaved forests. After this overview on nutrient dynamics let us now discuss specific aspects of nutrients in temperate forests.

A. Rhizosphere Biogeochemistry in Temperate Forests: Swedish Studies

Typically, rhizosphere of a tree root system is that portion of soil-plant continuum, wherein network of roots interact with microbes, solutes, gases and soil phases. Physically, it may extend from 0.2 to 0.4 mm cylindrical area surrounding each root segment. In spite of the small volume that a rhizosphere occupies in mineral soils, it plays a pivotal role in the maintenance of soil-plant interactive system in temperate forests. In nature, nutrients acquisitioned by the roots pass through the rhizosphere and then soil-root interface, wherein the impact of environmental stress factors on nutrient availability and uptake are perhaps most severely felt. Therefore, rhizosphere is actively involved in nutrient dynamics of the ecosystem. Our main understanding is that, fine roots produced by temperate trees and their associative organisms interact with each other to modify the rhizosphere chemistry. So that, appropriately higher level of nutrient availability is maintained in the rhizosphere, and in the vicinity of roots compared to bulk soils. We may then believe that rhizosphere interactions and the ensuing biogeochemical conditions have direct bearing on nutrient dynamics. Firstly, in the below-ground ecosystem of temperate tree roots, and consequently on the above ground portions due to better absorption of nutrients.

While assessing rhizosphere effects on nutrient dynamics it is imperative that diversity of tree species in the forest stand; the dominant species, its root architecture including depth and spread, fine root biomass, their density, physiological activity; and most importantly interacting soil physico-chemical and biological components are important. The spatial configuration of root system, with it the rhizosphere and its influence may also differ with time as roots grow, senesce or regenerate and occupy new portions in soil. To quote an example, nearly 50% of fine roots (1 to 2 mm thickness) distribution in a Norway spruce stand was confined to B horizon. Also, decline in root quantity with soil depth was a general pattern observed in many temperate locations in Northern Europe. External nutrient inputs too affect root distribution. Application of NH_4SO_4 enhanced fine root (<1.0 mm) formation and its density in the E horizon (Clegg et al. 1997). Clearly, several plant and soil related factors govern root formation and its proliferation, and indirectly influence rhizosphere structure, and its chemical functions.

Rhizosphere Mineralogy: Physical, chemical and mineralogical gradients are common between the rhizosphere of temperate tree roots and bulk soil. Models explaining substantial gradients that occur for soil pH, cation exchange equilibrium, mineral availability, mono-saccharides, organic acids, microflora etc., have been proposed (Courchesne and Gobran, 1997). Mineral composition in the rhizosphere, or near roots differs consistently from that encountered in bulk soils. In terms of nutrient release and availability, the physico-chemical modifications effected by rhizosphere may be crucial. Specific root and soil mineral associations are also important. Preferential dilutions of certain minerals on root surfaces can selectively change rhizosphere chemistry and nutrient availability. The X-ray diffraction analysis of rhizosphere mineralogy has revealed that, in most soil horizons, rhizospheres contained significantly lower levels of amphibolus and expandable phyllosilicates. Whereas, physico-chemical changes in the rhizospheres under Norway spruce had no influence on K-feldspars (Table 11.2). Obviously, the magnitude of changes in rhizosphere mineralogy was dependant on relative weathering stability of minerals, and on tree root activity (Gobran et al. 1998). Recently, Clegg et al. (1997) have reported detailed information on differences in physico-chemical characteristics of soil fractions derived from bulk soil

Table 11.2: Mineralogical differences between bulk and rhizosphere soil in a temperate forest stand, an example

Soil sample	Mineralogies (I/IQ_2)			
	Amphibole	Phyllosilicates	Plagioclase	K-feldspar
Bulk	0.12	1.14	2.29	1.29
Rhizosphere	0.03	0.54	1.73	1.28

Source: Courchesne and Gobran, 1997.
Note: I/IQ_2 = intensity of mineral divided by intensity of the 100 quartz pack (d = 0.426 mm); Soil mineralogy itself varies depending on location, parent material and external inputs if any.

rhizospheres, and soil-root interface, that occur in the below-ground portions of Norway spruce. They report that soil traits related to nutrient dynamics such as CEC, water soluble cations, pH, Ec, titerable Al, Ca:Al balance, extractable N, K, Ca, Mg, percent saturation of Ca, Mg, K, Na are enormously influenced by rhizosphere effect.

The mineral nutrient dynamics at the soil-root interface is definitely influenced by any of the variations in soil and plant components, and equally so by external fertilizer-based nutrient inputs. Quite often, the extent of nutrients recovered by tree crops might be pre-determined by nutrient dynamics at the soil-root interface. Clegg et al. (1997) have expressed that, generally the trends in physico-chemical transformation and concentrations of nutrients observed in rhizosphere are amplified at soil-root interface. For example, organic matter contents were enhanced 2 folds because of rhizosphere effect, and it increased by over 10 folds at soil-root interface (Table 11.3). Such effects on organic matter content persisted despite variations in nutrient supplements to trees.

Table 11.3: Organic matter content (%) in rhizosphere, soil-root interface and bulk soil at 0–30 cm depth under the *Picea abies* growth planted in 1996 in southwest Sweden (56°33'N, 13°13'E)

Nitrogen treatment	Rhizosphere	Soil-root interface	Bulk soil
Control	14.2	81.3	6.5
Ammonium sulphate	21.2	75.3	5.9

Source: Excerpted from Clegg et al. 1997

Amplification effects at soil-root interface were observed for several other factors such as CEC, BS (base saturation), water soluble cations, Ec, exchangeable K, Na, Mg, Ca, Ca:Al balance and pH. To summarize, regardless of external treatments such as drought, irrigation, nutrient deficiency, or excess, the general trend noticed for concentration, and for activities of several other soil characters were bulk soil < rhizosphere < soil-root interface (Clegg et al. 1997).

Sometimes rhizosphere exerts subtle influences on nutrient transformations, and it even affects ratio of different chemical forms of a nutrient element. For example, tree roots may preferentially support accumulation of specific nitrogenous compounds, a particular organic acid or a single ionic form of a nutrient element. In this regard, Clegg and Gobran (1997) have reported certain interesting observations on differences in P fractions in the rhizospheres of Norway spruce stands in Sweden. They suggest that ratio of inorganic P (Pi) versus organic P (Po) in the rhizosphere is a key factor during tree development. Drought, irrigation, or fertilizer inputs often alter the Pi/Po ratio in the rhizosphere. Despite higher Pi concentration in the rhizosphere than in bulk soil, it is Po activity (and its concentration) that dominated rhizosphere. The Pi accounted for only 16% of total P (P_t) in the rhizosphere. The Pi/Pt ratio was not influenced much by the soil horizon and depth, but definitely by the rhizosphere effect. Based on these above observations, Clegg and Gobran (1997) have remarked that biological processes buffer P transformation in the rhizosphere. Firstly, they cause accumulation of large Po fraction, which is mineralizable, and a smaller possibly easily absorbable Pi fraction. Clearly, maintenance of optimum Po pool in rhizosphere is crucial to its uptake by roots. Such rhizosphere mechanisms become operative to create optimum conditions for root growth and nutrient acquisition by Norway spruce.

Clegg *et al.* (1997) summarize that, if we desire to establish inter-relationships between soil chemistry and tree response in temperate forest ecosystem, then rhizosphere soil and its chemical nature should be analyzed carefully. Presently, the trend is to base much of the inferences on chemical nature of bulk soil. It may not be apt. Furthermore, although a simple relationship and continuum may exist in terms of soil solution chemistry between bulk soil and rhizosphere, it may actually involve complex physico-chemical interactions. Hence, a conceptual model that considers a range of specific soil factors and processes may appropriately explain the role of rhizosphere and its chemical dynamics in temperate tree nutrition (see Gobran and Clegg, 1996; Gobran *et al.* 1998). In this regard, a conceptual model that explains nutrient availability in temperate forest soils has been proposed by Gobran and Clegg (1996). Their working hypothesis is that, nutrient concentrations in most soils follow the order: bulk < rhizosphere < soil-root interface, and the three fractions are interlinked through physico-chemical and microbial phenomena. Ulrich (1987) has pointed out that morphologically roots, soil or micro-organisms could be delineated clearly, but in nature, functionally such demarcations are not possible. Obviously, within their model, Gobran and Clegg (1996) have identified three types of continuous streams of interaction between soil and roots, namely gaseous phase, solution phase and surface phase. Such interactions determine nutrient availability. They emphasize that root interactions with both organic and mineral layer can be equally important in terms of nutrient cycling and availability in the rhizosphere. Accumulation of organic matter in rhizosphere may induce mineralization and enhance exchange sites. Microbial activity including mycorrhizas, which release of organic acids etc. may enhance P and K supply to roots. It is believed that ultimately, interactions between soil, microbes and roots may create a mutually supportive system which is crucial to the maintenance of nutrient availability. Within such a system, mass flow and diffusion are also important, both in terms of nutrient transport and availability to roots. However, magnitude of nutrient translocation through such physico-chemical processes may be variable. Similarly, nutrient transport via mycorrhizas is also variable (Binkley, 1986; Finley and Soderstorm, 1989). Finally, Gobran and Clegg (1996) profess that temperate forest trees invest energy towards modifying the rhizosphere, then creating favorable conditions for functioning of beneficial soil microbes, and in maintaining interactions within rhizosphere. So that, it enables optimum nutrient availability to roots.

Soil Acidification in Temperate Forests: Forest decline is severe in North America, Scandinavia, central Europe and in several other temperate locations. It could be due to various causes, among which impaired nutrient dynamics is definitely a major factor. Soil acidity related maladies increase Al toxicity, leaching losses of cations, imbalance in Ca:Al ratios etc. Actually within the soil ecosystem, effects of acidic depositions are generally manifested through a long chain of processes that alter soil solution chemistry and nutrient equilibrium (Gobran and Bosatta, 1988). Several antagonistic interactions due to acidity and Al^{+3} accumulation affect cation uptake, carbohydrate partitioning, root growth and physiological functions. In the recent times, excessive N inputs into temperate forests has become detrimental by altering soil acidity and causing nutritional imbalances (Roberts *et al*. 1989; Zottl, 1990; Gobran *et al*. 1993; Galloway, 1995). On the whole, decreased temperate forest growth has often been attributed to soil acidification caused by high soluble Al^{++} and low base saturation. Application of Ca can suppress such detrimental effects of Al^{+3}, but it may interfere with acquisition of other cations such as Mg^{++}. Hence, Ca may be a useful detoxifying agent for Al^{+++}, only in soils where Mg^{++} occurs in significantly higher levels. Essentially, then interactions between Ca, Mg and Al must be carefully weighed out, before embarking on a forest revitalization program that combats nutrient imbalances in temperate soils (Ericsson *et al*. 1998). According to Ulrich (1987), the Ca and Al balance (i.e., CAB = $\log_{10} Ca^{+2}/Al^{+3}$ or Ca + Mg + K/Al) in soil medium plays an important role in plant nutrition. Still, it may not provide us with deeper understanding on physiological impacts of Ca and Al (Ericsson *et al*. 1998). In certain tree species, Al^{+3} related negative effects may confine to roots and their nutrient acquisition functions. But, in others, detrimental influence is seen in both above ground and below ground portions. Gobran *et al*. (1993) summarize that Ca^{++} dependent amelioration of soil acidity/Al^{++} toxicity may involve a combination of mechanisms, namely:

i) Effects of Ca^{+2} on reduced adsorption of Mn^{+2} and Al^{+2};
ii) Induction of NH_4^+ and P absorption by Ca^{+2} which restores N and P recovery levels; and/or
iii) Stimulation of tree growth in general.

In summary, soil acidity can cause wide ranging changes in nutrient recovery rates. It should be noted that, if soil acidity/Al^{+3} toxicity alters nutrient balance and causes physiological impairment in temperate trees, then Ca^{+2} applied into the soil system ameliorates and restores normal N and P recovery, as well as forest productivity.

B. Nutrient Loading in Tree Seedlings in Nursery—Canadian Examples

Nutrient status of containerized seedlings during their growth in the forest nursery, and at the time of planting in the out field are important. Mainly because, it affects regulation of nutrient dynamics immediately after transplantation into out fields, and controls successful plantation establishment in a variety of soils which may be deficient in nutrients (Fang, 1992; Xu and Timmer, 1998). Sometimes, despite several advantages attached with use of containerized seedlings, they turnout less competitive, due to small size and insufficiency of nutrients loaded into seedlings in the nursery. Hence, more effective preconditioning of seedlings by altering nutrient inputs and their schedules are necessitated. The steady state nutrition suggested by the Canadian group (see McAlister and Timmer, 1998) involves fertilization, where nutrients are scheduled at exponential rates that is commensurate with the rapid growth rate of seedlings, its nutrient consumption rates and required mineral nutrient concentration in tissues (Fig. 11.2).

Firstly, such nutrient loading definitely alters nutrient dynamics within the plant, right at the nursery stage. It is supposed to ensure relatively stable nutrient concentration in tissue, without causing dilutions (Munson and Bernier, 1993). Most strikingly, nutrient loading imparts greater competitive edge to seedlings to survive within forest out-plant sites (Timmer, 1997; Xu and Timmer, 1999). Ultimately, the forest plantation growth and productivity itself might partially depend on seedling nutrient status and its competitive ability to flourish in the out field. Furthermore, Xu and Timmer (1998) explain that such luxury nutrient consumption and reserves ensures proper remobilization and re-translocation of minerals. Therefore, it aids in overcoming nutrient dearth at, and immediately after transplantation.

While establishing a forest stand, knowledge about nutrient status of the seedlings used for transplantation is crucial. At times, foresters might resort to authentic predictions of the nutrient status of seedlings, based on analysis of previous experimental data and/or, using only a small test sample. In this context, a vector diagnosis is useful in judging interactions among nutrients, their concentrate and biomass has been proposed (Imo and Timmer, 1997; Quereshi and Timmer, 1998; Imo and Timmer, 1999; Quereshi and Timmer, 2000). This vector diagnosis is a graphical approach that helps in interpreting seedling nutrient status in relation to nutritional factors that normally influence growth and nutrition. Recently, it has

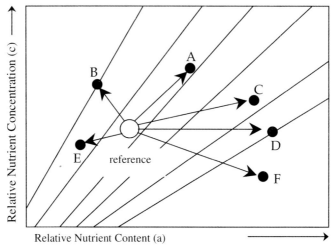

Vector Direction with time (t)	Change in			Interpretation	d(c)/dt	Possible Diagnosis
	m	c	a			
A	+	–	+	Dilution	<0	Growth Dilution
B	+	0	+	Sufficiency	=0	Steady state
C	+	+	+	Deficiency	<1, >0	Deficiency
D	0	+	+	Luxury	=1	Accumulation
E	–	++	±	Excess	>1	Toxic accum.
F	–	–	–	Excess	<0	Antagonism

Fig. 11.2 Diagramatic reperesentation of relative changes in nutrient contents of seedlings grown in forest nursery. (m = Biomass)
Source: Prof. Victor Timmer, University of Toronto, Canada.

been applicable on a variety of temperate forestry species, such as *Picea, Pinus, Prosopis, Alnus,* etc.

A comprehensive theoretical depiction of temporal variations in seedling nutrient status possible in nature or nursery, equivalent interpretations and causes are explained in Fig. 11.2 (Quereshi and Timmer, 2000). In this case, the vector diagnosis includes the time (t) variable. It therefore allows study of changes in nutrient concentration in relation to time factor. According to Imo and Timmer (1997), such changes in nutrient concentrations can then be expressed by the function:

$$dc/dt = d(a/m)/dt,$$

where c is nutrient concentration derived as a product of 'a/m, 'a' is nutrient uptake and 'm' is mass.

Therefore, a steady state nutrient concentration is possible when $dc/dt = 0$. That happens whenever nutrient uptake keeps pace with biomass accumulation, allowing no shift in nutrient concentration. Growth dilution occurs when $dc/dt < 0$, wherein dry mass and nutrient uptake increases, but still results in declining nutrient concentrations in the tissue.

To a certain extent, the temperate forest regeneration depends on the ability of transplanted seedlings to withstand severe competition for nutrients due to natural vegetation. For example, Malik and Timmer (1995) have remarked that in parts of Ontario, in Canada, Black spruce (*Picia mariana*) may get replaced by the invading hard wood trees which garner nutrients better. In case of conifers, managing neighboring vegetation through herbicides has been helpful, but not supported due to public resistance to use of herbicides, costs and at times unsure effectiveness. In this regard, Malik and Timmer (1995) suggest that use of nutrient loaded seedlings is a useful proposition. Their evaluations indicate that nutrient loaded seedlings outperformed similar sized conventionally fertilized seedlings by 28 to 44% in terms of growth and biomass. It ultimately resulted in 27% reduction in neighboring vegetation. According to them, loading nutrients into seedlings could stimulate nutrient absorption after planting in outfield. Also, translocation of more nutrients to actively growing parts from reserves built-up during nursery pre-conditioning could be the reason. Obviously, nutrient re-translocation and biomass partitioning are key to seedling competitive ability. In yet another study, Malik and Timmer (1998) evaluated nutrient loaded and unloaded Black spruce seedlings that were similar in biomass and height. They found that N re-translocation in seedlings was key to their competitive ability. Thus, internal nutrient reserves at the time of transplantation seems crucial. In the outfield, alterations in internal nutrient dynamics i.e., re-translocations were mainly driven by current growth rate, nutrient uptake and nutrient reserves. In conifers, nutrient loading favored higher needle production and root growth. Higher levels of available N apparently enhanced carbon fixation by increasing needle number, chlorophyll and carotinoid synthesis. Accelerated root turnover in nutrient loaded seedlings may increase nutrient uptake (Nadelhoffer *et al.* 1985). Nutrient loading of seedlings actually increases the proportion of easily available nutrients. In contrast to such nutrient loaded seedlings, non-nutrient loaded seedlings seem to conserve more of their nutrients, particularly N in structural parts, hence not easily available for re-translocation.

C. Nitrogen Saturation in Temperate Forests

In many North American and European locations, incessant long-term fertilizer N inputs into forest stands and chronic N depositions from atmosphere has resulted in excessive N in the ecosystem or overly saturated situations. For example, atmospheric N depositions in northeastern USA is currently 10 to 20 times above historic levels (Magill *et al.* 1997). The European temperate forests also face similar high N depositions. This is inspite of regulations on N_2O emissions. Hence, concerns regarding terrestrial systems ability to absorb and accumulate N have been expressed frequently. Firstly, such N saturation leads to greater nitrate mobility, then leaching losses accompanied by losses of cations, and enhanced soil acidity. Nutrient imbalances that ensue due to N saturation reduces/alters net photosynthesis, root and foliar chemistry, N retention, N-use efficiency, and finally forest growth (McNulty and Aber, 1993; Cronan and Grigal, 1995).

The concept of nitrogen saturation itself can be defined in a few different ways. Within the context of temperate forestry, nitrogen saturation generally refers to temporal changes in ecosystem functions in response to alleviation of N limitations, more aptly excessive N applications on plant, microbial and perhaps abiotic soil processes (Aber *et al.*, 1995). During the past decade, several long-term evaluations of N saturation have been made both in North America and Europe (Fig. 11.3). Through such studies, a wide range of inferences related to N dynamics and forest productivity have been possible. Aber *et al.* (1995) state that overall, chronic N additions into temperate forests causes (i) initial and often large increase in net N mineralization followed by decreases; (ii) increases in net nitrification; (iii) increase in N concentration in foliage and decreased Mg:N and Ca:Al ratios.

Fig. 11.3 A research team investigating aspects of long term nutrient dynamics within the pine plantations at the Austin Carey Forest Experimental Station, University of Florida, Gainesville, Florida, USA (Photo courtesy: Dr KR Krishna, Soil and Water Science Division, University of Florida, Florida, USA)

Aber *et al.* (1998) have tried to hypothesize and explain the consequences of such N saturation on various N transformation steps and forest productivity parameters. These enunciations might be helpful in gaining better insights into N dynamics in a present-day temperate forest ecosystem. During the evolution of a N saturated situation in forest soils, there occurs firstly a pretreatment period, which presumes a strong N limitation on forest growth. Under such low N input conditions, it is poor N availability that limits tree growth. Whereas limitations of carbon (energy) affects metabolism of free-living soil microbes. Next, a high N retention phase, and stimulatory effect of fertilizer N on growth are discernible in stage 1. With inorganic N additions, at least initially the N uptake and growth increase. Since plants garner N sufficiently, very little N is spared for immobilization by microbes, which already suffer C and energy limitations. Therefore, Aber *et al.* (1998) infer that N retention efficiency at this stage (i.e., as N saturation begins) should be mainly influenced by N uptake (recovery) by tree species, and very less by microbial activities. Induction of nitrification and nitrate leaching characterize stage 2. During stage 3, tree growth rate declines, nitrification and nitrate leaching losses continue, and fraction of NH_4^+ that gets nitrified increases. In summary, the hypothesis and explanations proposed are that plant growth physiology is affected first by nitrogen addition (that leads to N saturation). Then, soil microbe aided transformations, namely nitrogen mineralization, nitrification and denitrification follow in response to litter deposition and its chemical quality. However, nitrogen level at which plant species perceive the 'N saturation' might be crucial.

Experimental observations on forest stand in Europe (NITREX) and those in northeastern USA supported the view that N saturation leads to decline in forest tree growth (Schulze, 1989; Aber *et al.* 1998). Tree mortality too was higher, despite sufficient foliar N concentrations which could support better photosynthetic rates. In actuality, it was partly attributable to alteration in soil nutrient ratios, such as N : Mg or Ca : Al ratios. Sometimes, fine root physiology could be negatively affected by excessive N, which gets corrected if N is deleted from sites where soil N availability is high (Gunderson *et al.* 1998). Nitrogen saturation, clearly increased foliar N concentrations of pine stands and other tree species in several temperate locations. The decrease in Mg : N and Ca : Al ratios in foliar tissue were linked to N-saturation related forest decline in several European and North American sites (Shortle and Smith, 1988). At the same time, we may realize that increased foliar N achieved through N saturation can also affect the carbon balance in the forest ecosystem because of enhanced net photosynthetic rates (Reich *et al.* 1995). Increase in above-ground biomass may supply additional C and energy to induce excessive, rapid underground biomass. Ectomycorrhizas, the symbiotic fungi of pine and deciduous species may receive benefits from increased C availability/energy. Aber et al. (1996, 1998) say that increased foliar N due to N saturating gave a net C energy advantage equivalent to 150 gm^{-2} yr^{-1} over pines fed low N. Similarly, N saturation provided 26 gm^{-2} yr^{-1} C advantage with deciduous species.

The average dissolved inorganic N (DIN) retention was 70% in northeastern USA, and similar levels of retention are known to occur in European temperate forests (Aber et al. 1998). Nitrogen inputs as NH_4^+ salts supported better retention efficiency than NO_3–N. In addition to DIN loss through leaching, dissolved organic N (DON) losses too could be significant. Now, regarding the fate of high N inputs, Aber et al. (1998) inferred that nearly 66% of such high N inputs or deposits from atmosphere, resides in the soil organic matter. This fact was confirmed using ^{15}N traces analysis, then by examination of C : N ratios in the forest litter, ecosystem N budgets as well as soil nitrogen budgets. In addition, they believe that greater fraction of such high N inputs may not have passed through above-ground biomass. Certain other investigations revealed that nearly 85 to 99% of added N was retained by both pine-based or deciduous forests. Nitrogen cycling rates measured indicate that enhanced uptake of N by plant, and then litter production through fine roots accounts for major fraction of N incorporation. Forest floor leachetes from high N plots contained 40 to 60% higher DON (Currie et al. 1996). Magill et al. (1997) opine that, at a mere 4 to 5 kg N ha^{-1} yr^{-1}, N loss as DON may be significant compared to 15 to 50 fold higher losses via NO_3–N and NH_4^+–N leaching in high N plots. Data in Table 11.4 depicts the influence of low N and N saturation on N leaching losses, N retention, change in N storage and forest productivity. The three major mechanisms suggested, that may account for incorporation of high amounts mineral N into soil organic matter (SOM) are (a) immobilization by free-living microbes through biomass production; (b) conversion to organic nitrogen form by mycorrhizae and (c) abiotic incorporation into existing soil inorganic matter through complex chemical reactions. In summary, 'N saturation' is a situation that can occur once plants' N demand is satiated. Clearly, 'N saturation' in plants and soil could be influenced by several interacting factors.

Litter Turnover and Nitrogen Dynamics: Litter decomposition is an important aspect of nutrient dynamics in temperate forest ecosystems. The quality of litter, decided mainly by lignin, nitrogen and carbon contents (C:N ratio) influences decomposition rates. Several other environmental parameters also alter litter decomposition rates and nutrient recycling. Experts suggest that during recent decades, excessive nitrogen deposition into North American and European temperate forests has become a conspicuous factor affecting litter decomposition and nutrient

Table 11.4: An example of nitrogen dynamics and productivity of temperate forest stand[1] experimented for 4 years under low N and higher N (N saturation) at Beddington, Maine, Northeastern USA

	Control	Low N	High N
N inputs (kg NO_3–N ha^{-1})			
Atmospheric deposition	20	20	20
Fertilization	0	99	223
Total	20	119	243
N losses (kg ha^{-1})			
Gaseous	<1	<1	<1
Leaching	1	3	18
Total	1	3	18
N retention			
% of N inputs	96	97	93
Changes in N storage (kg ha^{-1})			
Soil extractable N	−3	−6	−23
Woody biomass	41	43	46
Foliage	−1	3	3
Soil organic matter	−18	76	199
Retention by soil pool			
% N retrieved	–	66	88
% of N added	–	64	82
Productivity (kg ha^{-1} $year^{-1}$)			
Mean wood production	5385	5625	6132
Mean litter production	3143	3804	4014
Mean above ground net pricing productivity	8928	9430	10146
Cumulative wood production	16154	16875	18395
Cumulative biomass change	14529	13027	12736
Cumulative tree mortality	1625	3547	3659

[1] Forest stand comprised American beach, Red spruce, Yellow birch, Red maple, Sugar maple, Striped maple.
Source: Excerpted from Magill et al. 1996.

dynamics (Galloway, 1995; Vitousek *et al.* 1997; Magill and Aber, 1998). Their reports suggest that during the initial stages of litter decomposition, higher N availabilities may induce rapid decay. However, during later stages, N inputs generally depressed decomposition rates in many types of forest stands, such as Red Pine, Red Maple, Yellow Birch, Black Oak, Alnus etc. The nitrogen concentration in the decaying litter heaps is itself regulated through immobilization and mineralization processs. Actually, total nitrogen contents that increases in the very early stages after litter deposition on forest floor has been attributed to immobilization of N derived from soil solution (McClaugherty *et al.* 1985). However, the initial phase of N immobilization is dependent on litter quality, that is its C : N ratio. For example, Fenn *et al.* (1991) reported that high C : N (< 70 : 1) found with litter under Scots pine (*Pinus sylvestrus*) immobilize N applied as urea. Therefore, priming litter with N inputs may not necessarily induce decay rates (Magill and Aber, 1998). Such inconsistencies have been observed even in long term evaluation of litter decomposition. Sometimes, higher concentrations in older litter and humus inhibits decomposition at later stages, because random bonding retards efficiency of extra cellular enzymes. In addition, lignolytic enzymes too could be depressed by N inputs in litter (Tein and Myer, 1990). Magill and Aber (1998) suggest that cellulose degradation in foliar litter could be due to layers of lignin polymer that shields cellulose. According to Berg and Staff (1980) three vivid phases of N dynamics are observable during litter decomposition in temperate forest ecosystems, and they are: (i) initial leaching phase; (ii) a period of N immobilization, therefore N accumulation into litter; and (iii) a net mineralization phase that is final.

In temperate forests with acidic soils, ammonia volatilization from urea applied to forest floor litter (FFL) can be significant. According to Fenn et al. (1991), such NH_3 losses range between 20 and 30%, and most of it occurs between 2nd and 3rd week after urea input. They further suggest that, NH_3 losses could be small for initial N inputs, but with repeated applications of urea, volatilization increases. Such a phenomenon is attributed to occurrence of microbes, one that proliferate rapidly, and the other groups which possess a lag period. Such a lag period is removable by supplying readily available energy substrate. In general, lowered pH in FFL reduces urea hydrolysis, hence it reduces NH_3 losses from applied urea. However, leaching loss of urea may occur. Incorporation of urea along with $CaCl_2$ mixture can reduce urea hydrolysis (NH_3 loss). Firstly, $CaCl_2$ addition seems to decrease urea mobility, perhaps through double salt formation. The second possibility suggested by Fenn et al. (1991) is that microbes proliferate so much that NH_3 release and immobilization rates balanced out.

3. Nutrient Dynamics in Boreal and Peat Land Forests

Peat lands are a common occurrence in the frosted areas of temperate sub-Arctic belt. The northern peat lands that occur in the boreal and sub-arctic zones of Eastern Europe, Scandinavia and British isles, occupy around 350 m ha (Laiho et al. 1993; Laine, 1989). Their formation in nature is augmented by excessive moisture and organic matter deposits. Such peat land soils may contain >75% organic matter, which is mainly derived from plants. In virgin state, the Finnish peat land and boreal locations support an abundance of vegetation comprising Scots pine (*Pinus sylvastrus*) and Birch (*Batula pubescence*). Tall sedge pines are also common. Whereas, hummocks are predominated by mire dwarfs such as *Batula nava*, *Ledium* sp and *Vaccinium uligirosum*. The oligo-ombrotrophic

Sphagna (*Sphagnum felox, S. angustifolia*), mesotrophic *S. subsedence* and *S. riparians* are frequent in Finnish peat lands (Laiho and Laine, 1994a, b, c). They are important organic C contributors. These boreal forests exhibit comparatively low productivity, mainly due to short period within a year when temperatures favor optimum biological activity. Actually, both photosynthesis and nutrient acquisition functions may get curbed. Weathering is minimal at such low temperatures and atmospheric inputs are feeble or unlikely from anthropogenic processes. Precipitations allow minor quantities of NH_4^+ to be absorbed by mosses. While, algae, which may be free-living or in symbiosis with mosses or lichens fix atmospheric nitrogen. A very large fraction of carbon and nutrients are stored in organic matter, because mineralization proceeds at very slow rate. The calculated nutrient turnover in certain sites in Alaska and European zones run into thousands of years. In the peat land forests greatest fraction of nutrients (N, P, K, Ca and Mg), more than 90% was harboured within the 0 to 50 cms of peat and litter layers. In the surface peat (0 to 50 cm), concentration of N, P and Ca were <1%, that of Mg 1 to 4% and K upto 8%. Laiho (1997) reports that relative proportions of nutrients accumulated in the tree stand, varied with those in peat surface. Most striking was N in relation to K found in peat compared with tree stand (Table 11.5). Clearly, organic matter decomposition is a key process in the release, turnover and cycling of nutrients (Kurka *et al.* 2000).

Table 11.5: Accumulation of nutrients (gm^{-2}) in tall sludge pice food vegetation, its components and, litter layer at an intensive study site in a peat land zone of southern Finland (peat land soils were drained 55 years ago)

Site components	N	P	K	Ca	Mg
Tree stand (canopy, above ground)	22.4	1.4	5.7	9.7	2.3
Ground vegetation	8.7	0.6	2.2	2.4	0.7
Stumps + coarse roots	13.4	0.5	2.5	3.4	0.8
Litter layer	7.4	0.4	0.6	2.4	0.3
Surface peat (0-15 mm including fine roots)	1355.8	51.8	12.2	84.1	12.6

Source: Data excerpted from site 6 in southern Finnish peat lands as reported by Laiho, 1997.

A. Drainage and Nutrient Dynamics in Peat Land

The Finnish temperate forests discussed here as an example were actually developed after a large scale drainage of mires, which extend into 10 to 15 m ha (Paavilainen and Taihonen, 1988). Such drainage has important physico-chemical consequences on peat/soil surface, nutrient transformations and their availability, tree growth, the general nutrient dynamics of the ecosystem and productivity of temperate forests. Drainage firstly affects the nutrient distribution in peat land soil profile.

For example, in oligo-ombrotrophic locations, N increased only in surface layers of peat (0-10 cm) following drainage. Whereas, enhanced P levels were discernible in all layers. In meso-oligo-ombrotrophic sites N and P changes, if any due to drainage were feeble and insignificant. However, peats of oldest age class tend to possess higher amounts of N and P compared to younger ones. There was a general decrease in K concentration in response to drainage. Secondary nutrients, Ca and Mg, decreased rather conspicuously at soil depths from 10 to 35 cm, in either types of soil, i.e., in oligo- and mesotrophic locations. Based on these estimates, Laiho *et al.* (1999) suggested that drainage of peat lands may have feeble effects on major nutrient recovery by a tree stand, but it severely depletes Ca and Mg in the root zone.

Time lapse after a drainage event in the peat land is important, because it affects the nutrient dynamics. On several drained meso-oligotrophic peat land sites, relative proportions of N increased only slightly with duration after drainage event. Some of the oldest drained peat lands in Finland stored nearly 20 to 30 times higher N and/or P than tree stands. Other mineral elements such as K, Ca, and micronutrients Mn, Fe, Cn, Mo were in higher quantities in peat compared to tree stand, but were not altered with age after drainage. Calcium and Fe were the only two essential elements that decreased in peat with increasing period after drainage.

In undrained boreal forests of Finland, or even elsewhere, most of the organic matter deposited as peat is derived from general vegetation. However, drainage results in a decline of peat-farming species such as *Carex* and *Sphagnum* species (Laine *et al.* 1995). Drainage of a mire (peat land bogs) lowers water table. Consequently, thickness of the aerated portion of peat layer increases. Thus, organic matter is increasingly exposed to aerobic microbial degradation. It results in loss of carbon through respiration and loss of nutrients from the organic matrix (Silvola, 1988; Laiho and Laine, 1994c). In general, redox changes in peat caused due to drainage enhances aeration, accelerates rate of mineralization and release of available nutrients (Freeman *et al.* 1998). Transitorily, increased runoff after drainage is known to enhance leaching losses of mineralized and available nutrients, particularly N (Salaantus, 1992a).

The increased tree growth after drainage is partly attributable to better oxygenation of peat layer, which aids rapid root growth. Hence, it leads to higher nutrient acquisition and growth. Actually, pine tree stands in undrained locations produce less needles and branches, and possess more dead branches. Unfavorable conditions due to poor oxygenation, that limits growth of fine roots are the reasons (Laiho and Finer, 1996). Therefore, favorable effects of drainage, first appear as increased production of needles and revived tree growth. Proportion of dead branches decreases rapidly as a result of drainage.

B. Carbon, Other Nutrients and Biomass

Accumulation of carbon in boreal forests is primarily due to low temperatures. Within Finnish peat land forests, on an average 0.5 kg C m^{-2} stored could be attributed to fine roots produced by trees aged between 22 and 55 years after drainage. Actually, fine root production in boreal peat lands increases and reaches a constant as canopy closure occurs (Vogt et al. 1987). Laiho (1997) attributes this phenomenon to functional balance, which regulates proportion of fine roots, foliage and above ground portion depending on the location. Accordingly, the carbon storage in fine roots will alter. The carbon storage in ground vegetation seems unrelated to drainage, or the years lapsed thereafter. However, carbon stored in ground vegetation depends on tree stand size. Larger tree stands may allow sparse ground vegetation, hence low levels of C will be stored (Braekke, 1998).

Drainage increases oxidation rates in peat. Under such conditions C efflux from soil/peat as CO_2 could vary between 50 and 70 g C m^{-2} yr^{-1} (Silvola et al. 1996a; Silvola et al. 1996b). Consequent to redox changes in peat, root biomass, root senescence and litter accumulation rates are affected. Finer and Laine (1996) state that nearly 40% increase in C efflux was due to root and root associated respiration. Laiho and Laine (1994c) suggest that litter quality that increases after drainage, and the resultant enhanced decomposition may account for the remaining portion of enhanced C efflux from peat lands. Carbon loss from the system through leaching is marginal after drainage. It could be as little as 1.0 g m^{-2} yr^{-1} (Sallantus, 1992b).

Laiho (1997) and Laine et al. (1995) argue that to compensate for enhanced C efflux from peat bog, the proliferation of Sphagna, and carbon inputs through its litter seems quite sufficient to maintain the carbon balance. Thus, whatever C accumulation occurs through photosynthesis in tree stand may represent net gain of C into the boreal ecosystem. However, in a forest plantation managed for commercial purpose, C storage and its equilibrium in a tree stand is dependent on length of rotation, regularity of cutting and soil management strategies.

Laiho (1997) remarks that in Finnish locations, the tree stand biomass, i.e., C storage is much higher in peat lands than uplands. In the southern boreal sites, Scots pine stands contribute high levels of stored C through stump plus coarse root system. Generally, the carbon storage in tree stand increased after drainage. On sites that were 50 years old, the C levels measured were approximately 10 folds greater than undrained sites (Laiho, 1997). We may therefore summarize that post-drainage carbon balance, or even the net ecosystem productivity largely depends on relation between carbon loss from the soil due to oxidation, and the carbon input through tree stand photosynthesis (Laiho and Laine, 1998; Laine and

Minkinen, 1996). The carbon sequestration potential of boreal peat lands is indeed an important factor that influences C cycle (Karjalinen, 1996).

The average yearly post-drainage accumulation of nutrients were 0.6 g N m^{-2}, 0.04 g P m^{-2}, 0.2 g K m^{-2}, 0.3 g Ca m^{-2} and 0.06 g Mg m^{-2}. Drainage, initially increased nutrient concentrations in foliage, which continued at such relatively higher rates. However, there was a transitory drop in nutrients such as N, Ca and Mg in leaves and stem, which later raised to parity. After drainage, increasing amounts of N and P were bound up in the tree stand (Table 11.6). Much of N and P sequestered were located in crowns > stems > stumps + roots > fine roots in that order. Crown and stumps accounted for over 70% of N and P stored in the tree system. Therefore, Laiho and Laine (1994a) suggested that stem wood harvesting may remove, generally less than 10 to 20 kg P ha^{-1} and 100 to 150 kg N ha^{-1} from any well developed boreal forest site. However, whole tree harvest will remove away sizeable quantities of N and P from the peat land ecosystem. Estimates by Laiho and Laine (1994b) indicate that some of the oldest tree stands in meso-oligotrophic, i.e., nearly 55 years after drainage, the trees had accumulated 200 kg Ca ha^{-1}, 90 kg K ha^{-1} and 40 kg Mg ha^{-1}. A similar trend but slightly reduced in quantity was found in oligo-ombrotroptic sites. Incidentally, amount of K in tree stand and that in peat were almost equal. Whereas, Ca and Mg levels sequestered were only 20% that estimated in peat. Packham et al. (1992) put forth the view that, since mineralization, nutrient turnover and availability are reduced in boreal locations, trees tend to become highly efficient in nutrient use. For example, they quote Cole and Rapp (1981) who reported that nutrient use efficiency of a boreal conifer site was 236 kg ha^{-1} yr^{-1}. Whereas it was only 138 and 184 kg ha^{-1} yr^{-1} for temperate conifers and temperate broad-leaved species respectively. Extreme longevity of needles, stems and trees which is generally, over 25 to 30 years is suggested as another example for nutrient efficient mechanisms in boreal stands. It allows for immobilization of carbon and other nutrients.

Table 11.6: Nitrogen and phosphorus stores (kg ha^{-1}) in northern peat land forest tree stands (meso-oligotrophic sites)

Years since drainage		Stems	Crowns	Stumps + coarse roots	Fine roots
Undrained	N	4	15	4	4
	P	1	2	<1	<1
1-20	N	32	92	27	22
	P	3	8	2	2
21-40	N	68	475	55	42
	P	5	14	4	3
41-55	N	73	226	61	60
	P	6	21	6	5

Source: Excerpted from a data compilation by Laiho and Laine, 1994b

4. Concluding Remarks

The temperate and boreal forests of the world provide a range of useful products and influence global geo-chemical cycles. They sequester large quantities of C and N, and emit gases such as CH_4, NO_2, and N_2 which affects atmospheric composition. Nutrient dynamics and productivity of temperate forests are influenced by a range of factors related to soils, tree species and environment, mainly temperature and moisture. The physico-chemical nature of a tree rhizosphere, nutrient transformations and the ecology of soil microflora, firstly are vital aspects that regulate nutrient availability and recovery rates in temperate forest ecosystem. Hence, for an appropriate interpretation in soil-plant relationship, the physico-chemical aspects of rhizosphere should be duly considered. Generally, nutrient concentrations and physico-chemical activity are higher at soil-root interphase compared with bulk. Models, simulations, and concepts that depict such rhizosphere phenomena are available. They could be useful in understanding tree nutrition better. We may also realize that inferences based solely on analysis of bulk soils could be, at times misleading. Soil nutrient dynamics, and its relevance to growth of temperate forests, begins earnestly as seedlings begin to grown in the nursery. The seedling growth rates, nutrients concentrations in the tissue, and rapidity of establishment in the outfield are crucial aspects. Nutrient loaded seedlings generally establish better. Nutrient loading, as a mechanism, involves feeding seedlings with higher levels of nutrient to match its growth without allowing dilutions.

Excessive N deposits and fertilizer based inputs into temperate forests have resulted in saturation of nutrients in the soil ecosystem. Nutrient leaching, enhanced acidity and gaseous emission (N_2O, NO_2, CH_4) can be detrimental to forest environment. Developing further understanding on N and C dynamics, and devising methods to regulate them seem very important. Peat lands and boreal forests/vegetation are unique in their own way. The nutrient transformations, turnover rates and growth of boreal forests stands are influenced by different factors, among which the peaty soils, moisture and temperate play an important role. These peat land forests sequester large quantities of C, N and other nutrients.

REFERENCES

Aber, J.D. (2001). Temperate Forests. Harvard University, Massachusetts, USA, website-http//www,harvard,edu/harvard forests.

Aber, J.D., McDowell, W., Nadehoffer, K., Magil, A., Bernston, G., Kamakea, M., McNulty, S., Curric, W., Rustad and Fernandez, I. (1998). Nitrogen saturation in temperate forest ecosystems—Hypothesis revisited. *BioScience,* **48:** 921-934.

Aber, J.D., Magill, A., McNulty, S.G., Boone, R.D., Knute, J.D., Downe, M. and Hallett, R. (1995). Forest biogeochemistry and primary production altered by nitrogen saturation. *Water, Air and Soil Pollution,* **83:** 1665-1670.

Aber, J.D., Reich, P.A., and Golden, M.L. (1996). Extrapolating leaf CO_2 exchange to canopy: A generalized model of forest photosynthesis validated by adding correlation. *Oecologia,* **106:** 257-265.

Berg, B. and Staaf, H. (1980). Decomposition rate and chemical changes in decomposing needle litter of Scots pine: II. Influence of chemical composition. *Ecol. Bull. (Stockholm),* **32:** 373-390.

Binkley, D. (1986). *Forest Nutrition Management.* John Wiley and Sons, New York, 290 pp.

Braekke, F.H. (1988). Nutrient relationships in forest stands: Field vegetation and bottom layer litter in peat land. *Comm. of the Norwegian Forest Research Institute,* **40:** 1-20.

Clegg, S. and Gobran, G.R. (1997). Rhizospheric P and K in forest soil manipulated with a minimum sulphate and water. *Canadian J. of Soil Science,* **77:** 525-533.

Clegg, S., Gobran, G.R. and Guan, X. (1997). Rhizosphere chemistry in an ammonium sulphate and water manipulated Norway spruce (*Picea abies* (L.) Karsta) forest. *Canadian J. of Soil Science,* **77:** 515-523.

Cole, D.W. (1995). Soil nutrient supply in natural and managed forests. *Plant and Soil,* **168/167:** 167172.

Cole C.W. and Rapp, S. (1981). Cited in functional ecology of woods and forests J.R. Packham, D.J.L. Harding, G.M. Hilton and R.A. Stutl and (eds.) (1992). Chapman and Hall, London, 407 pp.

Courchesne, F. and Gobran, G.R. (1997). Mineralogical variations of bulk and rhizosphere soils from a Norway spruce stand. *Soil Science Society of America Journal,* **61:** 1245-1249.

Cronan, C.R. and Grigal, D.F. (1995). Use of Ca/Al ratios as indicators of stress in forest ecosystems. *J. of Environmental Quality,* **24:** 209–226.

Crawley, M.C. (1986). *Plant Ecology.* Blackwell Publishers, Oxford, England, pp. 345-375.

Currie, W.S., Aber, J.D., McDowell, W.H., Boone, R.D. and Magill, A.H. (1996). Vertical transport of dissolved organic C and N under long-term N amendments in pine and hardwood forests. *Biogeochemistry,* **35:** 471-505.

Ericsson, T., Goransson, A. and Gobran, G. (1998). Effects of aluminium on growth and nutrition in birch seedlings under magnesium or calcium limiting growth conditions. *Z. Pflanzenernater. Bodenk.,* **161:** 653-660.

Fang, Q. (1992). Silviculture technique for preventing soil fertility decline in continuously cropped Chinese fir plantations. *In:* Research on site degradation of timber plantations. Division of Forest Ecology, Chinese Society of Forestry, Beijing, pp. 251-256.

Fenn, L.M., Gobran, G.R. and Agren, G.I. (1991). Nitrogen changes in forest litter after fertilization with calcium-urea. *Soil Science Society of America J.,* **55:** 509-514.

Finer, L. and Laine, J. (1996). Fine root production at drained peat land sites. *In:* Northern peat lands in global climatic change. Laiho, R., Laine, J. de Vasander, H. (eds.). Publications of the Academy of Finland, Edita, Helsinki, 1/96. pp. 40-46.

Finley, R.D. and Soderstorm, B. (1989). Mycorrhizal mycelia and their role in soil and plant communities. *In:* Ecology of arable land. M. Clarbolm and L. Bergstorm (eds.). Kluwer Academic Publishers, Dordrecht, The Netherlands, pp. 139-149.

Freeman, C., Lock, M.A. and Reynolds, B. (1998). Climatic change and the release of immobilized nutrients from Walsh riparian wetlands. *Ecol. Eng.* **2:** 369-373.

Galloway, J.N. (1995). Acid deposition: Perspectives in time and space. *Water, Air, and Soil Pollution,* **85:** 15-24.

Gobran, G.R. and Bosatta, E. (1988). Cation depletion rate as a measure of soil sensitivity to acidic deposition: Theory. *Ecological Modelling,* **40:** 25-36.

Gobran, G.R. and Clegg, S. (1996). Conceptual model for nutrient availability in the soil-root system. *Canadian J. of Soil Science.*, **76**: 125-131.

Gobran, G.R., Clegg, S. and Courchesne, F. (1998). Rhizospheric process influencing the biogeochemistry of forest ecosystems. *Biogeochemistry* **42**: 107-120.

Gobran, G.R., Fenn, L.B., Persson, H. and Winde A. (1993). Nutrition response of Norway spruce and willow at varying levels of calcium and aluminium. *Fertilizer Research,* **34**: 181-189.

Gundreson, P., Emmet, B.A., Kjonass, O.J., Koopmans, C.J. and Tietema, A. (1998). Impact of nitrogen deposition on nitrogen cycling in forests: A synthesis of NITREX data. *Forest Ecology and Management,* **101**: 37-56.

Imo, M. and Timmer, V.R. (1997). Vector diagnosis of nutrient dynamics in mesquite seedlings. Forest. *Science,* **43**: 1-6.

Imo, M. and Timmer, V.R. (1999). Vector competition analysis of Black spruce seedlings responses to nutrient loading and vegetation control. *Canada J. of Forest Research,* **29**: 474-486.

Karjalainen, T. (1996). Dynamics and potential of carbon sequestration in managed stands and wood products in Finland under changing climatic conditions. *Forest Ecology and Management,* **80**: 113-132.

Kurka, A.M., Starr, M., Heikinheimo, M. and Salkinajo-Salonen, M. (2000). Decomposition of cellulose strips in relation to climate, litterfall nitrogen, phosphorus and C:N ratio in natural forests. *Plant and Soil,* **219**: 91-101.

Laiho, R. (1997). Plant biomass dynamics in drained pine mires in southern Finland: Implication for carbon and nutrient balance. *The Finnish Forest Research Institute, Research papers,* 631, 20 pp.

Laiho, R. and Finer, L. (1996). Changes in root biomass after water-level draw down on pine mires in southern Finland. *Scandinavian J. of Forest Research,* **11**: 251-260.

Laiho, R. and Laine, J. (1994a). Nitrogen and phosphorus stores in peat lands drained for forestry in Finland. *Scandinavian J. of Forest Research,* **9**: 251-260.

Laiho, R. and Laine, J. (1994b). Changes in mineral element concentrations in peat soils drained for foresting in Finland. *Scandinavian J. of Forest Research,* **10**: 218-214.

Laiho, R. and Laine, J. (1994e). Plant biomass and carbon store after water-level draw down of pine mires. *In:* Northern peat lands in global climatic change. R. Laiho, J. Laine and H. Vasander (eds.). Publication of the Academy of Finland, Edita, Helsinki, 1/96, 54-57.

Laiho, R. and Laine, J. (1998). Tree stand biomass and carbon content in an age sequence of drained pine mires in southern Finland. *Forest Ecology and Management,* **90**: 17-25.

Laine, J. (1989). Classification of peat lands drained for forestry. *SWO.,* **40**: 37-51.

Laine, J. and Minkinen, K. (1996). Effect of forest drainage on the carbon balance of mire: A case study. *Scandinavian J. of Forest Research,* **11**: 307-312.

Laiho, R., Sallantaus, T. and Laine, J. (1999). The effect of forestry drainage on vertical distribution of major plant nutrients in peat soils. *Plant and Soil,* **207**: 169-181.

Laine, J., Vasander, H. and Laiho, R. (1995). Long-term effects of water level draw down on the vegetation of drained pine mires in southern Finland. *J. of Applied Ecology,* **32**: 785-802.

Margill, A.H. and Aber, J.D. (1998). Long-term effects of experimental nitrogen additions on foliar litter decay and humus formation in forest ecosystem. *Plant and Soil,* **203**: 301-311.

Magill, A.H., Aber, J.D., Hendricks, J.J., Bowden, R.D., Melillo, J.M. and Stendler, P.A. (1997). Biogeochemical response of forest ecosystem to simulated chronic nitrogen deposition. *Ecological Applications,* **7**: 403-415.

Magill, A.H., Downs, R.H., Nadelhaffer, K.J., Halleit, R.A. and Aber, J.D. (1996). Forest ecosystem response to four years of chronic nitrate and sulfate additions at Bear Boots watershed, Maine, USA. *Forest Ecology and Management,* **84:** 29-37.

Malik, V. and Timmer, V.R. (1995). Interaction of nutrient loaded Black spruce seedlings with neighboring vegetation in greenhouse environments. *Scandinavian J. of Forest Research,* **25:** 1017-1023.

Malik, V. and Timmer, V.R. (1998). Biomass partitioning and nitrogen retranslocation in black spruce seedlings a competitive mired wood site: A bio-assay study. *Scandinavian J. of Forest Research,* **28:** 206-215.

Mathews, E., Payne, R., Rohweder, M. and Murray, B. (2000). Forest ecosystems. World Resources Institute/International Food Policy Research Institute, Washington, D.C. pp. 114.

Mc Alister, J.A. and Timmer, V.R. (1998). Nutrient enrichment of White spruce seedlings during nursery culture and initial plantation establishment. *Tree Physiology,* **18:** 195-202.

McClaugherty, C.A., Pastor, J., Aber, J.D., and Melillo, J.M. (1985). Forest litter decomposition in relation to soil nitrogen dynamics and litter quality. *Ecology,* **16:** 266-275.

McNulty, S.G. and Aber, J.D. (1993). Effects of chronic N additions on nitrogen cycling in a high-elevation spruce for stand. *Scandinavian J. of Forest Research,* **21:** 1689-1693.

Munson, A.D. and Bernier, P.Y. (1993). Comparing natural and planted spruce seedlings II. Nutrient uptake and efficiency of use. *Scandinavian J. of Forest Research,* **23:** 2435-2442.

Nadelhoffer, K.J., Aber, J.D. and Metillo, J.M. (1985). Fine roots, net primary production and soil nitrogen availability: a mean hypothesis. *Ecology,* **66:** 1377-1390.

Oren, R. and Schulze, E.D. (1989). Nutritional disharmony and forest decline. *In:* Air pollution and forest decline. Shulze, E.D., Lange, O.L. and Oren, R. (eds.). Springer Verlag-Heidelberg-New York, pp. 425-443

Packham, J.R., Harding, D.J.L., Hilton, G.M. and Stuttand, R.A. (1992). *Functional Ecology of Woodlands and Forests.* Chapman and Hall, London, 407 pp.

Paavilainen, E. and Taihonen, P. (1988). Peat land forests in Finland in 1951-1984. *Folia Forestry, Helsinki,* **714:** 1-29.

Qurashi, A.M. and Timmer, V.R. (1998). Experimental fertilization increases nutrient uptake and ectomycorrhizal development of Black spruce seedlings. *Scandinavian J. of Forest Research,* **28:** 674-682.

Qureshi, A.M. and Timmer, V.R. (2000). Growth, nutrient dynamics, and ectomycorrhizal development of container-grown *Picea mariana* seedlings in response to exponential nutrient loading. *Scandinavian J. of Forest Research,* **30:** 199-201.

Reich, P.B., Koeppal, B., Ellsworth, D.S. and Walters, M.B. (1995). Different photosynthesis-nitrogen relationships in deciduous hardwood and evergreen coniferous tree species. *Oecologia,* **104:** 24-30.

Roberts, J. (1999). Plants and water in forests and woodlands. *In:* Eco-hydrology. A.J. Baird and R.L. Wilby (eds.). Routledge Publishers, London and NewYork, pp. 181-236.

Roberts, T.M., Skaffingon, R.A. and Blank, L.W. (1989). Causes of type 1 Spruce decline in Europe. *Forestry,* **62:** 180-222.

Sallantas, T. (1992a). Leaching in the material balance of peat lands—preliminary results. *Suo,* **43:** 253-258.

Sallantus, T. (1992b). The rate of leaching in the material balance of peat lands. *In:* The Finnish research program on climate change—progress report. Publications of the Academy of Finland, Painatuskeskes, Helsinki, 3/92, 239-242.

Schulze, E.D. (1989). Air pollution and forest decline in a spruce (*Picea abies*). *Forest Science,* **244:** 776-783.

Shortle, W.C. and Smith. K.T. (1998). Aluminium induced Ca deficiency syndrome in declaining Red spruce. *Science,* **246:** 239-240.

Silvola, J. (1988). Effect of drainage and fertilization on carbon output and nutrient mineralization of peat. *Suo,* **39:** 27-37.

Silvola, J., Alm, J., Ahlobriz U., Nykanen, H. and Marikainen, P.J. (1996a). CO_2 fluxes from peat in boreal mires under varying temperature and moisture conditions. *J. of Ecology,* **84:** 219-228.

Silvola, J., Alm, J., Atilohm, U., Nykanen, H. and Mastikainen, P.J. (1996b). The contribution of plant roots to CO_2 fluxes from organic soils. *Biology and Fertility of Soils,* **23:** 126-131.

Tien, H. and Myer, S.B. (1990). Selection and characterization of nutrients of *Phanearucheata chrysospermum* exhibiting lignolytic activity under nutrient rich conditions. *Appl. Environ. Microbial.,* **56:** 2540-2544.

Timmer, V.R. (1997). Exponential nutrient loading; a new fertilization technique to improve seedling performance on competitive sites. *New Forests,* **13:** 279-299.

Ulrich, B. (1987). Stability, elasticity and resilience of terrestrial ecosystems with respect to matter balance. *In:* Ecological studies. E.D. Shultze and H. Zolfer (eds.). Springer Verlag, Berlin, Heidelberg, pp. 11-49.

Vitouek, P.M., Aber, J.D., Howarth, R.W., Likens, G.E., Matson, P.A., Schindler, D.W., Schliringer, W.H. and Tilman, D. (1997). Human alteration of the global nitrogen cycle: sources and consequences. *Ecol. Appl.,* **7:** 737-750.

Vogt, K.A., Meore, E.E., Vogt, D.J., Radlin, M.J. and Edmonds, R.L. (1987). Conifer fine roots and mycorrhizal root biomass within the forest flows of Douglass—for stands of different ages and site productions. *Scandinavian J. of Forest Research,* **13:** 429-437.

Xu, X. and Timmer, V.R. (1998). Biomass and nutrient dynamics of Chinese fir seedlings under conventional and exponential fertilization regumes. *Plant and Soil,* **203:** 313-322.

Xu, X. and Timmer, V.R. (1999). Growth and nitrogen nutrition of Chinese fir seedlings exposed to nutrient loading and fertilization. *Plant and Soil,* **216:** 83-91.

Zottl, H.W. (1990). Remarks on the effects of nitrogen deposition to forest ecosystem. *Plant and Soil,* **128:** 83-89.

CHAPTER 12

Epilogue

The Agrosphere is a conglomerate of several agroecosystems, which are crop-based, food-generating ecological entities. Each of these agroecosystems is a product of several interactive, natural and man-made factors. However, its development, sustenance, and productivity to a great extent are guided through human ingenuity and preferences. Wood *et al.* (2000) opine that agrosphere, that is cropping activity within it is the dominant influence on global landscape. For this to happen agrosphere has expanded consistently, and throughout all the continents. Such an expansion has been more rapid during the past century. At present, agrosphere occupies 13,500 m km^2 which is equivalent to a third of the global land surface. Clearly, large enough to be deemed an ecosphere comparable to lithosphere, hydrosphere or atmosphere. However, there could be limits to such an expansion of agrosphere, particularly the individual agroecosystems. Although a debatable issue, it is generally accepted that, at least with a few agroecosystems such as the intensive rice culture in Southeast Asia expansion is difficult.

Overall, food production from agrosphere has kept pace with the global population growth and concomitant demand for food grains. Wood *et al.* (2000) report that percapita food supply has increased by 24% since 1967, and food commodities have costed 40% less in real terms. During this period human population increased from 3 to 6 billion. However, there is unequivocal data suggesting that the capacity of different agroecosystems to enhance their productivity is declining. In this regard, Gruhn *et al.* (2000) report an interesting phenomenon termed *'slowing yield growth rates'*. For example, cereal growth rates declined from 2.2% in 1970's to 1.1% in 1990's. More specifically, wheat yields increased by 4.3% annually in India during 1970's, but only by 0.7% in 1990's. Similarly, rice yield grew by only 1.5% in 1990's compared with 2.4% in 1970's. However, we do not have absolute knowledge about limits to productivity of any agroecosystem. Such yield increases since 1961, which now seems to stagnate in certain a'grozones, have actually forestalled conversion of an additional 3.3 billion ha of natural habitat into agricultural belts. Goklany (2000) warns that if productivity of major cereal/legume agroecosystems does not increase during the next 50 years, then extra 1.7 billion ha of natural habitat will have to be lost to agrosphere, so that anticipated 9.6

billion people by the year 2050 are nourished. Perhaps, ecologically not a good situation. However, a mere 1% annual increase in global food grain productivity will reduce the need for expansion of agrosphere to 400 m ha. Obviously, intensification of different agroecosystems, wherever feasible, is an important method to curb loss of natural vegetation to agrosphere. At this juncture, we lack precise knowledge regarding limits to expansion and/or intensification of agroecosystems. Even recently, during the past two decades, agrosphere has expanded at 13,000 km^2 yr^{-1}, mainly at the expense of natural forests and grassland. Expanses of unproductive land have also been converted into fields with good yield potential using a package of standardized procedures involving crop varieties, soil amendments and agronomic measures (Borlaug and Dowswell, 1994). Overall grain productivity increases are attributable, firstly to area expansion. Then, due to intensification effected through enhanced nutrient and water inputs, and improved (high yielding) crop genetic stocks, including those resistant to yield retardants, biotic and abiotic stress factors.

Now, in tune with the theme of this book let us critically summarize the significance of nutrients, and their dynamics to agrosphere productivity. Firstly, there is no doubt that nutrients and their dynamics are among the few key factors that determine productivity of different agroecosystems. At present, nearly 55% of the 128 million metric tons fertilizer based nutrients impinged annually into agrosphere is garnered by cereals, 12% by oilseed crops, 11% by pastures, and rest by horticultural crops (Harris, 1998). It is not a lopsided nutrient distribution, considering that we depend greatest on cereal grains to satisfy carbohydrate and energy requirements. On a unit area basis vegetables, sugar crops and tubers are fed with high dozes of nutrients averaging 200 to 250 kg ha^{-1}. Nutrients are also replenished into agroecosystems through recycling of organic residues, stover, FYM and green manure. Natural weathering of soil parent materials, atmospheric deposits, and biological nitrogen fixation processes may also contribute, at times sizeable amounts of nutrients to individual agroecosystems. Unlike agroecosystems nutrient inputs into natural vegetation are derived chiefly through natural weathering processes and atmospheric depositions.

In the absence of appropriate nutrient replenishment schedules, firstly 'nutrient mining' that is depletion sets in, then nutrient imbalances may appear. 'Nutrient mining' or depletion of available nutrients is a sort of scurge experienced during subsistence farming, and in low-nutrient input/ turnover based agroecosystems. For example, subsistence farming zones in WANA, or Indian subcontinent, or in pearl millet agroecosystem in West Africa, nutrient mining firstly leads to soil exhaustion and impaired nutrient dynamics, then disturbs ecosystematic functions and reduces productivity. Suggested remedies include, nutrient inputs, adoption of fallows, changing cropping pattern/sequence with species (or genotypes) that demand less nutrients (e.g. low-N or low-P tolerant genotypes), conserving nutrients by avoiding erosion, leaching and emissions etc.

Nutrient loss from any agroecosystem could be rampant, and variable depending on the soils, extent of inputs, the cropping pattern and environmental factors. No doubt, adoption of as many agronomic procedures that reduce losses of applied nutrients are important. Mainly because, they may directly determine the efficiency of the fertilizer technology and productivity of the given field/agroecosystem. Major avenues of nutrient loss from an agroecosystem are soil erosion, leaching, volatilization (NH_3) and emissions, immobilization, chemical fixation and retention in lower horizons of soil. Nutrient loss due to erosion can be severe. Rough estimates indicate that globally nutrient loss due to soil erosion and degradative processes can cause 13% yield loss from crop land and 4% from pastures.

Volatilization is a major threat to fertilizer N-use efficiency in most agroecosystems ranging from sandy, low fertility zones in West Africa, to wetland rice ecosystem in East Asia. Loss of N due to volatilization may range between 8 and 42% of applied urea fertilizer. However, deep placement of fertilizer granules, using urease inhibitors, splitting, adding FYM may all reduce nutrient loss and provide higher efficiency. Chemical fixation of applied P fertilizer again is a rampant problem in many agroecosystems. Overall, fertilizer N-use efficiency fluctuates between 20 and 48%. Similarly, fertilizer P-use efficiency varies between 8 and 20% depending on the agroecosystems. Clearly, there exists opportunities to enhance fertilizer-use efficiency, avoid unduly high rates of inputs and delay soil deterioration due to accumulation, and reduce atmospheric contamination. At this juncture, obtaining greater insights into nutrient transformations, and their significance to nutrient availability may be highly relevant. Mainly because, they may hold the key to achieving higher nutrient-use efficiency.

Foremost, the fertilizer recommendations should be based on accurate measurements of available nutrients in soil, including that derivable through mineralization, and accounting for loss due to leaching, chemical fixation, volatilization/emission. It is believed that nearly 30% reduction in fertilizer inputs could be effected by accurate fertilizer recommendations. Appropriate fertilizer formulation and delivery methods such as banding, deep placement, splitting dosages, fertigation etc. can enhance nutrient recovery and fertilizer-use efficiency. Hence, leading us towards better control over nutrient dynamics. A good example is shift to fertigation in the citrus grooves of Florida. Firstly, it enhances fertilizer N recovery and use efficiency. Then, it localizes nutrients nearer to roots, and in the upper horizons. Therefore, it avoids leaching of nutrients low into ground water sources/aquifers. Similarly, splitting N fertilizer application enhances recovery from 50 to 75%, improves fertilizer N-use efficiency and avoids emissions or accumulation in wetland rice soils.

Currently, a combination of field measurements, predictive models, simulations and precision farming methods are being tested for greater use in different agroecosystems. Precision farming techniques may, in future, enhance fertilizer-use efficiency and bring about conspicuous changes in soil nutrient dynamics. Essentially, any number techniques could be devised, tested and utilized, but all of these will have to aim at accurately matching soil nutrient dynamics, particularly nutrient availability with crop demand/recovery rates in time and space—really easier said than achieved.

Definitely, nutrient mis-management that occurs in parts of agrosphere has to be avoided/reduced. Excessive nutrients, common with intensively cultivated agroecosystems may not be congenial ecologically. It may affect soil physico-chemical transformations, nutrient equilibria, as well as many of the functions of the biotic component. Over application of nutrients, either into individual field, or on a agroecosystem basis is wasteful, even if it is inexpensive. Actually, positive effects of nutrients plateau off after reaching a certain level of saturation concentration, and further accumulation could be detrimental. World Bank (1996) reports suggest that North America, Europe and East Asia utilize enormous quantities of fertilizer-based nutrients. As a consequence nutrient storage in these soils have already risen to 2000 kg N, 700 kg P and 1000 kg K per ha. It may lead us to uncongenial soil conditions, if the trend continues. Fertilizer management programs that regulate soil mineral nutrient accumulation are required. Ecological consequences of excessive nutrient inputs can be severe. For example, excessive N inputs in parts of Great Plains, and in the citrus belt Florida has lead to NO_3 contamination of ground water that exceeds acceptable limits. Similarly, excessive CH_4 and NO_2 emissions from rice fields of Southeast Asia are detrimental to atmosphere and ozone layer.

Soil organic carbon is an important component of nutrient dynamics in any agroecosystem. In general, land conversion to agriculture is the primary cause of SOM decline. Decreased litter formation/residue recycling rates, and rapid oxidation due to incessant tillage induce SOM loss. To quote examples, sandy soils held in oxidized state allows very low SOM accumulation in the pearl millet agroecosystem of West Africa. Stubble burning is a preferred method to enhance mineral status of soils in Indo-Gangetic Plains, but it reduces C recycling. In general, if compared with tropical forest vegetation, the value of agroecosystems (field crops) for C sequestration is limited. Mainly because, biomass is regularly harvested/removed without being totally recycled. Plus, tillage induces rapid SOM oxidation. However, there are several methods employed depending on the agroecosytem, which reduce SOM loss, enhance C sequestration and improve soil quality. Any cropping system strategy that induces higher biomass/litter formation involves less tillage and burning, contains deep-rooted crops grown in sequence or as intercrops, and conserves soil moisture will definitely enhance C sequestration (Lal, 1998). For example, Houghton *et al.* (1999) estimate that no-tillage (or reduced-till) practices, and other United States of America government sponsored measures increased C storage in the Great Plains by 138 mt C during 1980's. Strategies that reduce C emissions as CH_4 need to be encouraged because methane is 20 times more potent than other emissions causing global warming. Presently, the agrosphere and farming activity within it is known to contribute 44% of anthropogenic CH_4 emissions. In particular, rice farming is one of the larger anthropogenic sources of CH_4 emission and global warming. Added to it, the intensive rice belt in East Asia which consumes > 200 to 250 kg N ha per year, emits high levels of NH_3 and N_2O. They contaminate the atmosphere, and N_2O in particular affects ozone layer. Hence, methods that sequester C, and reduce emission of N_2O and NH_3 are to be stringently followed during intensification of any agroecosytem.

Water resources can influence both nutrient dynamics and crop productivity. Soil nutrient transformations, their absorption and recovery by crops, partitioning are all dependent on moisture levels. In the dryland and semi-arid agroecosystems, quite often it is water first, or interactive effects of water and nutrients that controls productivity. Whereas, in humid tropics or wetland cropping systems, since water is not limiting, nutrient dynamics holds the key to productivity, and any intended intensification. Global estimates of dryland ecosystems indicate that they fare better by achieving 54 to 58% irrigation efficiency, compared with less than 30% with intensive wetland rice ecosystem in India and China (Seckler, 1998). Clearly, there exists opportunity to enhance irrigation use efficiency through trial and research. It might be a wise proposition considering that further intensification of different agroecosystem will need greater water inputs. Even in the past, intensification of cropping systems in U.S.A, Europe, China, and India have been possible, because, in addition to other inputs, water resources have been augmented. However, it is believed that such excessive use of water has resulted in decreased river flows, and falling ground water level, leading to strain in irrigation capacity. Enhanced precipitation-use efficiency, achievable through various agronomic measures such as reduced or no-tillage systems, or using stubble mulches could be useful propositions in many agroecosystems. Similarly, drip irrigation/fertigation feasible with certain agroecosystem is known to enhance both water and nutrient-use efficiency. Overall, we should realize that water resources can influence nutrient dynamics, and are key components wherever/whenever intensification of an agroecosystem is intended.

Crop improvement is a very important aspect of global crop production. Biomass and grain/fruit harvest increases in almost all agroecosystems of the agrosphere, to a certain extent is attributable to high yielding genetic stocks. Obviously, to yield more crops require proportionately greater nutrient inputs, so that sizeable photosynthates and minerals are partitioned into grains/harvestable product. In view of the above reason, introduction of each and every new crop species or its genotype will generate changes in nutrient dynamics. Immaterial whether farmers/researchers perceive it or not, nutrient dynamics and functional equilibrium of agroecosystem would be altered proportionately. Notable examples are the higher nutrient inputs/outputs that occurred due to introduction of high yielding semi-dwarfs of wheat or rice in different parts of the world. Such changes in nutrient dynamics were not authentically forecasted at that time when new genotypes were introduced. The phenomenon just occurred. However at present, with the advent of computers, simulation and modeling crop growth, soil nutrient dynamics and weather patterns are possible. Therefore, more accurate forecasts about altered nutrient dynamics due to introduction and spread of a genotype should be possible. Conversely, given a set of nutrient dynamics related parameters, we should be able to forecast the performance of crop genotype within an agroecosystem. This will be a very useful exercise to research administrators and policy makers while arriving at decisions. Let us consider a situation that may be unfolding in the Southeast Asian rice belt. Firstly, nutrient dynamics changed immensely as semi-dwarfs were introduced in 1960s and 70s. Such semi-dwarfs were provided 80-120 kg N, 25-40 kg P_2O_5 and 80-100 kg K_2O to yield 4 to 9 t grain and 7-12 t stover per ha. Now, with the anticipated spread of rice super hybrids that yield 10-15 t grain and 15-22 t stover per ha, requirements of NPK will increase enormously. We have no idea of its ecological consequence, at either field or agroecosystem level. Firstly, nutrient dynamics will change, which perhaps can be forecasted using an amalgamation of computer-based models simulations on soil nutrient dynamics, crop growth and weather patterns. Such drastic changes in nutrient dynamics due to introduction of a new genotype can occur irrespective of crop or agroecosystem, but it needs to be anticipated and appropriate measures adopted. With regard to subsistence farming zones, or agroecosystems with inherently low soil fertility, there seems to be opportunity to overcome nutrient paucity by breeding crop genotypes that tolerate low

N or low P in soil. Overall, irrespective of the type of agroecosystem, crop genetics will definitely influence the nutrient dynamics of the cropping zone. Again, intensification achieved through crop improvement may have limits imposed by crop physiological, soil and environmental factors. However, genetic potential for higher yields for most crops is much greater than that presently achieved by farmers in their fields. That again leaves us with vast opportunity to enhance productivity. As a word of caution, we must realize that despite excellent management of soil, nutrients, water and crops, the nutrient dynamics as well as productivity, both at times could be severely influenced by pestilence and disease. It is pertinent to analyze these aspects. However, they have not been dealt in this book.

Now, let us restate the initial sentences from the Preface of this book. In summary, the bounties bestowed by nature namely, crops, soil and their minerals, water, and environmental factors have so far ensured a certain level of productivity from the several agroecosystems that encompass agrosphere. Let us maintain it so, without disturbing the ecosystematic equilibria, but aiming at higher productivity. The future should then be brighter for us.

REFERENCES

Borlaug, N.E. and Dowswell, C.R. (1994). Fertilizer: To nourish the infertile soil that feeds a fertile population that crowds a fragile world. *In:* Proceedings of the 3rd Annual International Agribusiness Management Association (IAMA). Symposium on Managing agriculture in a global economy. Texas A and M University, College Station, TX, USA.

Goklany, I.M. (2000). Richer is more resilient: Dealing with climate change and urgent problems—Earth Report 2000. *In:* Agroecosystems. S. Wood, S. Kate and S.J. Scherr (eds.). International Food Policy Research Institute, Washington, D.C.

Gruhn, P., Goletti, F. and Yendelman, M. (2000). Integrated management, soil fertility, and sustainable agriculture. International Food Policy Research Institute, Washington, D.C. 31 pp.

Harris, G. (1998). An analysis of global fertilizer application rates for major crops. Paper presented at the agroeconomics committee on fertilizer demand. International Fertilizer Association, Toronto.

Houghton, R.A., Hackler, J.L. and Lawrence, K.T. (1999). The US carbon budget: contributions from land use change. *Science,* **285**: 574-578.

Lal, R., Kimble, J.M., Follett, R.F. and Cole, C.V. (1998). The potential of US crop land to sequester carbon and mitigate the greenhouse effect. Chelsea: Ann Arbor Press, MI. USA.

Seckler, D., Amarsinghe, V., Molden, D., De Silva, R. and Becker, R. (1998). World water demand and supply—1990 to 2025: Scenarios and issues. Research report 19, International Water Management Institute, Colombo, Sri Lanka.

World Bank. (1996). Natural resources degradation in sub-Saharan Africa: Restoration of soil fertility: Concept Paper and Action Plan. Washington, D.C

Wood, S., Sebastian, K. and Scherr, S.J. (2000). Agroecosystems. International Food Policy Research Institute, Washington, D.C., 103 pp.

Index

Acacia, Dalbergia 199
Acacia tortilis 198
acidic deposition 326
acidification 261, 262
Acrisols 242
Actinomycetes 48
aerosol deposition 286
Aerosols 308
African tropical forests 281
Agricultural ecosystems 26
agroecosystem 2, 5, 10, 11, 15, 19, 25, 27, 28, 32, 66, 141, 242, 330-333, 335, 336
Agrosphere 1, 3, 4-7, 10, 22, 28, 79, 141, 330, 336
Al saturation 244
Al toxicity 243, 244
Al^{+3} accumulation 312
Alfisols 30, 143, 198, 226, 239, 260, 261, 262, 271, 305
alkaline 273
alkalinity 11, 163, 164, 170, 261, 274
alternate bearing habit 275
AM fungi 253, 272, 273
AM symbiosis 252
Amazonia 281, 295
American neo-tropical forest 281
Ammonia fixation 176
Ammonia volatilization 173, 278, 320
Ammoniation 258
Ammonification 114
ammonium polyphosphate 156
Anabaena azollae 161
Anabaena variabilis 162
anaerobic 110, 116, 297
anaerobic conditions 289
anaerobic microsites 291
Annual P balance 126
anthropogenic contribution of S 19
Arbuscular mycorrhizal fungi 162, 184, 193, 226, 231, 237, 251
Ardisols 305
Arenosols 197

argillic layer 271
Aridisols 30, 210
Aspergillus 160
asymbiotic N fixation 160, 184, 245
atmosphere 6, 8, 330
atmospheric depositions 331
atmospheric fluxes 301
Atmospheric N depositions 269
Atmospheric S 121
Australasia, humid tropical forests 281
Australian wheat agroecosystem 79
Australian wheat belt 82
Availability of Phosphorus 192
Available P Pools 155
Azolla 128, 161, 177
Azolla hybrids 136
Azolla-BGA complex 165
Azolla-blue green algae 167
Azospirillum 224, 225
Azotobacter 224, 225
Azotobactor croococum 161

Bacillus 233
Bacillus polymixa 162, 226
Badia or Steppe-type pastures 186
banding 45, 196
Banding nutrients 41
beans 241
below-ground ecosystem 308
bio-temperature (BT) 281
Biofertilizers 127, 178, 204, 224, 235, 237, 239
biomes 306
Biosphere 1, 3,4, 6, 9, 301
biotic and abiotic stress factors 331
boreal 306
Boreal and peat land forests 306, 320
boreal mires 329
Boron deficiencies 159
Boron toxicities 182
Brassica 92
Brazilian Cerrado 218, 234

Bread basket   59
break crop   83, 87
broad-leaved trees   305
Broadbalk   67, 72, 76, 79, 101
Burkina-Faso   221, 222
Burning   293

C : N : S ratios   229
C : N : S relationships   240
C : N ratio   36, 37, 67, 68, 95, 110, 150, 160, 201, 224, 306, 318, 319
C : P ratio   77
C emissions   334
C sequestration   13, 103, 334
C sink   13
Ca : Al ratios   316, 317
Ca and Al balance   312
Ca deficiency   261
Cajanas cajan   239, 240
calcareous   186, 271, 273, 274
Calciferous   182
Calciorthids   180
Calcium deficiency   279
Cambisols   60, 61, 180, 192
Canadian prairies   46
canopy   276, 283, 284, 307, 326
carbon balance   317
carbon budget   336
carbon cycle   216
carbon loss   323
carbon monoxide   292, 301
Carbon Sequestration   25, 47, 48, 55, 56, 66, 324, 327
carbon storage   26, 323
Carrizo Citrange   262
Cassava   241, 242, 245, 249, 251, 254, 255
cassava-based agroecosystem   242, 244
catch crop   85
Central European Plains   59
Central Florida   270
Centrosema   246
CERES-wheat simulation model   103
Cerrados   243
change point   76
chemical fixation   254, 332
chemisorption   149
chemo-denitrification   297
chemolithotrophs   290
Chernozem   59, 60, 61, 64
chickpea   226, 232
chronic N additions   328
chronic nitrogen deposition   327

Cicer aerietinum   178
citriculture   258
citrus agroecosystem   259
Citrus agroecosystem in Florida   257
citrus decline   258
Citrus grooves   261, 264, 266, 267, 270, 271, 274, 276, 332
clear cutting   285
Clearing and burning   280
Cleopatra Mandarin   262, 266
climax vegetation   281
coarse textured soils   81
Conifer   305, 315, 329
Conservation tillage   95
controlled release fertilizers (CRP)   264
conventional plough tillage   61
Conventional Tillage   31, 37
copper effects   277
Corrective fertilization   214
Corsican pine   306
cow pea   237
critical rainfall level   216
Critical-P in soil   252
crop genetic stocks   331
Crop improvement   335
Crop load   275
Cropping Sequence   85, 91, 94
Crotalaria juncia   128
Cu replenishment   274
Cyanobacteria   161, 175, 246
Cyanodon dactylon   92, 168
Cyperus rotundus   168

deamination   114
deforestation   293, 294
Denitrification   17, 35, 70, 74, 105, 110, 115, 116, 150, 166, 171, 189, 212, 221, 266, 296, 297, 317
denitrifiers   290
desorption   122
Desorption of P   75
dew component   284
dissolved inorganic N (DIN)   318
dissolved organic N (DON)   318
disturbed ecosystems   3
Drainage   321-324
drought stress   210, 277
dry granular fertilizers (DGF)   268
dryland agroecosystem   179, 183, 190
Dryland Cropping Environment   181
Dryland crops   12
Drylands of South Asia   197

duplex soils 82
Dwarfing genes 217

East-coast lagoons 258
ecological stability 64
ecospheres 6, 7
ectomycorrhizal 328
Ectomycorrhizas 306, 317
edaphic factors 282
emissions 103, 332
Emissions and Deposits 288
Entic Haplustolls 92
Entisols 259, 263, 278
erosive runoff 242
euphotic zones 283
evapotranspiration 91, 199, 304
evapotranspirational losses 81
exchangeable sodium percentage (ESP) 163
exchangeable K 121, 157, 194, 229
exchangeable Zn 158
exudations 183

Fallows in Humid Tropics 250
Fe^{+3}–Fe^{+2} equilibrium 146
Ferralsols 242
fertigation 22, 260, 279, 332, 335
Fertile Crescent 58
Fertilizer N Recovery 221
fertilizer-N efficiency 87
fertilizer-N immobilized 74
Fibrous root 264, 277
fine litter fall 287
Finnish temperate forests 321
Flat Pampas 92
Flooded, deep water rice ecosystem 107
Flooded Rice Soils 119
flooded soils 289
Flooding 146
Flux 295, 286
Fluxes 286, 287, 290, 291, 296, 307
forest canopy 296
forest decline 328
Forgotten nutrient 194
Forms of N in Wetland Soils 111
Fourth Major Element 127
Full point 263
Fungi 48

G. calospora 272
G. desserticola 271
G. intraradicis 271
G. mossea 271
Ganges 143
Gangetic plains 144, 149, 151, 168
Gaseous Emission 293
Gaseous emission 280, 296, 325
Gaseous Loss of N 248
Gaseous N losses 270
gaseous phase 311
gene banks 10
Genetic progress 56
genotypes 34, 50, 72, 88, 89, 114, 125, 165, 218, 225, 226, 253, 335
German loess zone 71
germplasm 6
global carbon cycle 291
Global warming 13, 24
Glomus fasciculatum 271
Glomus occultum 253
Glyricidia 201
grassland 331
Great Plains 19, 28, 29, 31, 33, 34, 35, 40, 42, 43, 45, 47, 48, 50, 52, 54, 91, 333
Green manures 127, 170, 201, 246, 247, 250, 331
Green manuring 120, 155, 200
ground flora 308
ground layer 305
Ground vegetation 321, 323
ground water contamination 264
groundnut 226, 227, 230, 233, 235, 247
growing season 304
growing season precipitation (GSP) 32, 42
growth index 218
gypsiferous soils 183, 198

Hapludolls 151
Hardpan 144, 197, 271
herb layer 305
hidden hunger 200
high quality' or fine litter 287
high yielding genotypes 221
Hole-in-the-pipe conceptual model 294
Human ingenuity 5, 6, 7, 10, 28, 105, 141
humid Andes 248
humid tropical forests 286
Humid tropics 241-243, 249, 251, 252, 293
humified residues 111
Hungarian lowlands 60
Hybrid rice 140
hybrid sorghum 220
hydrosphere 6, 7, 330

ICRISAT   212
IFDC   212, 221
Immobilization   9, 16, 68, 69, 87, 96, 115, 122, 149, 152, 160, 188, 319, 324, 332
Inceptisol   119, 143, 185, 227, 238
Indo-Gangetic plains   12, 21, 141, 142, 145-148, 158, 162, 165, 168, 170, 333
Indo-Malaysian forest   281
Indus plains   142
infertile acid soils   252
Insular P sources   18
Integrated management   126, 336
Integrated Nutrient Management   123, 137, 138, 165, 175, 208, 222
integrated plant nutrient system (IPNS)   146, 165
Integrated weed management   235
intensification   22, 36, 196, 202, 210, 227, 234, 331
intensive cropping systems   56
Intensive cultivation   59, 118
intensive farming belts   21
intensive management strategies   31
intensive rice ecosystems   105
Irrigated rice-wheat   147
Isoprene   291-293, 299, 301, 302

K deficiency   224
K fixation   195
K$^+$/Na ratio   164
kankar   197
Kaolinite   186, 284
Kazakhstan   59, 60
Khejri   201
Kursk area   64

labile P   178
Land use change   280, 293, 294, 298, 300, 302
leaching   21, 221
leaching losses   119
Leaf fall   251
Legume N fixation   231
legumes Ecosystem   208
Legumes ecosystem in SAT   226
Lens   43
Leucana   201
Leucana leucociphala   246
lichens   305
Liebig's 'law of minimum   180
lignolytic enzymes   320

Lithosphere   8, 330
litter   286, 307
litter decomposition   319
litter fall   296
litter formation   334
Litter layer   321
Litter Turnover   318
Litter turnover ratios   287
local landraces   221
Long-term Nutrient Balance   58, 78
Low land irrigated rice ecosystem   107
Low land rainfed rice ecosystem   107
low-P tolerant genotypes   332
Lower montane (LM)   282
Lowland Rice-Wheat   147
lupin-wheat sequence   86

Managed fallows   250
mature stage   187
Medicago   43
Mediterranean Europe   61
Mehlich's extract   271
Mellitus   43
meso-oligo-ombrotrophic   322
meso-oligotrophic sites   324
Mesopotamian Pampas   92, 93
mesotrophic   321
metal ligands   261
Methane   289, 295, 301
Methane Emission/Utilization   289
Methane emissions   15
Methane fluxes   290, 302
methanogenesis   67, 289
methanogens   289
micro-irrigation   260
micro-sprinkler   277
microbial immobilization   20, 46
micronutrient balance   215
Micronutrients   157
microporosity   145
Mineralization   16, 68, 69, 84, 114, 183, 267
Mineralization-Immobilization   105, 115, 147, 222
mineralization/immobilization turnover (MIT)   191
Minimum Tillage   31, 33, 35, 37
minimum-till   32
Mn toxicity   245
Mollisols   30, 92, 171, 180, 185, 305
Molybdenum   159
Montane   282

Movement of N 112
mulch 62, 235
mung bean 240

N balance 232, 250
N budget 85, 270, 318
N credits 16, 74, 86, 200, 225, 230, 232, 234, 247, 254
N depositions 315
N deposits 325
N dynamics 75, 85, 267
N fixation 43, 168, 190, 231, 246, 306
N immobilization 147, 289
N inputs 16
N leaching 17, 85, 263
N losses 69, 117, 150, 152, 213, 249, 268, 288
N mineralization 43, 110, 203
N recovery 73
N removal 275
N saturation 288, 318
N sinks 288
N volatilization 149
N-fixation 62, 78, 86
N-leaching 36
N-mineralization 52
N-use efficiency (NUE) 43, 46, 51, 72, 74, 96, 116, 118, 119, 130, 191, 216, 222, 267, 275, 316, 332, 333
N_2O emission 71, 140, 288, 315
Natural Pine Forest 304
natural succession 5
needle litter 326
NH_3 volatilization 17
NH_3 volatilized 150
Nitrate leaching 35, 69, 84, 103, 317
nitric oxide emissions 18, 300
nitrification 9, 266, 296, 297, 302, 317
Nitrification inhibitors 175
Nitrification–denitrification 116-118, 150
nitrifiers 290
Nitrobacter 48, 116, 160, 183
Nitrogen Allocation 131
Nitrogen Availability 150
Nitrogen balance 124, 277
Nitrogen budget 86, 88
Nitrogen Budgets in Citrus Groves 269
nitrogen credits 231
nitrogen cycle 329
Nitrogen cycling 318, 327
Nitrogen deposition 100, 288

nitrogen dynamics 74, 83, 95, 111, 265, 319
Nitrogen Emissions 149
Nitrogen fixation 9, 17, 98, 177, 255, 256, 269, 298, 307
Nitrogen Losses 68
Nitrogen mineralization 36, 37, 68, 101, 188, 266
Nitrogen pools 100
Nitrogen Recovery 84, 151
Nitrogen Recovery and Losses 188
Nitrogen Removal 72
Nitrogen Saturation 315-317, 326
Nitrogen use efficiency 56, 106, 123, 132, 190, 206
Nitrogen Volatilization 266
nitrogen-fixing 52
Nitrosomonas, Nitrococcus 183
Nitrosomonas 48, 116, 160
Nitrosomonas/Nitrobacter 9
nitrous oxide 103
Nitrous oxide emissions 23, 70, 103
Nitrous oxide fluxes 290, 295, 301
NO emissions 249, 289, 296, 297
NO fluxes 293
No tillage 31, 37, 61, 63, 82, 93, 100
no-till 32, 34, 37, 41, 44, 45, 47, 54, 55, 56, 63, 68
no-till system 33, 35
NO_2 emission 290, 291, 333
NO_3 leaching 98
NO_3–N contamination 267
non-bearing 275
non-exchangeable K 157, 225
North Central Florida 257
Norway spruce 309, 310
Nostoc muscorum 162
NO_x emissions 300
Nutrient 298
Nutrient Balance 220
nutrient buffering 182
nutrient cycling 196, 206, 212, 286
nutrient depletion 250
Nutrient deposits 288
Nutrient disequilibria 26
Nutrient Dynamics 241
nutrient equilibria 259
nutrient imbalances 166
Nutrient loading 313, 315, 328
Nutrient Losses 248
nutrient mining 331
nutrient partitioning ratio 217

Nutrient pumping 298
Nutrient Ratios 166
Nutrient Recovery 220
nutrient recycling 200
nutrient turnover 324
Nutrients versus Water 216

octo-calcium phosphates 183
'off' years 276
oligo-ombrotrophic 320
on years 276
Orange splitting 259
organic carbon 65
organic hardpan 260
organic matrix 322
organic recycling 159
organic S 122, 222, 229
organic–N 74, 225
Orlando tangelos 275
ORSTOM 212, 213
Oryza sativa 137, 138, 173
Oxisols 227, 243, 284
oxygenation 322
ozone 296

P availability 242
P balance 193
P diffusion 120
P efficiency 167
P leaching 76
P losses 76
P retention 270
P sufficiency 49
P transport 226, 252, 272
P–use efficiency 133, 156, 221
P-fixation 11
Paecilomyces 160
Pampa Humeda 94
Pampas in Argentina 58, 90, 91
Panamenian tropical forests 292
pCO_2 163
pearl millet 12, 19, 21, 210, 215, 234, 235, 236, 238, 331
Pearl Millet Agro-environment 210
Peat Land Forests 303
Peat lands 320, 322, 325
pedogenesis 284
pedosphere 6, 8, 22
Peuraria 250
Phaseolus 43
phenological stage 88, 89
phenology 191

Phosphate Physiology 133
Phosphate rock 214, 251-254
Phosphate Solubilizers 162
phosphogypsum 230
Phosphogypsum urea 124
Phosphorus dynamics 18, 28, 49, 88, 192
Phosphorus fixation 111
Phosphorus in Citrus Plantation 270
Phosphorus leaching 75, 271
Phosphorus Management 125
physiognomy 281
Physiological Genetics of Rice 129
Picea 314
Picea abies 310, 326
Picia mariana 315
Pigeon pea 227, 230, 231, 233, 237
Pilot analysis of global ecosystems 256
Pinus 314
pioneer crops 91
Pisum 43
Plant growth promoting rhizobacteria (PGPR) 232
Plugging N leaks 69
Podzols 60, 61
Polygonium 92
post-freeze season 276
Potassium deficiency 273
Potassium Fixation 194, 203
Potassium in Rice-Wheat Sequence 156
Potassium in WANA Drylands 194
Potassium Recovery 195
potentially mineralizable nitrogen (PMN) 267
prairies 2, 3
Precipitation 55, 242, 283, 284, 286, 304
precipitation-use efficiency 33, 334
Precipitations 321
Precision farming techniques 333
Premontane (P) 282
Prilled Urea 124, 152
Prosopis junifer 198
Pseudomonads 233
Pseudomonas striata 226
Pseudotsuga 304
puddled 118
Puddling 105, 108, 135, 144, 145, 171
puddling index 108
pulverizing 145

re-oxidation 120
re-translocation 315
recycling stover 247

redox   117, 145, 149, 154, 291
redox potential (Eh)   110, 145
reduced tillage   82
Refill point   263
remobilization   313
residual N   39, 191, 247
residual nitrate-N   35
Residual Nutrients   200
residual P   168
residue burning   46
Restricted tillage   63, 93
retranslocation   130
Rhizobium   231
rhizosphere   145, 160, 162, 225, 246, 308, 310, 311, 312, 325, 326
Rhizosphere bio-geochemistry   303
Rhizosphere Mineralogy   309
rhizospheres   238, 309
Rhizospheric   327
Rice Agroenvironment   107
Rice-Wheat-based cropping sequence   141, 146
rice-based cropping systems   107
rice-rhizosphere   117
rice-wheat cropping   174
rice-wheat cropping system   135
Rice-Wheat Cropping Zones   143
rice-wheat ecosystem   150
rice-wheat sequence   153, 155, 158, 163
Rolling and southern Pampas   91
Rolling Pampas   92, 93, 104
root exudates   48
rustic crop   245

S deposition   77
S-free fertilizers   20, 78
Sahel   218, 234
Sahelian West Africa   208, 209, 213
salinity   11
salt-affected soils   162
sapling layer   305
scenarios   66, 104
Scots pine (*Pinus sylvestrus*)   320
secondary forests   294
secondary rainforests   302
semi-arid tropics (SAT)   208
semi-dwarf   50, 59, 133, 217, 335
semi-dwarf rice   106
Semi-dwarfing   72
sequestering   306
sequestering C   45
Sesbania   128, 161, 246
shallow soils   222

shifting cultivation   244
shrub layer   305
Sinks   24, 66, 67, 302, 307
Sitka spruce   306
Slash and burn   300
slowing yield growth rates   330
small litter fall   287
Sodic   162
sodic soil   172, 173
Sodicity   11, 142, 164, 165, 219
Sodicity versus Nutrient Dynamics   162
Soil acidification   44, 278, 312
soil aerobicity   291
soil deterioration   170
Soil erosion   254, 255
Soil Exhaustion   249, 331
soil fertility   62
Soil fertility change   169
Soil microbes and Nutrient Dynamics   225
Soil N Pools   73, 98
soil organic carbon (SOC)   45, 54, 62, 65, 94
soil organic matter (SOM)   30, 53, 63, 83, 215, 318
Soil P Fractions   154
Soil P Transformations   154
Soil Quality   31, 34, 52, 55, 170
soil-N buffering   52
Soil-Plant N Buffering   39
soil-root interface   310, 311
Sorghum   236, 238
Sorghum-based cropping System   208, 218
Sorption   122
Sorption-desorption   271, 278
sour or sevile orange (*Citrus aurentium*)   258
Southern Great Plains   37
Southern Pampas   92, 93, 100
Southern plains   30
Soybean   226
Sphagna   323
Sphagnum   322
Sphagnum felox   321
spodosol   259, 260, 261, 262, 274, 278, 284, 305
Stem flow   283, 288
structural K   184, 194
stubble mulches   38
stubble mulching   33
Stylosanthes   246
sub-Saharan Africa   220, 221, 235
Submerged Soils   121
Submergence   110, 112, 119, 122, 147, 164, 174
subsistence farming   331

sulphate-S   215
sulphur   26
Sulphur dynamics   77
Sulphur in Wetland Paddy Soils   121
Sulphur   19
SUNDIAL   74
super hybrids   131
super varieties   106
surface flow   271
surface runoff   68
sweet orange (*C. sinensis*)   258
Sweet potato   241, 247
symbiosis   232

Taiga   303
temperate forests   303, 306
Temperate Wheat   58
Terrestrial Ecosystem Model (TEM)   13
Thar Desert   197
the forgotten element   204
Throughfall   284, 287, 288
tiechoic acid-P   76
Tirs   186
to reduced tillage   61
Tolypothrix timus   162
Top dressing   199
traditional clearing methods   285
Traditional system   199
Trans-Gangetic plains   142
Transgenic crops   10
Transpiration rates   303
Tree physiology   257, 274
Tree rhizosphere   298
tree stratum   305
Trifolium   43
Trifolium subtarraneum   80
tripartite relations   253
tripartite symbiosis   232
Triticum aestivum   173, 175
Tropical forests   280, 299
Troyer Citrange   272

Ultisols   284, 305
understory   304
undisturbed ecosystem   3-6
Untimely nitrate^{-1} 69
Upland dry rice ecosystem   107

urea hydrolysis   164
urea mobility   320
Urea super granules   124
Urease activity   149, 173
Ustalfs   210

Vadose zone   36, 259
Valencia citrus   271
Valencia orange   277
Vector competition analysis   327
vector diagnosis   313
Vertic Inceptisol   108, 182, 198, 219
Vertical leaching   89
Vertisols   119, 219, 223, 227, 237-240
Vetch   190
Vicia   43
Volatile Organic Compounds   291
Volatilization   35, 44, 46, 70, 113, 114, 118, 119, 164, 166, 189, 212, 213, 221, 234, 248, 268, 283, 295, 332
Volga territory   60

WANA drylands   195
Water Deficits   183
Water resources   334
water soluble K   229
water soluble K$^+$ 157
water-use efficiency (WUE)   33, 148, 203, 222, 227, 230
Weeds   233
West Asia and North Africa (WANA)   179-181, 184
Western Pampas   92
Wheat based rotation   65
Wheat-based Cropping Sequence   31
wheat-fallow (WF)   32
WUE   34

xerolchrepts   193
xerosols   192

yam   241

Zadocks' decimal scale   51
Zn deficiency   273
Zn use efficiency   158